Studies in Human Ecology and Adaptation

More information about this series at http://www.springer.com/series/6877

Erick Robinson • Frédéric Sellet
Editors

Lithic Technological Organization and Paleoenvironmental Change

Global and Diachronic Perspectives

Springer

Editors
Erick Robinson
Department of Anthropology
University of Wyoming
Laramie, WY, USA

Frédéric Sellet
Department of Anthropology
University of Kansas
Lawrence, KS, USA

ISSN 1574-0501
Studies in Human Ecology and Adaptation
ISBN 978-3-319-64405-9 ISBN 978-3-319-64407-3 (eBook)
DOI 10.1007/978-3-319-64407-3

Library of Congress Control Number: 2017955029

Printed on acid-free paper

This Springer imprint is published by Springer Nature
The registered company is Springer International Publishing AG
The registered company address is: Gewerbestrasse 11, 6330 Cham, Switzerland

Contents

1 **Lithic Technological Organization and Paleoenvironmental Change** 1
Erick Robinson and Frédéric Sellet

2 **Provisioning Responses to Environmental Change in South Africa's Winter Rainfall Zone: MIS 5-2** 13
Alex Mackay, Emily Hallinan, and Teresa E. Steele

3 **The South Asian Microlithic: *Homo sapiens* Dispersal or Adaptive Response?** 37
Chris Clarkson, Michael Petraglia, Clair Harris, Ceri Shipton, and Kasih Norman

4 **The Hoabinhian of Southeast Asia and its Relationship to Regional Pleistocene Lithic Technologies** 63
Ben Marwick

5 **Horizons of Change: Entanglement of Paleoenvironment and Cultural Dynamics in Australian Lithic Technology** 79
Peter Hiscock

6 **Human Adaptive Responses to Environmental Change During the Pleistocene-Holocene Transition in the Japanese Archipelago** 91
Kazuki Morisaki, Masami Izuho, and Hiroyuki Sato

7 **Isolation, Exploration or Seasonal Migration? Investigating Technological Organization in the Late Middle Palaeolithic of Britain During Marine Isotope Stage 3** 123
Rebecca M. Wragg Sykes

8 **Proto-Aurignacian Lithic Technology, Mobility, and Human Niche Construction: A Case Study from Riparo Bombrini, Italy** ... 163
Julien Riel-Salvatore and Fabio Negrino

9 **Environmental Change and Technological Convergence in Southern Germany** ... 189
Michael Jochim

10 **Technology and Human Response to Environmental Change at the Pleistocene-Holocene Boundary in Eastern Beringia: A View from Owl Ridge, Central Alaska** ... 203
Angela K. Gore and Kelly E. Graf

11 **Technological Change from the Terminal Pleistocene Through Early Holocene in the Eastern Great Basin, USA: The Record from Bonneville Estates Rockshelter** ... 235
Ted Goebel, Aria Holmes, Joshua L. Keene, and Marion M. Coe

12 **My Flute Is Bigger Than Yours: Nature and Causes of Technological Changes on the American Great Plains at the End of the Pleistocene** ... 263
Frédéric Sellet

13 **The Peopling of Southeastern South America: Cultural Diversity, Paleoenvironmental Conditions, and Lithic Technological Organization During the Late Pleistocene and Early Holocene** ... 281
Rafael Suárez

14 **Changes in the Technological Organization and Human Use of Space in Southern Patagonia (Argentina) During the Late Holocene** ... 301
Nora Viviana Franco, Luis Alberto Borrero, George A. Brook, and María Virginia Mancini

15 **The Costs and Benefits of Technological Organization: Hunter-Gatherer Lithic Industries and Beyond** ... 321
Michael J. Shott

Index ... 335

Contributors

Luis Alberto Borrero CONICET-University of Buenos Aires, Buenos Aires, Argentina

George A. Brook Department of Geography, University of Georgia, Athens, GA, USA

Chris Clarkson Archaeology, School of Social Sciences, The University of Queensland, Brisbane, QLD, Australia

Marion M. Coe Center for the Study of the First Americans, Texas A&M University, College Station, TX, USA

Nora Viviana Franco CONICET-University of Buenos Aires, Buenos Aires, Argentina

Ted Goebel Center for the Study of the First Americans, Department of Anthropology, Texas A&M University, College Station, TX, USA

Angela K. Gore Center for the Study of the First Americans, Department of Anthropology, Texas A&M University, College Station, TX, USA

Kelly E. Graf Center for the Study of the First Americans, Department of Anthropology, Texas A&M University, College Station, TX, USA

Emily Hallinan Division of Archaeology, University of Cambridge, Cambridge, UK

Claire Harris School of Archaeology, Research Laboratory for Archaeology and the History of Art, University of Oxford, Oxford, UK

Peter Hiscock Department of Archaeology, University of Sydney, Sydney, NSW, Australia

Aria Holmes Department of Anthropology, University of New Mexico, Albuquerque, NM, USA

Masami Izuho Archaeology Laboratory, Faculty of Social Sciences and Humanities, Tokyo Metropolitan University, Tokyo, Japan

Michael Jochim Department of Anthropology, University of California, Santa Barbara, CA, USA

Joshua L. Keene Center for the Study of the First Americans, Department of Anthropology, Texas A&M University, College Station, TX, USA

Alex Mackay Centre for Archaeological Science, University of Wollongong, Wollongong, NSW, Australia

María Virginia Mancini University of Mar del Plata, Mar del Plata, Argentina

Ben Marwick Department of Anthropology, University of Washington, Seattle, WA, USA

Center for Archaeological Science, University of Wollongong, Wollongong, NSW, Australia

Kazuki Morisaki Monuments and Sites Division, Agency for Cultural Affairs -Japan, Tokyo, Japan

Fabio Negrino Dipartimento di antichità, filosofia, storia e geografia, Università di Genova, Genova, Italy

Kasih Norman Archaeology, School of Social Sciences, The University of Queensland, Brisbane, QLD, Australia

Michael Petraglia Max Planck Institute for the Science of Human History, Jena, Germany

Julien Riel-Salvatore Département d'anthropologie, Université de Montréal, Montréal, Canada

Erick Robinson Department of Anthropology, University of Wyoming, Laramie, WY, USA

Hiroyuki Sato Department of Archaeology, Faculty of Letters, The University of Tokyo, Tokyo, Japan

Frédéric Sellet Department of Anthropology, University of Kansas, Lawrence, KS, USA

Ceri Shipton McDonald Institute for Archaeological Research, University of Cambridge, Cambridge, UK

Michael J. Shott Department of Anthropology and Classical Studies, University of Akron, Akron, OH, USA

Teresa E. Steele Department of Anthropology, University of California, Davis, CA, USA

Rafael Suárez Depto. Arqueología – FHCE, Universidad de la República and Sistema Nacional de Investigadores – ANII, Montevideo, Uruguay

Rebecca M. Wragg-Sykes PACEA UMR-5199, Université de Bordeaux, Bordeaux, France

About the Editors

Erick Robinson is a postdoctoral researcher in the Department of Anthropology at the University of Wyoming. He specializes in geochronology, lithic technology, and paleodemography, focusing on hunter-gatherer adaptations to terminal Pleistocene and Holocene environmental change and transitions to agriculture in northern Europe and Western US. He is currently working to develop a comprehensive archaeological radiocarbon database for the US that will be used to reconstruct prehistoric population growth and migration throughout the Holocene.

Frédéric Sellet is an Associate Professor of Anthropology at the University of Kansas. His research revolves mainly around the study of stone tools in small-scale prehistoric societies, with a focus on issues such as technological organization, planning depth, and human mobility. He has been conducting fieldwork in the American Plains and in the Rockies for the past 25 years, primarily exploring the archaeology of the First Americans. He is an affiliate member of the French National Center for Scientific Research (CNRS) and the Smithsonian Institution.

Chapter 1
Lithic Technological Organization and Paleoenvironmental Change

Erick Robinson and Frédéric Sellet

"Forty-Seven Trips" After Forty Years

Forty years ago, Lewis Binford (1977) published a paper that set the stage for considering how the organization of lithic technologies facilitated hominin adaptations to environmental changes throughout prehistory. In this paper, Binford presented findings from his ethnoarchaeological research with the Nunamiut, specifically how these findings had implications for the formation of the archaeological record and the interpretation of past cultural systems. He noted how the organizational dynamics of cultural systems were based on the relationships between the procurement of material items and their discard ("discharge or entropy"). The now widely used terms of "curation" and "expediency" were introduced to describe the relative organizational intensity of different cultural systems. Highly curated technologies, such as those of the Nunamiut, would not lead to archaeologists finding "butchering tools at butchering locations, or wood cutting tools at wood collecting locations" (Binford 1977: 34). In these kinds of organizational systems, most of the materials left in the field were the "immediate by-products of consumption, such as bones from dry meat or shell cases from firing guns" (Binford 1977: 34), not tools that we would expect to be related to a specific activity. With this observation, Binford introduced the importance of moving beyond a sole focus on stone tools toward the investigation of the by-products of tool production, or debitage:

E. Robinson (✉)
Department of Anthropology, University of Wyoming, Laramie, WY, USA
e-mail: Erick.Robinson@uwyo.edu

F. Sellet
Department of Anthropology, University of Kansas, Lawrence, KS, USA
e-mail: fsellet@ku.edu

E. Robinson, F. Sellet (eds.), *Lithic Technological Organization and Paleoenvironmental Change*, Studies in Human Ecology and Adaptation 9,
DOI 10.1007/978-3-319-64407-3_1

> We need to give more analytical attention to the relationships between features and mundane remains such as fire-cracked rock, bone fragments, chipping debris, and garbage, not as a means to categorical types of generalization about diet, etc., but as a means to understanding the behavioral context in terms of which the site was produced. (Binford 1977: 34)

This better understanding of behavioral context was based on his proposition that, "…regardless of the degree of curation and maintenance of the tool assemblage, we can anticipate variability in the immediate by-products of consumption to relate to seasonal or situational variation in the activities performed…" (Binford 1977: 36).

These various insights from working with the Nunamiut culminated in Binford's 1980 publication of "Willow Smoke and Dogs' Tails," in which he outlined a theory for relating hunter-gatherer settlement systems, adaptation, and archaeological site formation processes. In this paper, Binford posed two questions that the volume before you attempts to address with global and diachronic archaeological case studies:

> Can we now begin the important task of building an explanation for the variability presented? Can we begin to understand the particular adaptive conditions which human groups differently face by virtue of coping with different environments? (Binford 1980: 13)

At the same time that Binford was producing this pioneering work, the field of Quaternary science was going through a similar period of important and foundational change. The CLIMAP project was developed in 1971, with the aim of bringing together scientists from around the world to reconstruct global climate over the past million years from deep-ocean sedimentary records (CLIMAP Project Members 1976). The first publication from this project was produced in 1976 and was titled "The Surface of Ice-Age Earth" (CLIMAP Project Members 1976). The paper starts with a statement that still has major relevance in contemporary Quaternary science, and also problems highlighted by the various case studies in this volume:

> Over the last few million years, the earth's climate has alternated between ice ages and warmer intervals. The cause of these climatic fluctuations is an intriguing and as yet unsolved problem. The difficulty in understanding the cause lies in the complexity and global scope of the climate system, for changes in climate occur over a wide range of time and space scales and involve interactions within a planetary system that includes the ice sheets, the atmosphere, the surface of the land, and the entire world ocean. Thus, any strategy for attacking this problem empirically must be linked to a physical model of the global climate system. (CLIMAP Project Members 1976: 1131).

In the same year, Hays et al. (1976) published a synthesis of oxygen isotope data from marine cores that confirmed the Milankovitch hypothesis that changes in the orbital patterns of earth regulated ice ages over the past 500,000 years.

Thus, at the same period of time, both archaeology and the Quaternary sciences were attempting to develop models or theoretical frameworks to adequately understand the complexities of their respective records. How have these different models/frameworks been refined over the past 40 years? How have their refinements helped us to understand the relationships between technological organization (TO) and paleoenvironmental change?

Together, the thirteen case studies in this volume seek to answer these questions. This volume is the first to compile different case studies focusing on TO approaches to paleoenvironmental changes from around the world (Fig. 1.1), covering a broad

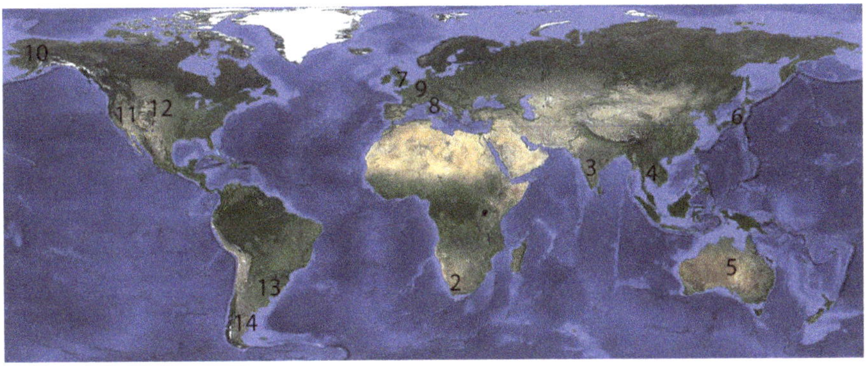

Fig. 1.1 Geographical locations of the different chapters in the volume

expanse of time (much of the second half of the Quaternary Period). Over the past 40 years, much has changed in both the use of TO approaches in archaeology, as well as in our knowledge of how regional ecosystems are linked to global-scale processes of climate change. However, despite these changes, researchers in 2017 face some similar problems that were confronted by the pioneering work of the 1970s. For example, Shott (Chap. 15) highlights how the use of the term "curation" continues to be treated as a type rather than a continuous variable to be measured. Likewise, while our knowledge of paleoclimates and their interactions with regional paleoenvironmental changes advanced tremendously over the past 40 years, the problems of time-transgressive ecosystem responses to paleoclimate changes remain a key challenge for researchers, just as it was when highlighted by Watson and Wright in 1980. Furthermore, beyond the problem of time-transgressive responses, researchers are also still trying to understand the forcing mechanisms of shorter millennial and centennial-scale climate change events, as indicated by recent papers on the Younger Dryas (Murton et al. 2010; Not and Hillaire-Marcel 2012; Muschitiello et al. 2015; Hogg et al. 2016).

Two refinements over the past 40 years have opened up greater challenges for researchers seeking to apply a TO approach to the investigation of human adaptations to paleoenvironmental changes. These challenges provide the impetus for bringing these different case studies together. The first challenge is that TO approaches have now been applied to myriad case studies around the world, from the Middle Paleolithic (Wragg-Sykes, Chap. 7) and Middle Stone Age (Mackay et al., Chap. 2) to the Late Holocene (Franco et al., Chap. 14). These studies have applied different concepts from TO to very different problems. After Binford's first three papers from 1977 to 1980 that introduced the general theoretical framework, a series of seminal papers were produced in the 1980s and 1990s that introduced different concepts to highlight different aspects of the TO approach (Torrence 1983; Bamforth 1986, 1991; Bleed 1986; Shott 1986; Kelly 1988; Nelson 1991; Kuhn 1994; Bamforth and Bleed 1997). This was the period when the TO approach was fully elaborated. However, this elaboration was carried out differently by different researchers, and there was still a lack of a single unified theoretical framework that could mitigate the potential for what Shott (Chap. 15) has called "interpretive ambiguity." The majority of these studies were focused on purely theoretical investigations, or the ethnographic record, rather than archaeological case studies. This means that by the turn of the millennium, there were still few widespread applications of TO concepts to different archaeological case studies (cf. Carr 1994). By bringing together different case studies throughout the world, this volume aims to highlight similar applications of TO concepts to vastly different archaeological problems. We believe that these similarities will help to establish a less ambiguous theoretical framework for the future and advance how TO concepts are operationalized, both qualitatively and quantitatively.

The second refinement over the past 40 years has been the increase of knowledge about the complexities of feedback relationships between different paleoclimate forcing mechanisms (orbital cycles glacier meltwater pulses) and regional paleoenvironmental changes. Regional ecosystem responses to paleoclimate change determined the various ecological constraints to which prehistoric cultures had to adapt (Birks et al. 2015). When Binford was formulating the TO approach, and when researchers were elaborating it in the 1980s and 1990s, paleoenvironmental reconstructions were often based on single proxies such as pollen, and at relatively low chronological resolution. In a synthesis paper of the TO approach, Nelson states:

> One of the greatest contributions of studies of technological organization has been in emphasizing the dynamics of technological behavior. Dynamics refers to the plans or strategies that guide the technological component of human behavior...Technological strategies weigh social and economic concerns with respect to environmental conditions and are implemented through design and activity distribution...Strategies are viewed as problem-solving processes that are responsive to conditions created by the interplay between humans and their environment. (Nelson 1991: 57–58)

Over the past 40 years, high-resolution, multiproxy paleoenvironmental reconstructions have advanced our knowledge of the complexity of ecosystem responses to change, and the temporal variability of changes. This makes it important for researchers to integrate regional paleoenvironmental records with interregional or hemispheric records such as ice or marine cores. In 2017, we therefore have more "environmental conditions" to consider.

In the following sections, we address three different aspects of this volume that highlight advances, or continuing problems, in the 40 years since "Forty-seven trips." We believe that these three areas provide a platform for continuing refinement of the TO approach and the power it has to advance our understanding of hominin adaptations to paleoenvironmental change.

Chronological Resolution

This volume confronts two different but interrelated problems of chronological resolution. The first problem is the chronological resolution of archaeological records. As noted in the previous section, the TO approach as originally developed by Binford

(1977) focused on the environment in terms of different seasonal constraints. While seasonal constraints are easily observable from the ethnographic record, most archaeological records only provide time-averaged data on centennial or millennial scales. The challenge therefore lies in adapting concepts developed in ethnographic contexts to archaeological records that often have gaps in evidence. For instance, one major problem that archaeologists will face in the coming years is how to interpret the impacts of abrupt, centennial-scale paleoclimate changes on prehistoric societies. However, different case studies in this volume show that similar TO changes were carried out in similar social and environmental contexts, regardless of the relative chronological resolution of regional archaeological records for different time periods. A good example of this comes from contributions highlighting similar raw material procurement and replacement patterns during the colonization of new landscapes, from the Middle Paleolithic (Wragg-Sykes, Chap. 7), to the terminal Pleistocene (Gore and Graf, Chap. 10; Goebel et al., Chap. 11), to the Late Holocene (Franco et al., Chap. 14).

A related problem with the chronological resolution of archaeological records concerns the various scales of analysis undertaken. Site level analyses can reveal important diachronic fluctuations in technological organization with higher chronological resolution, given that they are stratified. Three great examples are provided in the contributions of Riel-Salvatore and Negrino (Chap. 8), Goebel et al. (Chap. 11), and Gore and Graf (Chap. 10). Breaking down the different Proto-Aurignacian levels at Riparo Bombrini allowed Riel-Salvatore and Negrino to reveal greater internal variability within this culture than traditionally recognized. They were able to relate different Proto-Aurignacian levels to two separate environmental periods (one colder and one warmer), with more expedient technological organization (defined by ratios of total debitage volumetric density to retouch frequency) and logistical land use being associated with a colder period, and more curated and residential land use being associated with warmer conditions. The contributions by Goebel et al. and Gore and Graf provide important insights into the relationships between technological organization and the colonization of landscapes during the Pleistocene–Holocene transition from the perspective of a single site.

Of course, well-stratified sites with multiple occupation events are rare relative to the wealth of data derived from single-component or surface sites. While providing coarser chronological resolution, syntheses of these data provide important information on different strategies of technological organization through time throughout a region, or between multiple regions. For example, despite having coarse resolution for the proliferation of backed artifacts in Australia, Hiscock (Chap. 5) is able to delineate the variable timing of proliferation between arid and more humid regions. Using regional-scale approaches has enabled several contributors to investigate the relationships between different types of TO strategies and varying resource predictability (Mackay et al., Chap. 2; Clarkson et al., Chap. 3; Morisaki et al., Chap. 6; Jochim, Chap. 9; Sellet, Chap. 12).

The second problem concerns the lack of high-resolution paleoenvironmental records at regional scales, which is noted by several chapters (Mackay et al., Chap. 2; Clarkson et al., Chap. 3; Marwick, Chap. 4; Hiscock, Chap. 5; Sellet, Chap. 12). As mentioned above and in the contribution by Sellet, different kinds of paleoenvironmental changes occurred at different scales. This requires that we obtain regional records that are dated at high enough resolution to delineate these scales and their relative impacts on prehistoric societies. The contributions by Suarez (Chap. 13) and Franco et al. (Chap. 14) highlight what our ultimate goal should be in the coming years to overcome these problems: paleoenvironmental records that are as near as possible to the archaeological sites of interest. As our knowledge of the complexities of regional ecosystem responses to hemispheric and global processes of paleoclimate change has increased, the legacy records that we rely on to reconstruct regional environments will become less and less useful unless we can return to these records to redate and reanalyze them at higher resolution. Marwick (Chap. 4) points out that future work will rely on more interdisciplinary collaborations that can produce the necessary high-resolution paleoenvironmental data to address this problem.

Linking Technological Parameters to Environmental Conditions

In his discussion chapter, Shott notes how TO approaches still suffer from a lack of theoretical specification that can qualitatively and quantitatively "detail the strength and direction of relationships between technological parameters and organizational constraints" (Shott, Chap. 15: 306). Our intention in assembling case studies from different time periods and continents was to cast a wide enough net that would provide a foundation for linking similar technological parameters to similar environmental constraints. However, as the contributions to this volume illustrate, contradicting technological parameters and environmental constraints can also provide an important foundation for future studies. While TO still lacks a unified theory with clear qualitative and quantitative expectations (Andrefsky 2009), this volume establishes relationships that can be tested by future work. Ideally, future work can rely on more formalized approaches such as those from evolutionary ecology (cf. Sellet, Chap. 12; Kuhn 1994; Andrefsky 2009; Surovell 2009; Kuhn and Miller 2015), which produce clear and testable hypotheses that can be compared between different case studies. Andrefsky (2009) has provided a positive outlook for this future, where studies in data patterning can help to develop testable predictions that can feed into a more unified theory of TO. The studies in this volume provide the greatest time depth and context variability yet to isolate patterns in the data.

The first data pattern that can help us move forward concerns the relationships between raw material procurement and the colonization of new landscapes (Wragg-Sykes, Chap. 7; Gore and Graf, Chap. 10; Goebel et al., Chap. 11; Suarez, Chap. 13; Franco et al., Chap. 14). Despite ranging from Neanderthal recolonization of Britain (Wragg-Sykes) to the Late Holocene of Patagonia (Franco et al.), these contributions highlight how hunter-gatherers colonizing new landscapes in uncertain and dynamic environments will utilize high-quality nonlocal raw materials, eventually incorporating lower-quality, local raw materials through time.

The second pattern arising from the contributions pertains to the relationships between curation and increased residential mobility (Morisaki et al., Chap. 6; Wragg-Sykes, Chap. 7; Riel-Salvatore and Negrino, Chap. 8; Goebel et al., Chap. 11; Suarez, Chap. 13). Not all are explicit about curation being a continuous variable comparing realized to maximum tool utility (Shott 1996), but they unanimously emphasize how curated tools will have higher retouch, resharpening, and/or recycling. These data can be converted into continuous variables in further studies. These studies relate curated technologies to increased residential mobility and link them with worse, dynamic, and/or uncertain environments (Wragg-Sykes, Morisaki et al., Goebel et al., Suarez). The lone exception to this comes from Riel-Salvatore and Negrino, who associate curation and increased residential mobility with better environments. This contradiction highlights a further problem to be investigated in the future: what makes an environment "good" or "bad" for a hunter-gatherer society? A "good" or "bad" environment is based on the relative resource variability in a given ecosystem and the specific feedback relationships maintaining resource abundance, in combination with the adaptive options that are regionally available to a particular hunter-gatherer group. These problems can be examined with future comparative work employing high-resolution regional paleoenvironmental records that assess multiple proxies, preferably proxies that would have had direct impacts on hunter-gatherers.

The third pattern identified in the volume deals with the relationships between resource predictability and blade/microblade systems. The contributions present an interesting contradiction that will hopefully be the focus of further study. The chapters by Clarkson et al. (Chap. 3), Hiscock (Chap. 5), and Morisaki et al. (Chap. 6) note how blade/microblade industries enhance adaptations to periods of environmental change characterized by low resource productivity, whereas chapters by Mackay et al. (Chap. 2), Jochim (Chap. 9), and Gore and Graf (Chap. 10) associate them with periods of high resource predictability. This contradiction is even more interesting considering that Marwick proposes that their lack of appearance in SE Asia proves they provide adaptive advantages in specific environmental contexts. The chapters here can provide a springboard for further testing of this observation.

Recording Strategies, Debitage, and Time-Series Data

The development of a more unified approach to TO—one that establishes clear links between technological parameters and environmental constraints—can be facilitated by increasing the comparability of lithic recording strategies and reporting. The various concepts developed under the TO umbrella ("curation," "expediency," "reliability," "maintainability," "flexibility") do not provide prescriptions for how lithic assemblages should be recorded, which increases the risk for the "interpretive ambiguity" that Shott (Chap. 15) describes. Unified recording and reporting strategies would enhance our ability to compare not only different archaeological case studies but also archaeological and paleoenvironmental data. Lithic assemblages have traditionally not been recorded with considerations for how that data might be compared with paleoenvironmental data. Paleoenvironmental data are usually arrayed as time-series data, often including ratio scales of measurement. Thus, archaeological data arrayed as ratio scale time-series data would facilitate more robust statistical tests of hypotheses articulating TO within a regional paleoenvironmental framework.

In an insightful but unfortunately undercited paper, Binford (1983) investigated the intensity of use at the stratified site of Sudden Shelter by arraying total debitage concentrations versus the different stratified levels of the site. These observations were not compared to local environmental data, but the way they were arrayed in a continuous time-series could easily facilitate this. A further example where ratio-scale time-series are used to compare different lithic assemblage measurements to environmental proxies is Surovell's (2009: Fig. 3.11) comparison of the Puntutjarpa Rockshelter data to regional lake-level records. Four examples in this volume similarly emphasize the value of these approaches.

Clarkson et al. (Fig. 3.14) display time-series data comparing the frequency of microlithic ages to varying cycles of Late Pleistocene monsoon strength on the Indian subcontinent. Morisaki et al. (Fig. 6.4) compare different technological systems to different periods of environmental change. The "curation continuum" employed by Riel-Salvatore and Negrino (Fig. 8.4) could be converted into a ratio that would enable these data to be contrasted with regional environmental data. Finally, Franco et al. (Fig. 14.3) highlight the power that these time-series arrays could have for comparing different lithic-ratio measurements with multiple paleoenvironmental proxies. Comparisons with different paleoenvironmental proxies are especially important for testing hypotheses about what spatial and temporal scales of paleoclimate and paleoenvironmental change were most impactful on hunter-gatherer societies.

Working debitage data into these time-series arrays is critical for testing hypotheses concerning the relationships between TO, land use, and paleoenvironmental change. Figure 1.2 illustrates this. Hypothetical ratios of local/nonlocal raw materials and debitage/tools are compared against regional moisture records from Wyoming (Shuman et al. 2010) and the oxygen isotope record from the GRIP ice core in Greenland (Bond et al. 2001). These kinds of time-series arrays could help test hypotheses about varying TO strategies and land-use intensities during periods of drought versus more humid periods. They could also enable measuring whether temperature or precipitation had a greater impact on hunter-gatherer adaptations. Ideally, the data points for the lithics should come from the same stratified archaeological site. But, as mentioned above, this is rarely the case. This hypothetical data could also be generated from well-dated components of multiple sites within a given region and then synthesized to develop regional time-series of various ratio-scale data. Both chapters by Clarkson et al. and Franco et al. also illustrate how regional data can be synthesized and compared in time-series to different paleoenvironmental records.

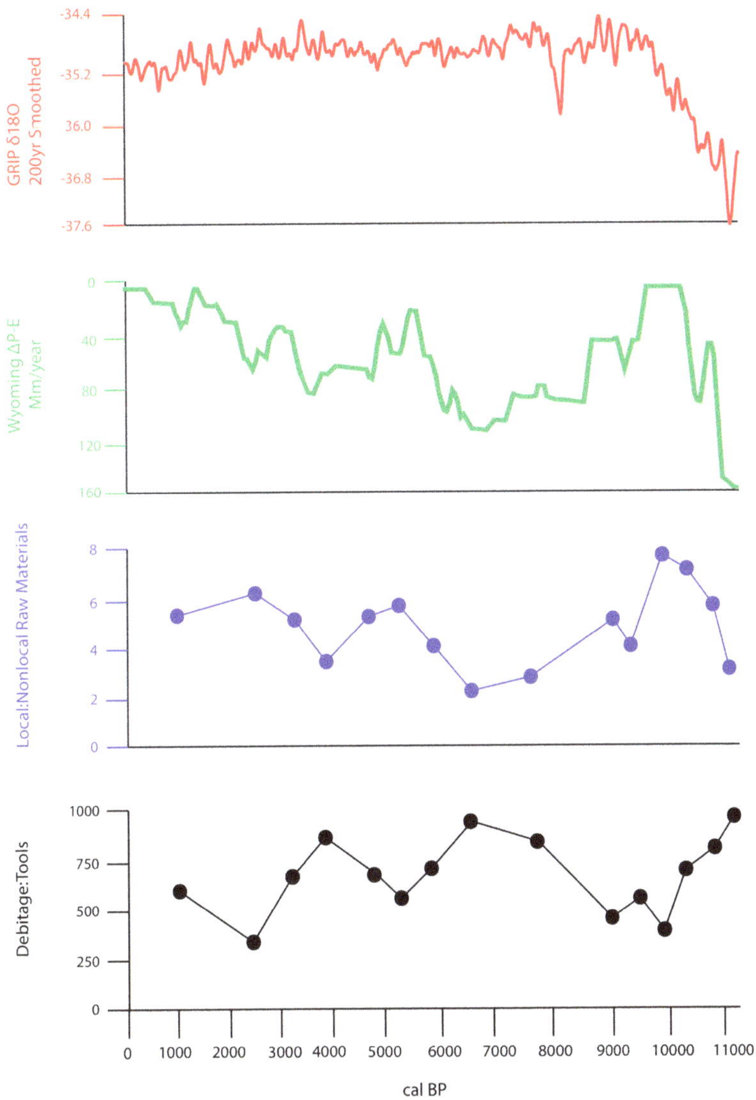

Fig. 1.2 Hypothetical debitage:tool and local:nonlocal raw material ratios compared to a regional moisture record from Wyoming (Shuman et al. 2010) and a hemispheric-scale paleoclimate record from GRIP ice core in Greenland (Bond et al. 2001)

Conclusion

The chapters in this volume provide the first global and diachronic study of the application of the TO approach to the investigation of hominin adaptations to paleoenvironmental change. In this preamble, we have discussed changes to the TO approach over the past 40 years and isolated some of the new challenges that confront researchers. While TO can provide important insights into how prehistoric societies changed their lithic technological organization strategies to adapt to different kinds of paleoclimate and paleoenvironmental changes, some improvements need to be made in the future to develop a more robust, hypothesis-driven framework. The chapters in this volume provide a foundation for these improvements by presenting case studies from which different data patterns can be isolated and tested by further research. We believe the stage is set for lithic TO studies to advance our knowledge on the entire prehistory of human adaptations to paleoenvironmental change.

Acknowledgments We thank all of the contributors for their strong contributions to this volume and for their patience in its production. We also thank all of the peer reviewers for enhancing the quality of this volume. Finally, we would like to thank Hana Nagdimov and Teresa Krauss of Springer Press for their support, guidance, and patience throughout the production and editing of this volume.

References

Andrefsky, W. (2009). The analysis of stone tool procurement, production, and maintenance. *Journal of Archaeological Research, 17*, 65–103.

Bamforth, D. (1986). Technological efficiency and stone tool curation. *American Antiquity, 51*, 38–50.

Bamforth, D. (1991). Technological organization and hunter-gatherer land use: A California example. *American Antiquity, 56*, 216–234.

Bamforth, D., & Bleed, P. (1997). Technology, flaked stone technology, and risk. *Archaeological Papers of the American Anthropological Association, 7*(1), 109–139.

Binford, L. R. (1977). Forty-seven trips: A case study in the character of archaeological formation processes. In R. V. S. Wright (Ed.), *Stone tools as cultural markers: Change, evolution and complexity* (pp. 24–36). Canberra: Australian Institute of Aboriginal Studies.

Binford, L. R. (1980). Willow smoke and dogs tails: Hunter-gatherer settlement systems and archaeological site formation. *American Antiquity, 45*(1), 4–20.

Binford, L. R. (1983). Long-term land-use patterning: Some implications for archaeology. In L. R. Binford (Ed.), *Working at archaeology* (pp. 379–386). New York: Academic Press.

Birks, H. H., Gelorini, V., Robinson, E., & Hoek, W. Z. (2015). Impacts of paleoclimate change 60,000-8000 years ago on humans and their environments in Europe: Integrating paleoenvironmental and archaeological data. *Quaternary International, 378*, 119–133.

Bleed, P. (1986). The optimal design of hunting weapons: Maintainability or reliability. *American Antiquity, 51*, 737–747.

Bond, G., Kromer, B., Beer, J., Muscheler, R., Evans, M. N., Showers, W., Hoffman, S., Lotti-Bond, R., Hajdas, I., & Bonani, G. (2001). Persistent solar influence on North Atlantic climate during the Holocene. *Science, 294*, 2130–2135.

Carr, P. J. (Ed.). (1994). *The organization of North American prehistoric chipped stone tool technologies*. Ann Arbor: International Monographs in Prehistory.

CLIMAP Project Members. (1976). The surface of Ice-age earth. *Science, 191*(4232), 1131–1137.

Hays, J. D., Imbrie, J., & Shackleton, N. J. (1976). Variations in the earth's orbit: Pacemaker of the Ice Ages. *Science, 194*(4270), 1121–1132.

Hogg, A., Southon, J., Turney, C., Palmer, J., Bronk Ramsey, C., Fenwick, P., Boswijk, G., Friedrich, M., Helle, G., Hughen, K., Jones, R., Kromer, B., Noronha, A., Reynard, L., Staff,

R., & Wacker, L. (2016). Punctuated shutdown of Atlantic Meridional overturning circultation during Greenland Stadial 1. *Scientific Reports, 6*, 25902.
Kelly, R. L. (1988). The three sides of a biface. *American Antiquity, 53*, 717–734.
Kuhn, S. L. (1994). A formal approach to the design and assembly of mobile toolkits. *American Antiquity, 59*, 426–442.
Kuhn, S. L., & Miller, D. S. (2015). Artifacts as patches: The marginal value theorem and stone tool life histories. In N. Goodale & W. Andrefsky (Eds.), *Lithic technological systems and evolutionary theory* (pp. 172–197). Cambridge: Cambridge University Press.
Murton, J. B., Bateman, M. D., Dallimore, S. R., Teller, J. T., & Yang, Z. (2010). Identification of Younger Dryas outburst flood path from Lake Agassiz to the Arctic Ocean. *Nature, 464*, 740–743.
Muschitiello, F., Pausata, F. S. R., Watson, J. E., Smittenberg, R. H., Salih, A. A. M., Brooks, S. J., Whitehouse, N. J., Karlatou-Charalampopoulou, A., & Wohlfarth, B. (2015). Fennoscandian freshwater control on Greenland hydroclimate shifts at the onset of the Younger Dryas. *Nature Communications, 6*(8939). doi:10.1038/ncomms9939.
Nelson, M. (1991). The study of technological organization. In M. B. Schiffer (Ed.), *Archaeological method and theory* (Vol. 3, pp. 57–100). Tucson: University of Arizona Press.
Not, C., & Hillaire-Marcel, C. (2012). Enhanced sea-ice export from the Arctic during the Younger Dryas. *Nature Communications, 3*(647). doi:10.1038/ncomms1658.
Shott, M. J. (1986). Technological organization and settlement mobility: An ethnographic examination. *American Antiquity, 42*(1), 15–51.
Shott, M. J. (1996). An exegesis of the curation concept. *Journal of Anthropological Research, 52*, 259–280.
Shuman, B., Pribyl, P., Minckley, T. A., & Shinker, J. J. (2010). Rapid hydrologic shifts and prolonged droughts in Rocky Mountain headwaters during the Holocene. *Geophysical Research Letters, 37*, L06701.
Surovell, T. A. (2009). *Toward a behavioral ecology of lithic technology: Cases from Paleoindian archaeology*. Tucson: University of Arizona Press.
Torrence, R. (1983). Time budgeting and hunter-gatherer technology. In G. Bailey (Ed.), *Hunter-gatherer economy in prehistory: A European perspective* (pp. 11–22). Cambridge: Cambridge University Press.
Watson, R. A., & Wright, H. E. (1980). The end of the Pleistocene: A general critique of chronostratigraphic classification. *Boreas, 9*, 153–163.

Chapter 2
Provisioning Responses to Environmental Change in South Africa's Winter Rainfall Zone: MIS 5-2

Alex Mackay, Emily Hallinan, and Teresa E. Steele

Introduction

The later Middle Stone Age (MSA) archaeological record of southern Africa is known for the diversity and antiquity of its complex lithic technologies, including early examples of the manufacture of blades and bladelets, bifacial points, and backed artifacts (or microliths) (Volman 1980; Mitchell 1988; Soriano et al. 2007; Wadley 2007; Villa et al. 2009, 2010; Högberg and Larsson 2011; Porraz et al. 2013a, b; Wurz 2013). Associated production techniques include the use of pressure flaking, marginal percussion with soft stone, and heat treatment (Brown et al. 2009; Soriano et al. 2009; Mourre et al. 2010; Schmidt et al. 2012; Porraz et al. 2013b). These lithic systems underwent turnover on the order of 2–10 kyr throughout the Late Pleistocene (Jacobs et al. 2008a, Tribolo et al. 2012; Porraz et al. 2013a).

There has been considerable discussion of the drivers underlying technological variability through this period (Ambrose and Lorenz 1990; Wurz 1997; Ambrose 2002; McCall 2007; Mackay 2009; Powell et al. 2009; McCall and Thomas 2012; Ziegler et al. 2013; Mackay et al. 2014a, Soriano et al. 2015). While the focus of these arguments has largely been on the form of typical technological items, we consider evidence for changes in provisioning – effectively, the ways in which lithic technologies were delivered to their point of need – through the period from ~75 to ~50 ka. We focus on southern Africa's winter rainfall zone and the relationship between provisioning systems and environmental variation, as it is known.

A. Mackay (✉)
Centre for Archaeological Science, University of Wollongong, Wollongong, NSW, Australia
e-mail: amackay@uow.edu.au

E. Hallinan
Division of Archaeology, University of Cambridge, Cambridge, UK

T.E. Steele
Department of Anthropology, University of California, Davis, CA, USA

E. Robinson, F. Sellet (eds.), *Lithic Technological Organization and Paleoenvironmental Change*, Studies in Human Ecology and Adaptation 9,
DOI 10.1007/978-3-319-64407-3_2

The Archaeology and Environments of Southern Africa's Winter Rainfall Zone

Southern Africa is divisible into three broad climatic regions – the winter rainfall zone (WRZ), summer rainfall zone (SRZ), and year-round rainfall zone (YRZ) (Chase and Meadows 2007) (Fig. 2.1) – based on factors controlling precipitation. The WRZ is situated in the southwest of the subcontinent and receives most of its annual precipitation from the northward migration of westerly winds during the austral winter. Controls on the delivery of winter rain in the region are complex, but the bulk of the available theory and data suggest an expansion of the zone of westerly influence under cooler conditions associated with expansions of Antarctic sea ice (Nicholson and Flohn 1980; Chase and Meadows 2007; Toggweiler and Russell 2008; Chase 2010; Mills et al. 2012; Stager et al. 2012; Truc et al. 2013). In consequence, increases in the extent and duration of the rainy season are expected to accompany glacial conditions, resulting in greater regional humidity. This is most evident in marine isotope stage (MIS) 4 and MIS 2, with evidence for decreases in humidity into MIS 3 (Cowling et al. 1999; Klein et al. 1999; Shi et al. 2000; Stuut et al. 2002; Avery et al. 2008; Bruch et al. 2012).

The major vegetation community in the WRZ is fynbos, characterized by low shrublands and high rates of endemism (Mucina and Rutherford 2006). Fynbos taxa show considerable resilience to climatic change (Meadows and Sugden 1993; Valsecchi et al. 2013), possibly a consequence of edaphic controls arising from the

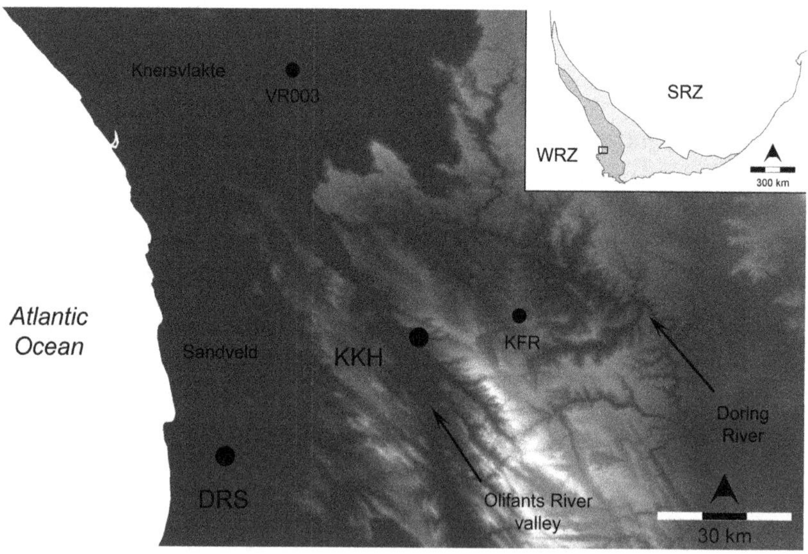

Fig. 2.1 Study area showing the major shelter sites (*KKH* Klein Kliphuis, *DRS* Diepkloof) and minor sites of relevance to the paper (*VR003* Varsche Rivier 003, *KFR* Klipfonteinrand), as well as the rainfall zones of southern Africa

low-nutrient soil related to underlying sandstone geology (Valsecchi et al. 2013; Carr et al. 2016). Increased humidity in the WRZ has been associated, in some cases at least, with increases in Afromontane forest taxa, though in higher-nutrient areas, it may be associated with expansions of grasslands (Klein 1976; Cowling et al. 1999; Cartwright 2013). Aridity clines to the north and northeast result in increased prevalence of arid-adapted plants, most notably in the form of succulents (Mucina and Rutherford 2006).

The later MSA archaeological sequence in the WRZ can be broken down into several components or industries based on the characteristics of flaked stone artifact assemblages. The earliest of these may be referred to as the earlier Middle Stone Age (MSA) which some researchers subdivide into multiple phases (Volman 1980; Wurz 2002). Generally dating to greater than 80 ka, the earlier MSA is outside the scope of interest of this paper and will receive little further discussion.

Chronologically, the earlier MSA is followed by the Still Bay industry in which bifacial points are common and the prevalence of fine-grained rocks often increases (Henshilwood et al. 2001; Högberg and Larsson 2011; Porraz et al. 2013b) (Fig. 2.2). Researchers have noted an underrepresentation of cores in Still Bay sites across southern Africa (Henshilwood et al. 2001; Wadley 2007; Porraz et al. 2013b), and some have suggested that this may have been a period in which residential mobility was emphasized, based largely on inferences drawn from technological systems design theory (Mackay 2009; McCall and Thomas 2012). Most ages place the Still Bay around the transition from MIS 5 to MIS 4 between 75 and 70 ka, in the context of rapidly cooling conditions and increasing humidity (Stuut et al. 2002). These ages, however, are contested (Jacobs et al. 2008a, Tribolo et al. 2009; Högberg and Larsson 2011; Tribolo et al. 2012; Jacobs et al. 2013). Conflicting OSL ages at the WRZ site of Diepkloof place the Still Bay either 75–71 ka or >100 ka. It should be noted that while bifacial points occur throughout MSA sequences at sites in the SRZ (e.g., Kaplan 1990; Wadley 2012; de la Pena et al. 2013; Porraz et al. 2015), this pattern has not been documented in the WRZ or YRZ (Mackay et al. 2014a). Bifacial points recovered from excavated contexts in these regions invariably occur as a single discrete temporal band within sequences (Henshilwood et al. 2001; Vogelsang et al. 2010; Porraz et al. 2013b, Will et al. 2015; Steele et al. 2016) and where dated exceed 70 ka.

The termination of the Still Bay at rock-shelter sites across southern Africa is commonly marked by an occupational hiatus (Jacobs et al. 2008a, b; Vogelsang et al. 2010; Högberg and Larsson 2011). Diepkloof in the WRZ is an exception, with apparently continuous occupation after the end of the Still Bay, into a period characterized by the production of small blades, backed artifacts, and pièces esquillées (Igreja and Porraz 2013; Porraz et al. 2013b). Porraz and colleagues refer to this industry as "early Howiesons Poort" as it shares many of the characteristics of the widespread later industry. An apparently similar early Howiesons Poort industry is also known from Pinnacle Point 5/6 in the YRZ (Brown et al. 2012). Both sites have OSL ages for the start of this industry at around 70 ka, though Diepkloof also has ages >100 ka for the same layers produced by a different team of analysts (Jacobs et al. 2008a; Tribolo et al. 2012). From 65 ka, "classic" Howiesons Poort assemblages are in evidence at

Fig. 2.2 Distinctive artifacts from the study period. Panel (**a**) Still Bay bifacial points (all from Hollow Rock Shelter); (**b**) cores, blades, backed artifacts, and notched flakes from the Howiesons Poort (from Klein Kliphuis, Putslaagte 8, and Varsche Rivier 003); (**c**) cores, retouched flakes, and unifacial points from the post-Howiesons Poort (all from Klein Kliphuis). Scale bars are 10 mm

numerous sites across southern Africa. These assemblages are marked by the production of blades, notched flakes, and backed artifacts (Fig. 2.2) and also, in many instances, by extremely dense accumulations of flaked stone (Volman 1980; Singer and Wymer 1982; Jacobs et al. 2008a; Mackay 2010; Vogelsang et al. 2010; Porraz et al. 2013a, b). In the WRZ, the classic or later Howiesons Poort is notable for the preferential selection of fine-grained rocks and specifically silcrete. Various researchers have argued for the Howiesons Poort as a time of, alternatively, increased residential mobility (Ambrose and Lorenz 1990; Porraz et al. 2013a) and decreased residential mobility (Mackay 2009; McCall and Thomas 2012) in response to the cool, humid conditions suggested by environmental archives (Chase 2010).

The Howiesons Poort terminates around the MIS 4/3 transition ~58–60 ka at most sites (Jacobs et al. 2008b; Piennar et al. 2008; Guerin et al. 2013). Various proxies suggest increased temperatures and decreased humidity at this time (Chase 2010). Technologically, blade production fades, and backed artifacts are replaced by unifacial points (morphologically equivalent to convergent scrapers) as the dominant implement type (Fig. 2.2) (Volman 1980; Soriano et al. 2007; Villa et al. 2010; Mackay 2011; Conard et al. 2012; Porraz et al. 2013b). In the WRZ, these "post-Howiesons Poort" assemblages also witness decreases both in artifact discard and in the prevalence of silcrete.

The archaeological signal in the WRZ becomes increasingly muted after 50 ka (Mackay 2010; Faith 2013). Few sites have ages in this range, and those that do are often associated with only ephemeral accumulations of flaked stone. The causes of this quietude are presently not well understood. Notably, it is not matched by a similar occupational absence in the SRZ where signals are relatively robust (Mackay et al. 2014a). After ~25 ka, the occupational signal in the region strengthens once again, but by this time, the MSA has ended and Later Stone Age (LSA) technologies dominate (Deacon 1978; Deacon et al. 1984; Orton 2006; Mackay 2010).

Provisioning, Mobility, and Population

Provisioning can be defined as the means by which hunter-gatherers overcame the spatial and temporal mismatch between opportunities to make stone tools and locations at which stone tools were deployed. Kuhn (1995) differentiates two idealized forms of provisioning: provisioning of individuals and provisioning of places. Provisioning of individuals involves foragers equipping themselves with tools sufficient to underwrite tasks encountered before the next opportunity to retool. Provisioning of places involves foragers equipping landscape nodes with tool stone, turning them into sources, and allowing them to function as gearing-up locations.

The influences on these provisioning systems, and their archaeological manifestations, are expected to be different. Individual provisioning can be deployed when impending tasks are difficult to anticipate, providing maintainable gear of general utility (Nelson 1991). In contrast to individual provisioning, place provisioning can function where extended or repeated occupation of a given location can be anticipated (Kuhn 1995) and therefore presumes some foreknowledge of the nature and duration of impending activities. Necessarily then, place provisioning is predicated on predictable (though not necessarily abundant) subsistence conditions. Where individual provisioning might emphasize the transport, use, and conservation of a small number of largely utilitarian items in maintainable toolkits (Shott 1986; Kuhn 1994), place provisioning is expected to result in on-site reduction of transported rocks – most likely cores – and the consequent accumulation of manufacturing debris at residential bases (Binford 1980; Nelson 1991; Kuhn 1995; Riel-Salvatore and Barton 2004; Mackay 2005).

A key difference between these provisioning systems is in the way in which gearing up occurs. Place provisioning allows groups to gear up at selected locations – presumably central places in a foraging range. Provisioned places can then function as nodes for logistical forays into the surrounding landscape. The viability of such an approach, as noted above, depends on the predictability of subsistence resources, and particularly water, at that occupational node (Kelly 1983, 1995; Read 2008). In the absence of place provisioning, however, gearing up needs to have occurred on the landscape at points where the foraging round intersected appropriate sources of usable stone.

While there has so far been little discussion of provisioning systems in the Late Pleistocene of southern Africa, there has been some discussion of changing patterns of mobility (Ambrose and Lorenz 1990; McCall 2007; McCall and Thomas 2012). Perhaps the most influential of these is the Ambrose and Lorenz (1990) model of mobility during the Howiesons Poort. In that model, cool and dry conditions presumed to be associated with the glacial MIS 4 were inferred to have resulted in decreased resource abundance. As a consequence, this increased the distance and frequency of movements required to fulfill minimum subsistence needs. In alternative arguments, Mackay (2009) and McCall and Thomas (2012) have argued that the Howiesons Poort was a time in which logistical movements from residential bases were emphasized.

The relative abundance of artifacts in Howiesons Poort assemblages, coupled with the apparent humidity of that time (Chase 2010), seem to argue against the Ambrose and Lorenz (1990) model of constant long-range population movement; however, it is still viable if there was significant change in a third factor – population. Recent arguments have attempted to link the perceived increases in technological and behavioral complexity during the Still Bay and Howiesons Poort periods to population increases across southern Africa (Powell et al. 2009). The absence of direct evidence for population increase in the Howiesons Poort notwithstanding (Mackay 2011), and allowing that only the Howiesons Poort seems to be distinctively abundant, it is possible to reconcile increases in mobility and relative abundance during the Howiesons Poort if there were dramatic increases in the number of individuals on the landscape at this time. Under these conditions, while people may have been moving further and more often, the increased population could conceivably have resulted in greater accumulations of artifacts given a sufficient number of visits to sites, however brief. (It is unclear why population increase should align with a period of resource duress in this scenario, but this is not important with respect to testing the model).

Taken together, these various principles and models allow us to develop some expectations for changes in provisioning systems through the Late Pleistocene in southern Africa. First, if humid conditions were those most favorable to place provisioning, then this is most likely to have occurred during the mid to late MIS 4 (following Stuut et al. 2002; Chase 2010). We would expect to see these periods associated with increases in numbers of artifacts and specifically cores, as tool-making potential was transported to specific sites. Assuming direct-to-site transport of cores, rather than their transportation and maintenance as part of a mobile toolkit, we would expect to see multiple stages of reduction on-site, potentially including initial or early decortication. Gearing up would also occur at these locations, and this may have resulted in the production and discard of complex tools if there were sufficient levels of subsistence risk (Bousman 1993; Collard et al. 2005; Read 2008). The other side of this inference is that beyond the selected landscape "nodes" which were provisioned, we would see little evidence of gearing up and consequently fewer sites around the landscape generally. Sites beyond the nodes in these periods would be represented by isolated artifacts from episodic discard events.

Second, if less humid conditions were less suited to place provisioning and instead encouraged a greater reliance on individual provisioning, then we would expect to see this during late MIS 5/early MIS 4 and potentially during MIS 3. Gearing up would be more likely to occur at landscape locations where tool stone was directly available. Otherwise, sites would show relatively small accumulations of artifacts with fewer cores and potentially more tools.

Third, if human populations were abundant, and regardless of the provisioning systems used, we might reasonably expect an abundance of archaeological material at the landscape scale. In effect, the entire signal of such periods should be particularly "loud" relative to others.

Study Area

Our study area for this paper is centered on the northern Cederberg mountains situated in the heart of the WRZ in the Western Cape Province of South Africa (Fig. 2.1). Rainfall arrives predominantly from northwesterly winds during winter storms, and there is little summer rain. In the middle of the study area is the Olifants River valley located between the low coastal mountain range to the west and the Cederberg mountains which rise to >2000 masl to the east. The Cederberg and coastal mountains are composed mainly of deeply incised Table Mountain Sandstone (Visser and Theron 1973). The Olifants River is the only permanently flowing watercourse in the study area. The river is presently dammed in the area of Clanwilliam, creating a 10-km-long body of standing water, the volume of which exhibits marked seasonal fluctuations. Further to the west is the Sandveld, a thick coastal sandsheet falling between the coastal range and the sea. The Sandveld is punctuated by sandstone ridges which are often widely spaced. Drainages in this area tend to be relatively ephemeral with short catchments rising to the west of the coastal range.

Rocks suitable for the manufacture of flaked stone artifacts show a variable distribution across the landscape. Quartzite and quartz are relatively abundant and widely distributed. Quartzite occurs as massive geological bands and associated scree and as river cobbles in the Olifants River. Quartz occurs as small pebbles eroding from conglomerate beds and is dispersed across land surfaces. The fine-grained sedimentary rock silcrete is less common than the other major rock types and tends to occur as discrete, infrequent outcrops (Roberts 2003; Porraz et al. 2013b). Cobbles of silcrete have not been observed in the Olifants River during surveys.

The archaeological sites for this study include both excavated rock-shelter deposits and numerous open artifact scatters. The two main rock-shelters are Klein Kliphuis in the Olifants River valley and Diepkloof in the Sandveld. Both sites have later Howiesons Poort and post-Howiesons Poort components, while Diepkloof additionally has an earlier Howiesons Poort and a Still Bay industry (Mackay 2009, 2010; Porraz et al. 2013b). The Diepkloof data presented here derive from the L6 column sequence (Mackay 2009).

Surveys for open sites have generally not been systematic with the exception of those conducted by one of us (EH) in the area surrounding Clanwilliam Dam (Hallinan 2013). In those surveys, temporally diagnostic artifact markers – bifacial points for the Still Bay, backed artifacts for the Howiesons Poort, and unifacial points for the post-Howiesons Poort – were used to characterize distributions for various industries rather than "sites" per se (see discussion in Hallinan 2013). In addition to these surveys, we will present some artifact distribution data from locations around other excavated contexts in the region, specifically around Varsche Rivier 003 immediately to the north of the present study area and along the Doring River to the east (Steele et al. 2012).

Methods

Our analytic methods for shelter sites largely follow Riel-Salvatore and Barton (2004) in using artifact density and proportions of retouched flakes to assess place and individual provisioning. Density of artifacts per unit volume of deposit is not presently available for Diepkloof, and artifact discard per unit time is complicated by the contested OSL ages. Consequently, we use numbers of artifacts per industry as a basic measure of discard and supplement this with discard per unit time within the available age constraints.

Caveats aside, we expect that where place provisioning has occurred, large numbers of flakes and cores will have been discarded with relatively low proportions of retouched flakes. The opposite should hold where individual provisioning dominates. The proportion of cores to retouched flakes should thus provide an approximation of shifts between place and individual provisioning.

In addition to these data, we also consider evidence for cortex proportions, particularly in the silcrete that is not locally available at either of the shelters under consideration. If cores were transported and reduced as part of toolkits, then we would expect to see evidence for reduction consistent with spatial distance to source. That is, there should be little cortex on silcrete flakes. Alternatively, if place provisioning occurred and cores were transported more or less directly from source to site, then distance to source should less effectively predict cortex prevalence. In that sense, it is pertinent to remember that spatial distance to source is relevant only insofar as it is a proxy for time. Where spatial distance to source is held constant, direct movements from source to discard location are expected to produce less reduction than where movements in the intervening space are tortuous.

For the open sites, our interest is in the distribution and abundance of implements relative to stone sources. We use the term "implement" in this sense to refer to morphologically regular retouched flakes and more specifically those that are distinctive of different industries, here, bifacial points, backed artifacts, and unifacial points. If individual provisioning dominates and assuming that, in any given industry, multiple implements are being made at procurement opportunities, we would expect to see clustering of these implements at or near sources of stone. Where place provisioning dominates, implements should cluster at selected and presumably central foraging nodes that are not predicted by the distribution of stone sources.

Results

We start our results by looking at the data from rock-shelter sequences and move on to open site distribution data.

Discard Rates, Cores, and Retouched Flakes

At Diepkloof and Klein Kliphuis, artifact numbers are exceptionally large in the later Howiesons Poort, with weaker discard during the Still Bay and post-Howiesons Poort and moderate discard during the earlier Howiesons Poort (Table 2.1). As noted, these data are based on discard per industry as opposed to discard per unit time. We can factor for time if we allow for multiple potential outcomes given the

Table 2.1 Numbers of artifacts and discard rates per industry for Diepkloof and Klein Kliphuis calculated using the Jacobs et al. (2008a, b) and Tribolo et al. (2012) chronologies separately

			Duration (kyr) (central age range)		Discard rate (artifacts / kyr)			
Site	Industry	*n* artifacts	Jacobs et al. (2008a, b)	Tribolo et al. 2013	Jacobs max	Jacobs central	Tribolo max	Tribolo central
DRS	Still Bay	1519	7–3	29–9	217	506	52	169
	Early HP	2852	12–8	39–20	238	357	73	143
	Later HP	9604	7–3	47–33	1372	3201	204	291
	Post-HP	614	11–8		56	77		
KKH	Later HP	26139	12–6		2178	4357		
	Post-HP	5060	7–3		723	1687		

Note that no discard rates are calculated for the post-Howiesons Poort at Diepkloof using the Tribolo et al. (2012) chronology because only a single central age is given and the duration of the industry cannot be ascertained

Table 2.2 Proportions of cores to retouched flakes and adjusted residuals, for all industries at Diepkloof and Klein Kliphuis

		Cores		Retouched flakes		
Site	Industry	*n*	Adj. res.	*n*	Adj. res.	Cores/Retouched flakes
DRS	SB	35	−3.4	77	3.4	0.45
	Early HP	54	2.1	41	−2.1	1.21
	Later HP	132	0.9	139	−0.9	0.95
	Post-HP	13	0.4	13	−0.4	1
KKH	Later HP	267	1.9	274	−1.9	0.97
	Post-HP	52	−3.0	97	3.0	0.53

discrepancies between the Jacobs et al. and Tribolo et al. chronologies and their inherent uncertainties (Table 2.1). The dates suggest that, regardless of the chronology used or whether discard rates are calculated for the maximum spread at one sigma error or just for the central ages, the later Howiesons Poort always returns the highest values. At Klein Kliphuis, discard rates are typically three times higher in the Howiesons Poort than the post-Howiesons Poort. At Diepkloof, later Howiesons Poort discard rates are roughly double those seen in the earlier Howiesons Poort and Still Bay and roughly 25–40 times higher than in the post-Howiesons Poort, albeit the post-Howiesons Poort in this part of the site may be truncated (Porraz et al. 2013b).

Core and retouch numbers show similar patterns in terms of absolute quantities to total artifact numbers; however, their relative proportions are quite variable (Table 2.2). Retouched flakes are proportionally most common in the Still Bay at Diepkloof and in the post-Howiesons Poort at Klein Kliphuis, while cores are most common in the early Howiesons Poort at Diepkloof. A χ^2 test for variance across both sites is significant at 0.05 ($p < 0.001$, df = 5, Cramer's $V = 0.145$). Adjusted residuals suggest that the variance is being driven principally by the oversupply of cores in the early Howiesons Poort and the oversupply of retouched flakes in the Still Bay (at Diepkloof) and post-Howiesons Poort (at Klein Kliphuis). With respect to the post-Howiesons Poort at Diepkloof, it should be noted that this has the smallest sample of any of the units considered.

It should also be noted that the Howiesons Poort industries (both early and later) include numerous backed artifacts which were probably not maintained tools; thus, even though retouched flake proportions are relatively low in these layers, they probably still overestimate the proportion of maintained tools in the assemblage.

Raw Material Proportions and Cortex

Considerable changes occur in the proportions of different rock types through the sequences (Table 2.3); only the major rock types - quartz, quartzite and silcrete - are included, as these combined account for more than 90% of artifacts in all units.

Table 2.3 Relative proportions of the major raw material types per industry for Diepkloof and Klein Kliphuis, artifacts >15 mm only

		Quartz		Quartzite		Silcrete	
Site	Industry	%	n	%	n	%	n
DRS	SB	18.3	178	54.9	533	26.8	260
	Early HP	49.6	555	38.2	427	12.2	137
	Later HP	32.5	745	16.4	377	51.1	1173
	Post-HP	18.6	41	10.0	22	71.5	158
KKH	Later HP	8.4	411	5.9	288	85.7	4200
	Post-HP	8.7	154	28.6	505	62.7	1109

Both sites have high proportions of silcrete in the later Howiesons Poort and post-Howiesons Poort. At Klein Kliphuis, silcrete peaks in the former; at Diepkloof, it peaks in the latter. The abundance of silcrete at both sites in the later Howiesons Poort industry, which also has the largest numbers of artifacts and cores, suggests that large quantities of that material, though likely not available in the immediate surrounds of the site, were being transported to sites in these periods.

Silcrete cores are also particularly common in these industries. There are 133 complete silcrete cores in the later Howiesons Poort at Klein Kliphuis, compared with 20 in the post-Howiesons Poort. At Diepkloof, there are 57 silcrete cores in the later Howiesons Poort; the remaining industries combined have only 16. Thus, there is a lot of silcrete, particularly cores, being transported to sites in the later Howiesons Poort. The remaining question is whether silcrete cores were being reduced in a way that suggests extensive transport and maintenance prior to arrival on-site.

Cortex data suggest that silcrete flakes with >50% cortical coverage on their dorsal surface are more common in Howiesons Poort than other units at both sites, though the patterns are statistically weak ($\chi^2_{\text{Diepkloof}}$: $p = 0.87$, df = 3, Cramer's $V = 0.113$; $\chi^2_{\text{KleinKiphuis}}$: $p = 0.81$, df = 1, Cramer's $V = 0.041$)[1]. Variance at Diepkloof is driven chiefly by the absence of cortical silcrete flakes in the Still Bay and their comparative abundance in the later Howiesons Poort. While typical "Howiesons Poort" cores (cf., Villa et al. 2010) retain a cortical lower surface throughout their reduction, flakes with 50% cortex likely relate to initial core setup and ongoing maintenance (Porraz et al. 2013b). Thus, while large numbers of silcrete cores were being transported to both sites in the Howiesons Poort, they appear often to have been in sufficiently early stages of reduction to have produced highly cortical flakes. This observation is not easily reconciled with consistently extensive reduction of cores prior to arrival on-site (Table 2.4).

While the later Howiesons Poort and Still Bay seem to have relatively clear signals with respect to provisioning systems, the post-Howiesons Poort and early Howiesons Poort are perhaps more enigmatic. The post-Howiesons Poort has very high percentages of silcrete in the context of systems that seem otherwise consistent

[1] These data are generated by site rather than for the whole sample given differences in the raw material availability in the Sandveld and in the Olifants River discussed earlier, where all data are considered together $p = 0.001$, df = 5, and Cramer's V = 0.093.

Table 2.4 Proportions of silcrete flakes with >50% and <50% cortex and adjusted residuals, for all industries at Diepkloof and Klein Kliphuis

		Cortex < = 50			Cortex > 50		
Site	Industry	%	*n*	Adj. res.	%	*n*	Adj. res.
DRS	SB	100	79	2.1	0	0	−2.1
	Early HP	95.7	45	0.1	4.3	2	0.1
	Later HP	93.8	316	−2.3	6.2	21	2.3
	Post-HP	98.0	50	1.0	2.0	1	−1.0
KKH	Later HP	90.2	1329	−1.7	9.8	145	1.7
	Post-HP	93.2	316	1.7	6.8	23	−1.7

with individual provisioning. In the early Howiesons Poort, cores are relatively common but artifact discard rates are moderate. Proportions of retouched flakes indicate low frequencies of flake maintenance, but silcrete artifacts are uncommon and silcrete cores – which might be taken to indicate place provisioning with high-quality rocks – are rare. The most abundant rocks in this industry are locally available quartz and quartzite. Based on this, the provisioning possibilities that arise are either episodic place provisioning with local rocks or individual provisioning principally with maintained cores. We are unable to resolve these possibilities with the available data.

Open Site Data

So far, the data presented suggest that the later Howiesons Poort has a particularly strong archaeological signal in the study area as measured by the abundance and discard rates of artifacts and that this may be a result of place provisioning. A further test of this proposition comes from the distribution of open sites. If the later Howiesons Poort is a consequence of place provisioning to allow gearing up at specific locations, then we may or may not find open site expressions of that industry. On the other hand, if the abundance of the later Howiesons Poort is simply a product of massive increases in the number of people using the landscape, then we would expect to find the signal of the later Howiesons Poort widely distributed in open site contexts. Furthermore, if the Still Bay and post-Howiesons Poort industries reflect individual provisioning, we expect to find evidence of that in gearing-up sites near sources of selected tool stone.

Clanwilliam Dam Surveys

Extensive surveys in the area surrounding Clanwilliam Dam were conducted by one of us (EH) with assistance from students at the University of Cape Town. These surveys resulted in the location of artifacts dating to the Earlier, Middle, and Later

Stone Ages. The MSA component included bifacial points likely indicative of the Still Bay, backed artifacts that may relate to the Howiesons Poort[2], and unifacial points likely related to the post-Howiesons Poort.

Bifacial points were identified at ten localities, distributed throughout the landscape (Fig. 2.3a). Most locations had between one and three points, but one site, known as Clanwilliam Dam East, has at least 50. Clanwilliam Dam East occurs in a small embayment on the edge of the dam where seasonal fluctuations in the water level have eroded a shallow body of sediment, which is exposed during the dry summer months and submerged through most of winter. The 50 points so far located represent a minimum for the site. Of the 37 analyzed points, ~95% are made of silcrete and ~46% of these retain cortex. The assemblage includes both heavily reduced and recycled points and those in the earliest stages of manufacture (Fig. 2.4). While no silcrete source has been identified at the site, a proximate source cannot be precluded due to the dam, which forms the western boundary of the site.

A single backed artifact was located during the surveys (Fig. 2.3b). The artifact was made of gray-green silcrete and measured 33 mm, making it comparable in size to the later Howiesons Poort backed artifacts from Klein Kliphuis (Mackay 2011). Interestingly, the artifact was found in the same embayment as the bifacial points (Fig. 2.4). Given the overlap of bifacial points and backed artifacts in the latest Still Bay phase at Diepkloof (cf., Porraz et al. 2013b), it is possible that this backed piece is not in fact Howiesons Poort. This aside, the surveys revealed no evidence for the concerted production of backed artifacts at any open site location around Clanwilliam Dam.

Unifacial points, like bifacial points, were widely distributed in the survey area (Fig. 2.3c). A total of 26 unifacial points was identified, spread over 9 localities. Three sites had five, three, and two points, respectively, and five were found as single points within larger assemblages. At Clanwilliam Dam East, 11 unifacial points were recorded, but given their context alongside a substantial number of bifacial points, these may represent a stage in bifacial point manufacture (Högberg and Larsson 2011). As with backed artifacts, there was no evidence for the concerted production of these implements at any location.

The Clanwilliam Dam survey data reveal some interesting patterns but they have limitations. The presence of the dam precludes survey of the margins of the Olifants River in this area, meaning that certain, specific locations remain unexplored. We can to some extent assess the validity of the patterns found there by considering partial survey data from other areas.

[2] Two points are made here. First, it is not presently possible to differentiate the earlier and later phases of the Howiesons Poort solely on the basis of implement form, and the two are therefore considered together here. Second, backed artifacts do recur in the LSA, but these are commonly smaller than their MSA counterparts. We return to this point later in this section.

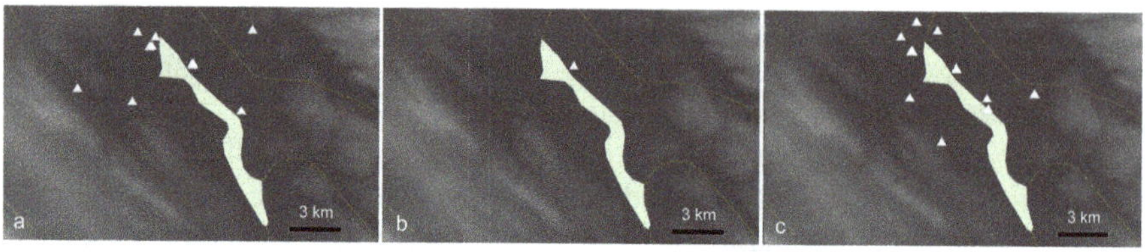

Fig. 2.3 Triangles showing the distribution of (**a**) bifacial points, (**b**) backed artifacts, and (**c**) unifacial points around Clanwilliam Dam

Fig. 2.4 Clanwilliam Dam East, looking east. *Black oval* shows the location of the concentration of bifacial points. Points on the right-hand side include early-stage (*bottom half*) and late-stage/reworked points (*top half*). The uppermost point image shows both faces of a recycled point which had been snapped at one end and subsequently flaked to remove blades from the margins

Doring River Surveys

To the east of the Olifants River valley is the Doring River. The two rivers run parallel for much of their course before merging and flowing to the sea together as the Olifants. The Doring River has remnant Late Pleistocene terraces along more than 50 km of its course, and these are often covered in MSA artifacts (Mackay et al. 2014b). Surveys so far have resulted in the identification of bifacial points at eight localities (Fig. 2.5a). These include two instances of isolated points, two instances of two points together, three instances of multiple points (six, six, and eight points), and one instance of >140 bifacial points of which 136 have been analyzed. This large assemblage of points occurs as a spatially discrete component of a much larger MSA scatter known as Kleinhoek 1, situated on a large terrace (Fig. 2.6). More than 95% of these points are made of relatively fine-grained blue-gray quartzite which outcrops on the ridge and scree slope that forms the western boundary of the site. The assemblage includes early-stage and late-stage points.

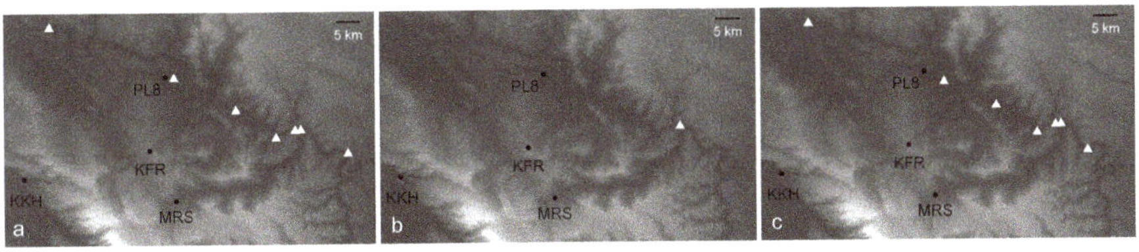

Fig. 2.5 Triangles showing the distribution of (**a**) bifacial point sites, (**b**) backed artifact sites, and (**c**) unifacial point sites along the Doring River. Rock-shelters (*circles*) are *PL8* Putslaagte 8, *KFR* Klipfonteinrand, and *MRS* Mertenhof

Fig. 2.6 Kleinhoek 1, looking northeast. The scatter covers the entire erosion feature; however, all of the ~140 points are constrained to the area covered by the *black oval*. The points on the right-hand side include examples of early-stage points (*bottom half*) and late-stage/reworked points (*top half*). Scale bars are 10 mm. Note that the scree slope at the bottom of the site photo includes the same raw material of which the points are made

So far, we have only identified two backed artifacts, both at a single locality – Uitspankraal 7 (Fig. 2.5b). These artifacts were both identified as isolated finds during repeated nonsystematic surface sampling of the site.

Unifacial points have been identified at seven localities, five of which also have bifacial points (Fig. 2.5c). Four included single, isolated points. Of the remainder, one included two unifacial points, and another included four points though these were in the same cluster as the bifacial points at Kleinhoek 1 and may therefore relate to bifacial point manufacture (Högberg and Larsson 2011). The final site, Uitspankraal 7, like Kleinhoek 1, is a large terrace with multiple archaeological components including MSA and LSA artifacts. In the center of the scatter is a dense concentration of heated silcrete cores and flakes and 18 unifacial points which we believe to be associated with the post-Howiesons Poort (Will et al. 2015). While there are also bifacial points at this site, they are located in a different area of the extensive scatter, and their distributions do not appear to overlap. Another similarly dense post-Howiesons Poort scatter has also been located on the Tankwa River – a tributary of the Doring River – a further 40 km southeast of Uitspankraal 7 under a different survey program (Hallinan and Shaw 2015).

The 16 artifact-bearing localities on terraces so far identified along the Doring River have yielded more than 160 bifacial points and 28 unifacial points, with only two backed artifacts. In spite of this, three of the four excavated rock-shelter sites in the Doring River catchment have Howiesons Poort components (Klipfonteinrand, Mertenhof, Putslaagte 8), and only two each have the Still Bay and post-Howiesons Poort (Högberg and Larsson 2011; Mackay et al. 2015; Will et al. 2015). Consistent with the patterns elsewhere, in the only site to contain all three of these industries (Mertenhof), the density of artifacts is greatest in the Howiesons Poort (Will et al. 2015).

Knersvlakte Surveys

The final open site data come from around the excavated site of Varsche River 003 (Steele et al. 2012; Steele et al. 2016), located in the Knersvlakte region of Namaqualand. Varsche River 003 includes Still Bay and Howiesons Poort industries but as yet no post-Howiesons Poort. Limited surveys were undertaken around the site during excavation, and full details will be published elsewhere. We note here only that we recorded bifacial points at five sites. Three of these were isolated, and one site included three points. At the fifth site, we identified 60 bifacial points during our initial analysis (Mackay et al. 2010) and have subsequently expanded this sample to 142, with three unifacial points. The site – Soutfontein – occurs on the banks of the Varsche River immediately below a large outcrop of quartz. Quartz accounts for ~67% of points. No other unifacial points have been identified in the surrounding area and no backed artifacts.

Discussion

The data presented here depict fairly clear patterns with respect to provisioning in some industries and less clear patterns in others. It seems probable that the commonly noted density of artifacts in the later Howiesons Poort is a consequence of place provisioning in the context of cool and humid conditions of mid to late MIS 4. Large numbers of cores from spatially restricted sources were transported to shelter sites in this industry and subsequently reduced from an initial state in which they still retained considerable cortical coverage. The fact that many of these retouched pieces were not maintainable suggests that they would have been suited to logistical and possibly task-specific trips (McCall 2007; Mackay 2009; McCall and Thomas

2012). In support of the artifact data, geoarchaeological evidence for site use suggests extended periods of fairly intensive occupation during the later Howiesons Poort (Goldberg et al. 2009; Miller et al. 2013; Karkanas et al. 2015), consistent with the use of sites as residential bases.

The near-complete absence of evidence for Howiesons Poort sites on the landscape provides important additional information. Site occupation was clearly very specific at this time, with gearing up occurring mainly if not exclusively at selected sites away from stone sources. Even though backed artifacts are small and perhaps more easily missed than the other implement types on which we focused, had there been instances of substantial backed artifact production sites in the open comparable to those in shelters, we expect we would have seen them. While other researchers have found Howiesons Poort backed artifacts in dunefield settings under perhaps more ideal survey conditions, it is notable that they too have failed to find clear open-air Howiesons Poort manufacturing sites (e.g., Dietl et al. 2005; Kandel and Conard 2012) (though note Carrion et al. 2000). Overall, it seems to us improbable that the signal we are witnessing is the result of very large numbers of highly mobile people moving across the landscape.

The Still Bay inverts many of these outcomes. Discard rates are generally low, cores are rare, and the proportion of retouched flakes is high. The dominant implement form – bifacial points – could be characterized as maintainable. While there was a reasonable amount of silcrete transported to Diepkloof in this period, the paucity of cortical flakes combined with the shortage of cores argues against place provisioning. We also found clear evidence for dedicated gearing-up locations out on the landscape. At least two such instances were in close proximity to sources of stone that comprised the majority of assemblages, and this could not be precluded at the third (Clanwilliam Dam East). Given that the Clanwilliam Dam East assemblage was dominated by a single stone type, included a mix of heavily reduced and early-stage points, and had a reasonably high proportion of points on which cortex was still visible, it is likely that a source of stone was located fairly close to the site. Overall, we believe that the Still Bay signal is consistent with gearing up at sources of stone to provision individuals with maintainable implements. This system of organization occurred in the context of rapidly cooling and perhaps increasingly humid conditions from MIS 5 to MIS 4 (accepting the Jacobs chronology) or under variable conditions of earlier MIS 5 (accepting the Tribolo chronology with its proportionally large uncertainties).

The earlier Howiesons Poort presents a less clear signal with the use of local rocks, moderate rates of discard, reasonable numbers of cores, and low proportions of retouched flakes, among which the dominant implement type is not considered maintainable. Either limited/episodic place provisioning with local rock or individual provisioning with a combination of cores and tools – or perhaps variable use of both – may explain this pattern. This issue requires further consideration. The Jacobs chronology places this industry in the context of cool and humid conditions of early to mid MIS 4. In the Epica Dome C southern hemisphere temperature record, this is the coldest period of that stage (Jouzel et al. 2007).

The post-Howiesons Poort, like the earlier Howiesons Poort, has a somewhat mixed signal. Most of its characteristics in shelter sites – low discard rates, few cores, high proportions of retouched flakes, and maintainable implements – suggest individual provisioning. However, this needs to be balanced against the abundance of silcrete which is not local to either of the sites studied. As Binford (1979) notes, where individuals are provisioned with a limited amount of transported gear, we expect them to make supplementary use of locally available rocks; yet the evidence for this at Diepkloof and Klein Kliphuis is weak. Furthermore, the clearest post-Howiesons Poort site identified on the landscape appears in many respects to reflect place provisioning. At Uitspankraal 7, we see preferential transportation and reduction of cores of nonlocal rock at a specific landscape location where multiple implements were made.

We suspect that the solution to this puzzle lies in the coarse nature of industries and their complex relationship with behavior. An industry is generally identified based on a limited number of artifact characteristics. In the case of our open site data, it was one variable alone that we used to determine industry affinity. Yet technologies are complex mixtures of mobility, provisioning, material selection, core reduction and discard, and implement production. These variables may change at different rates and in relation to different controls. The Howiesons Poort to post-Howiesons Poort transition, for example, has been shown to involve gradual decreases in blade production and increases in flake size (Soriano et al. 2007; Villa et al. 2010; Mackay 2011). In contrast, the change from backed artifacts to unifacial points seems to have been relatively rapid. At Klein Kliphuis, artifact densities remain quite high in the earliest post-Howiesons Poort (Mackay 2010). These are also the layers in which silcrete is abundant; thereafter, quartzite and eventually quartz come to dominate. Cores are also common in the earliest post-Howiesons Poort, but what is notable is that their size at discard increases through time from the later Howiesons Poort into the earlier post-Howiesons Poort before cores become altogether infrequent. It may be that while other elements of technology were changing across the Howiesons Poort to post-Howiesons Poort transition, the nature of provisioning initially remained much the same – that of large quantities of silcrete transported to site, sometimes as cores. However, the return on this provisioning investment, measured here as the reduction of cores prior to discard, may have begun to decrease as increased aridity resulted in shorter and perhaps less predictable durations of site use. Thus, the change from place to individual provisioning may have occurred some time after the transition from the Howiesons Poort to post-Howiesons Poort, resulting in the mixed provisioning signal for the later industry – something suggested by the core to retouched flake ratios in the earliest post-Howiesons Poort at Diepkloof. Viewed in this light, the open site post-Howiesons Poort occurrence at Uitspankraal 7 may be seen to document the continuation of place provisioning in the earliest post-Howiesons Poort but a spatial shift from selected and presumably preferred landscape locations in rock-shelters to a better-watered location on the Doring River. The value of the post-Howiesons Poort here may be in cautioning us against making an oversimplistic association between industry and behavior (Mackay 2016).

Conclusions

The objective of this paper was to explore changes in provisioning through the later MSA in southern Africa's WRZ and to compare these changes to environmental variability. The latter goal is complicated by chronological uncertainties and the paucity of terrestrial paleoenvironmental archives. Nevertheless, we feel it has been possible to identify some clear changes in provisioning and, where chronologies allow, to suggest that place provisioning occurred most notably in the context of cool humid conditions during the later Howiesons Poort. During the Still Bay, though environmental conditions are poorly resolved, we feel confident that individual provisioning dominated, with gearing up occurring at raw material sources. Provisioning in the other industries was less easily resolved, but we posit that at least part of our difficulty here arose from the concept of industries and their capacity to mask behaviorally meaningful variation.

Beyond provisioning, our paper also suggests that the noted abundance of archaeology during the later Howiesons Poort is biased by an occupational system focused on rock-shelters and an approach to archaeology which often does very much the same. Consideration of the rock-shelter record alone can thus be misleading. It seems quite likely that rock-shelters had specific conditions of use and that these conditions were only met periodically. An extension of this observation is that the lack of an observable archaeological record in the WRZ from 50 to 25 ka, on which others and we have commented (Mitchell 2008; Mackay 2010; Faith 2013), cannot necessarily be taken at face value. If occupation can focus on shelters to the exclusion of open sites under some conditions, it is plausible that the inverse could hold under others (Mackay et al. 2014b).

Acknowledgments AM's research has been supported by grants from the Australian Research Council (DP1092445, DE130100068). Fieldwork conducted by EH was funded by an NRF African Origins Platform grant awarded to John Parkington.

References

Ambrose, S. H. (2002). Small things remembered: Origins of early microlithic industries in sub-Saharan Africa. In R. G. Elston & S. L. Kuhn (Eds.), *Thinking small: Global perspectives on Microlithization, Archaeological Papers of the American Anthropological Association, no 12* (pp. 9–30). Arlington, VA: American Anthropological Association.

Ambrose, S. H., & Lorenz, K. G. (1990). Social and ecological models for the Middle Stone Age in Southern Africa. In P. Mellars (Ed.), *The emergence of modern humans* (pp. 3–33). Edinburgh: Edinburgh University Press.

Avery, G. D., Halkett, D., Orton, J., Steele, T. E., & Klein, R. G. (2008). The Ysterfontein 1 Middle Stone Age Rockshelter and the evolution of coastal foraging. *South African Archaeological Society Goodwin Series, 10*, 66–89.

Binford, L. R. (1979). Organization and formation processes: Looking at curated technologies. *Journal of Anthropological Research, 35*(3), 255–273.

Binford, L. R. (1980). Willow smoke and dogs tails: Hunter-gatherer settlements systems and archaeological site formation. *American Antiquity, 45*(1), 4–20.

Bousman, C. B. (1993). Hunter-gatherer adaptations, economic risk and tool design. *Lithic Technology, 18*(1&2), 59–86.

Brown, K. S., Marean, C. W., Herries, A. I. R., Jacobs, Z., Tribolo, C., Braun, D., Roberts, D. L., Meyer, M. C., & Bernatchez, J. (2009). Fire as an engineering tool of early modern humans. *Science, 325*(5942), 859–862.

Brown, K. S., Marean, C. W., Jacobs, Z., Schoville, B. J., Oestmo, S., Fisher, E. C., Bernatchez, J., Karkanas, P., & Matthews, T. (2012). An early and enduring advanced technology originating 71,000 years ago in South Africa. *Nature, 491*(7425), 590–593.

Bruch, A. A., Sievers, C., & Wadley, L. (2012). Quantification of climate and vegetation from Southern African Middle Stone Age sites: An application using late Pleistocene plant material from Sibudu, South Africa. *Quaternary Science Reviews, 45*, 7–17.

Carr, A. S., Chase, B. M., & Mackay, A. (2016). Mid to late quaternary landscape and environmental dynamics in the Middle Stone Age of Southern South Africa. In B. A. Stewart & S. Jones (Eds.), *Africa from MIS 6–2: Population dynamics and Paleoenvironments* (pp. 23–47). Dordrecht: Springer.

Carrion, J. S., Brink, J. S., Scott, L., & Binneman, J. N. F. (2000). Palynology and palaeoenvironment of Pleistocene hyaena coprolites from an open-air site at Oyster Bay. *South African Journal of Science, 96*, 449–453.

Cartwright, C. R. (2013). Identifying the woody resources of Diepkloof Rock Shelter (South Africa) using scanning electron microscopy of the MSA wood charcoal assemblages. *Journal of Archaeological Science, 40*(9), 3463–3474.

Chase, B. M. (2010). South African palaeoenvironments during marine oxygen isotope stage 4: A context for the Howiesons Poort and Still Bay industries. *Journal of Archaeological Science, 37*(6), 1359–1366.

Chase, B. M., & Meadows, M. E. (2007). Late quaternary dynamics of southern Africa's winter rainfall zone. *Earth-Science Reviews, 84*(3–4), 103–138.

Collard, M., Kemery, M., Banks, S., & S. (2005). Causes of toolkit variation among hunter-gatherers: A test of four competing hypotheses. *Canadian Journal of Archaeology, 29*, 1–19.

Conard, N. J., Porraz, G., & Wadley, L. (2012). What's in a name? Characterising the 'post-Howiesons Poort' at Sibudu. *South African Archaeological Bulletin, 67*(196), 180–199.

Cowling, R. M., Cartwright, C. R., Parkington, J. E., & Allsopp, J. C. (1999). Fossil wood charcoal assemblages from Elands Bay Cave, South Africa: Implications for late quaternary vegetation and climates in the winter-rainfall fynbos biome. *Journal of Biogeography, 26*, 367–378.

de la Pena, P., Wadley, L., & Lombard, M. (2013). Quartz bifacial points in the Howiesons Poort of Sibudu. *South African Archaeological Bulletin, 69*(198), 119–136.

Deacon, H. J., Deacon, J., Scholtz, A., Thackeray, J. F., Brink, J. S., & Vogel, J. C. (1984). Correlation of palaeoenvironmental data from the late Pleistocene and Holocene deposits at Boomplaas cave, southern cape. In J. C. Vogel (Ed.), *Late Cainozoic Palaeoclimates of the southern hemisphere* (pp. 339–352). Rotterdam: Balkema.

Deacon, J. (1978). Changing patterns in late Pleistocene/early Holocene prehistory in southern Africa as seen from the Nelson Bay Cave stone artifact sequence. *Quaternary Research, 10*, 84–111.

Dietl, H., Kandel, A. W., & Conard, N. J. (2005). Middle Stone Age settlement and land use at the open-air sites of Geelbek and Anyskop, South Africa. *Journal of African Archaeology, 3*(2), 231–242.

Faith, J. T. (2013). Taphonomic and paleoecological change in the large mammal sequence from Boomplaas Cave, Western Cape, South Africa. *Journal of Human Evolution, 65*(6), 715–730.

Goldberg, P., Miller, C. E., Schiegl, S., Ligouis, B., Berna, F., Conard, N. J., & Wadley, L. (2009). Bedding, hearths, and site maintenance in the Middle Stone Age of Sibudu Cave, KwaZulu-Natal, South Africa. *Archaeological and Anthropological Sciences, 1*(2), 95–122.

Guerin, G., Murray, A. S., Jain, M., Thomsen, K. J., & Mercier, N. (2013). How confident are we in the chronology of the transition between Howieson's Poort and Still Bay? *Journal of Human Evolution, 64*(4), 314–317.

Hallinan, E. (2013). *Stone age landscape use in the Olifants River Valley, Western Cape*. MPhil, University of Cape Town.

Hallinan, E., & Shaw, M. (2015). A new Middle Stone Age industry in the Tankwa Karoo, Northern Cape Province, South Africa. *Antiquity Project Gallery, 89*, 344.

Henshilwood, C. S., Sealy, J. C., Yates, R., Cruz-Uribe, K., Goldberg, P., Grine, F. E., Klein, R. G., Poggenpoel, C., van Niekerk, K., & Watts, I. (2001). Blombos Cave, Southern Cape, South Africa: Preliminary report on the 1992–1999 excavations of the Middle Stone Age levels. *Journal of Archaeological Science, 28*(4), 421–448.

Högberg, A., & Larsson, L. (2011). Lithic technology and behavioural modernity: New results from the Still Bay site, Hollow Rock Shelter, Western Cape Province, South Africa. *Journal of Human Evolution, 61*(2), 133–155.

Igreja, M., & Porraz, G. (2013). Functional insights into the innovative early Howiesons Poort technology at Diepkloof Rock Shelter (Western Cape, South Africa). *Journal of Archaeological Science, 40*(9), 3475–3491.

Jacobs, Z., Hayes, E. H., Roberts, R. G., Galbraith, R. F., & Henshilwood, C. S. (2013). An improved OSL chronology for the Still Bay layers at Blombos Cave, South Africa: Further tests of single-grain dating procedures and a re-evaluation of the timing of the Still Bay industry across southern Africa. *Journal of Archaeological Science, 40*(1), 579–594.

Jacobs, Z., Roberts, R. G., Galbraith, R. F., Deacon, H. J., Grun, R., Mackay, A., Mitchell, P., Vogelsang, R., & Wadley, L. (2008a). Ages for the Middle Stone Age of southern Africa: Implications for human behavior and dispersal. *Science, 322*(5902), 733–735.

Jacobs, Z., Wintle, A. G., Duller, G. A. T., Roberts, R. G., & Wadley, L. (2008b). New ages for the post-Howiesons Poort, late and final Middle Stone Age at Sibudu, South Africa. *Journal of Archaeological Science, 35*(7), 1790–1807.

Jouzel, J., Masson-Delmotte, V., Cattani, O., Dreyfus, G., Falourd, S., Hoffmann, G., Minster, B., Nouet, J., Barnola, J. M., Chappellaz, J., Fischer, H., Gallet, J. C., Johnsen, S., Leuenberger, M., Loulergue, L., Luethi, D., Oerter, H., Parrenin, F., Raisbeck, G., Raynaud, D., Schilt, A., Schwander, J., Selmo, E., Souchez, R., Spahni, R., Stauffer, B., Steffensen, J. P., Stenni, B., Stocker, T. F., Tison, J. L., Werner, M., & Wolff, E. W. (2007). Orbital and millennial Antarctic climate variability over the past 800,000 years. *Science, 317*(5839), 793–796.

Kandel, A. W., & Conard, N. J. (2012). Settlement patterns during the earlier and Middle Stone Age around Langebaan Lagoon, Western Cape (South Africa). *Quaternary International, 270*, 15–29.

Kaplan, J. (1990). The Umhlatuzana rock shelter sequence: 100 000 years of stone age history. *Natal Museum Journal of Humanities, 2*, 1–94.

Karkanas, P., Brown, K. S., Fisher, E. C., Jacobs, Z., & Marean, C. W. (2015). Interpreting human behavior from depositional rates and combustion features through the study of sedimentary microfacies at site Pinnacle Point 5-6, South Africa. *Journal of Human Evolution, 85*, 1–21.

Kelly, R. L. (1983). Hunter-gatherer mobility strategies. *Journal of Archaeological Research, 39*(3), 277–306.

Kelly, R. L. (1995). *The foraging spectrum: Diversity in hunter-gatherer Lifeways*. Washington D.C.: Smithsonian Institution Press.

Klein, R. G. (1976). The mammalian fauna of the Klasies River mouth sites, southern Cape Province, South Africa. *South African Archaeological Bulletin, 31*, 75–99.

Klein, R. G., Cruz-Uribe, K., Halkett, D., Hart, T., & Parkington, J. E. (1999). Paleoenvironmental and human behavioral implications of the Boegoeberg 1 late Pleistocene hyena den, Northern Cape Province, South Africa. *Quaternary Research, 52*, 393–403.

Kuhn, S. L. (1994). A formal approach to the design and assembly of mobile toolkits. *American Antiquity, 59*(3), 426–442.

Kuhn, S. L. (1995). *Mousterian lithic technology*. Princeton: Princeton University Press.

Mackay, A. (2005). Informal movements: Changing mobility patterns at Ngarrabullgan, Cape York Australia. In C. Clarkson & L. Lamb (Eds.), *Lithics down under: Recent Australian approaches to lithic reduction, use and classification* (pp. 95–108). Oxford: Archaeopress.

Mackay, A. (2009). *History and selection in the Late Pleistocene archaeology of the Western Cape, South Africa*. PhD, Australian National University.
Mackay, A. (2010). The late Pleistocene archaeology of Klein Kliphuis rockshelter, Western Cape, South Africa: 2006 excavations. *South African Archaeological Bulletin, 65*(192), 132–147.
Mackay, A. (2011). Nature and significance of the Howiesons Poort to post-Howiesons Poort transition at Klein Kliphuis rockshelter, South Africa. *Journal of Archaeological Science, 38*(7), 1430–1440.
Mackay, A. (2016). Technological change and the importance of variability: The Western Cape of South Africa from MIS 5 to MIS 2. In S. Jones & B. A. Stewart (Eds.), *Africa from MIS 6–2: Population dynamics and Paleoenvironments* (pp. 49–63). Dordrecht: Springer.
Mackay, A., Jacobs, Z., & Steele, T. E. (2015). Pleistocene archaeology and chronology of Putslaagte 8 (PL8) rockshelter, Western Cape, South Africa. *Journal of African Archaeology, 13*(1), 71–98.
Mackay, A., Orton, J., Schwortz, S., & Steele, T. E. (2010). Soutfontein (SFT)-001: Preliminary report on an open-air bifacial point-rich site in southern Namaqualand, South Africa. *South African Archaeological Bulletin, 65*, 84–95.
Mackay, A., Stewart, B. A., & Chase, B. M. (2014a). Coalescence and fragmentation in the late Pleistocene archaeology of southernmost Africa. *Journal of Human Evolution, 72*, 26–51.
Mackay, A., Sumner, A., Jacobs, Z., Marwick, B., Bluff, K., & Shaw, M. (2014b). Putslaagte 1 (PL1), the Doring River, and the later Middle Stone Age in southern Africa's Winter Rainfall Zone. *Quaternary International, 350*, 43–58.
McCall, G. S. (2007). Behavioral ecological models of lithic technological change during the later Middle Stone Age of South Africa. *Journal of Archaeological Science, 34*(10), 1738–1751.
McCall, G. S., & Thomas, J. T. (2012). Still Bay and Howiesons Poort foraging strategies: Recent research and models of culture change. *African Archaeological Review, 29*(1), 7–50.
Meadows, M. E., & Sugden, J. M. (1993). The late quaternary palaeoecology of a floristic kingdom: The southwestern Cape South Africa. *Palaeogeography, Palaeoclimatology, Palaeoecology, 101*, 271–281.
Miller, C. E., Goldberg, P., & Berna, F. (2013). Geoarchaeological investigations at Diepkloof Rock Shelter, Western Cape, South Africa. *Journal of Archaeological Science, 40*(9), 3432–3452.
Mills, S. C., Grab, S. W., Rea, B. R., Carr, S. J., & Farrow, A. (2012). Shifting westerlies and precipitation patterns during the late Pleistocene in southern Africa determined using glacier reconstruction and mass balance modelling. *Quaternary Science Reviews, 55*, 145–159.
Mitchell, P. J. (1988). Early microlithic assemblages of southern Africa. *World Archaeology, 20*(1), 27–39.
Mitchell, P. J. (2008). Developing the archaeology of marine isotope stage 3. *South African Archaeological Society Goodwin Series, 10*, 52–65.
Mourre, V., Villa, P., & Henshilwood, C. S. (2010). Early use of pressure flaking on lithic artifacts at Blombos Cave, South Africa. *Science, 330*(6004), 659–662.
Mucina, L., & Rutherford, M. C. (2006). *The vegetation of South Africa, Lesotho and Swaziland*. Pretoria: South African National Biodiversity Institute.
Nelson, M. (1991). The study of technological organization. *Archaeological Method and Theory, 3*, 57–100.
Nicholson, S. E., & Flohn, H. (1980). African environmental and climatic changes and the general atmospheric circulation in the late Pleistocene and Holocene. *Climatic Change, 2*, 313–348.
Orton, J. (2006). The Later Stone Age lithic sequence at Elands Bay, Western Cape, South Africa: Raw materials, artefacts and sporadic change. *Southern African Humanities, 18*(2), 1–28.
Piennar, M., Woodborne, S., & Wadley, L. (2008). Optically stimulated luminescence dating at rose cottage cave. *South African Journal of Science, 104*, 65–70.
Porraz, G., Parkington, J. E., Rigaud, J.-P., Miller, C. E., Poggenpoel, C., Tribolo, C., Archer, W., Cartwright, C. R., Charrié-Duhaut, A., Dayet, L., Igreja, M., Mercier, N., Schmidt, P., Verna, C., & Texier, P.-J. (2013a). The MSA sequence of Diepkloof and the history of southern African Late Pleistocene populations. *Journal of Archaeological Science, 40*(9), 3542–3552.

Porraz, G., Texier, P.-J., Archer, W., Piboule, M., Rigaud, J.-P., & Tribolo, C. (2013b). Technological successions in the Middle Stone Age sequence of Diepkloof Rock Shelter, Western Cape, South Africa. *Journal of Archaeological Science, 40*(9), 3376–3400.

Porraz, G., Val, A., Dayet, L., De la Pena, P., Douze, K., Miller, C., Murungi, M. L., Tribolo, C., Schmid, V., & Sievers, C. (2015). Bushman Rock Shelter (Limpopo, South Africa): A perspective from the edge of the highveld. *South African Archaeological Bulletin, 70*(202), 166–179.

Powell, A., Shennan, S., & Thomas, M. G. (2009). Late Pleistocene demography and the appearance of modern human behavior. *Science, 324*(5932), 1298–1301.

Read, D. W. (2008). An interaction model for resource implement complexity based on risk and number of annual moves. *American Antiquity, 73*, 599–625.

Riel-Salvatore, J., & Barton, C. M. (2004). Late Pleistocene technology, economic behavior, and land-use dynamics in Southern Italy. *American Antiquity, 69*(2), 257–274.

Roberts, D. L. (2003). *Age, genesis and significance of South African coastal belt silcretes*. Pretoria: Council for Geoscience.

Schmidt, P., Porraz, G., Slodczyk, A., Bellot-Gurlet, L., Archer, W., & Miller, C. E. (2012). Heat treatment in the South African Middle Stone Age: Temperature induced transformations of silcrete and their technological implications. *Journal of Archaeological Science, 40*(9), 3519–3531.

Shi, N., Dupont, L. M., Beug, H.-J., & Schneider, R. (2000). Correlation between vegetation in Southwestern Africa and oceanic upwelling in the past 21,000 years. *Quaternary Research, 54*(1), 72–80.

Shott, M. J. (1986). Technological organisation and settlement mobility. *Journal of Anthropological Research, 42*(1), 15–51.

Singer, R., & Wymer, J. (1982). *The Middle Stone Age at Klasies River Mouth in South Africa*. Chicago: The University of Chicago Press.

Soriano, S., Villa, P., Delagnes, A., Degano, I., Pollarolo, L., Lucejko, J. J., Henshilwood, C., & Wadley, L. (2015). The Still Bay and Howiesons Poort at Sibudu and Blombos: Understanding Middle Stone Age technologies. *PloS One, 10*(7), e0131127.

Soriano, S., Villa, P., & Wadley, L. (2007). Blade technology and tool forms in the Middle Stone Age of South Africa: The Howiesons Poort and post-Howiesons Poort at Rose Cottage Cave. *Journal of Archaeological Science, 34*(5), 681–703.

Soriano, S., Villa, P., & Wadley, L. (2009). Ochre for the toolmaker: Shaping the Still Bay points at Sibudu (KwaZulu-Natal, South Africa). *Journal of African Archaeology, 7*(1), 45–54.

Stager, J. C., Mayewski, P. A., White, J., Chase, B. M., Neumann, F. H., Meadows, M. E., King, C. D., & Dixon, D. A. (2012). Precipitation variability in the winter rainfall zone of South Africa during the last 1400 yr linked to the austral westerlies. *Climate of the Past, 8*(3), 877–887.

Steele, T. E., Mackay, A., Fitzsimmons, K., Igreja, M., Marwick, B., Orton, J., Schwortz, S., & Stahlschmidt, M. (2016). Varsche Rivier 003: A Middle Stone Age site with Still Bay and Howiesons Poort assemblages in southern Namaqualand, Western Cape, South Africa. *PaleoAnthropology, 2016*, 100–163.

Steele, T. E., Mackay, A., Orton, J., & Schwortz, S. (2012). Varsche Rivier 003, a new Middle Stone Age site in southern Namaqualand, South Africa. *South African Archaeological Bulletin, 67*(195), 108–119.

Stuut, J.-B., Prins, M. A., Schnieder, R. R., Weltje, G. J., Jansen, J. H. F., & Postma, G. (2002). A 300-kyr record of aridity and wind strength in southwestern Africa: Inferences from grain-size distributions of sediments on Walvis Ridge, SE Atlantic. *Marine Geology, 180*, 221–233.

Toggweiler, J. R., & Russell, J. (2008). Ocean circulation in a warming climate. *Nature, 451*, 286–288.

Tribolo, C., Mercier, N., Douville, E., Joron, J. L., Reyss, J. L., Rufer, D., Cantin, N., Lefrais, Y., Miller, C. E., Porraz, G., Parkington, J., Rigaud, J. P., & Texier, P. J. (2012). OSL and TL dating of the Middle Stone Age sequence at Diepkloof Rock Shelter (South Africa): A clarification. *Journal of Archaeological Science, 30*, 3401–3411.

Tribolo, C., Mercier, N., Valladas, H., Joron, J. L., Guibert, P., Lefrais, Y., Selo, M., Texier, P. J., Rigaud, J. P., Porraz, G., Poggenpoel, C., Parkington, J., Texier, J. P., & Lenoble, A. (2009).

Thermoluminescence dating of a Stillbay–Howiesons Poort sequence at Diepkloof Rock Shelter (Western Cape, South Africa). *Journal of Archaeological Science, 36*(3), 730–739.

Truc, L., Chevalier, M., Favier, C., Cheddadi, R., Meadows, M. E., Scott, L., Carr, A. S., Smith, G. F., & Chase, B. M. (2013). Quantification of climate change for the last 20,000 years from Wonderkrater, South Africa: Implications for the long-term dynamics of the Intertropical Convergence Zone. *Palaeogeography, Palaeoclimatology, Palaeoecology, 386*, 575–587.

Valsecchi, V., Chase, B. M., Slingsby, J. A., Carr, A. S., Quick, L. J., Meadows, M. E., Cheddadi, R., & Reimer, P. J. (2013). A high resolution 15,600-year pollen and microcharcoal record from the Cederberg Mountains, South Africa. *Palaeogeography, Palaeoclimatology, Palaeoecology, 387*, 6–16.

Villa, P., Soressi, M., Henshilwood, C. S., & Mourre, V. (2009). The Still Bay points of Blombos Cave (South Africa). *Journal of Archaeological Science, 36*(2), 441–460.

Villa, P., Soriano, S., Teyssandier, N., & Wurz, S. (2010). The Howiesons Poort and MSA III at Klasies River main site, Cave 1A. *Journal of Archaeological Science, 37*(3), 630–655.

Visser, H. N., & Theron, J. N. (1973). Clanwilliam. 3218, Geological series. Pretoria: Government Printer.

Vogelsang, R., Richter, J., Jacobs, Z., Eichhorn, B., Linseele, V., & Roberts, R. G. (2010). New excavations of Middle Stone Age deposits at Apollo 11 Rockshelter, Namibia: Stratigraphy, archaeology, chronology and past environments. *Journal of African Archaeology, 8*(2), 185–218.

Volman, T. P. (1980). *The Middle Stone Age in the Southern Cape*. PhD, University of Chicago.

Wadley, L. (2007). Announcing a Still Bay industry at Sibudu Cave, South Africa. *Journal of Human Evolution, 52*(6), 681–689.

Wadley, L. (2012). Two 'moments in time' during Middle Stone Age occupations of Sibudu, South Africa. *Southern African Humanities, 24*, 79–97.

Will, M., Mackay, A., & Phillips, N. (2015). Implications of Nubian-like Core reduction systems in Southern Africa for the identification of early modern human dispersals. *PloS One, 10*(6), e0131824.

Wurz, S. (1997). *The Howiesons Poort at Klasies River: from artefacts to cognition*. MA, University of Stellenbosch.

Wurz, S. (2002). Variability in the Middle Stone Age lithic sequence, 115,000–60,000 years ago at Klasies River, South Africa. *Journal of Archaeological Science, 29*(9), 1001–1015.

Wurz, S. (2013). Technological trends in the Middle Stone Age of South Africa between MIS 7 and MIS 3. *Current Anthropology, 54*(S8), S305–S319.

Ziegler, M., Simon, M. H., Hall, I. R., Barker, S., Stringer, C., & Zahn, R. (2013). Development of Middle Stone Age innovation linked to rapid climate change. *Nature Communications, 4*, 1905.

Chapter 3
The South Asian Microlithic: *Homo sapiens* Dispersal or Adaptive Response?

Chris Clarkson, Michael Petraglia, Clair Harris, Ceri Shipton, and Kasih Norman

Introduction

The question of the origins of the microlithic in South Asia has been one that is hotly debated in the last decade and a prominent topic within the scientific and public domain (Balter 2010; Appenzeller 2012). Two opposed models have invoked different mechanisms to explain the origins of the microlithic in India. These see the microlithic as either the signature of the arrival of *Homo sapiens* with essentially African Later Stone Age technology (Mellars 2006; Mellars et al. 2013; Mishra et al. 2013; Bar Yosef and Belfer Cohen 2013;) or the indigenous development of technologies suited to coping with increasing risk and uncertainty during periods of climatic variability and change (Clarkson et al. 2009; Petraglia et al. 2009a; Hiscock et al. 2011; Clarkson 2014).

The *Homo sapiens* dispersal argument would see microlithic technologies appearing in South Asia before 50 ka (Mellars 2006; Mellars et al. 2013) via a

C. Clarkson (✉) • K. Norman
Archaeology, School of Social Sciences, The University of Queensland, Brisbane, QLD, Australia
e-mail: c.clarkson@uq.edu.au

M. Petraglia
Max Planck Institute for the Science of Human History, Jena, Germany

C. Harris
School of Archaeology, Research Laboratory for Archaeology and the History of Art, University of Oxford, Oxford, UK

C. Shipton
McDonald Institute for Archaeological Research, University of Cambridge, Cambridge, UK

E. Robinson, F. Sellet (eds.), *Lithic Technological Organization and Paleoenvironmental Change*, Studies in Human Ecology and Adaptation 9,
DOI 10.1007/978-3-319-64407-3_3

coastal route. Under this scenario, the microlithic should appear as a clear break from the Middle Paleolithic, as new people brought a new technological and symbolic package with them.

The microlithic adaptation model, on the other hand, would expect to see the microlithic appear at a time of significant climatic downturn or as climatic variability increased. Furthermore, it would be expected that the microlithic would emerge out of existing technologies, and hence gradual changes leading up to the adoption of microlithic technology might be expected. We might also expect to see microlithic elements appearing during times of higher innovation, such as during demographic expansions, but becoming more common when conditions worsen, as is seen in Australia and Africa (Hiscock et al. 2011; Blinkhorn and Petraglia 2014). With increasingly well-resolved sequences in India and Sri Lanka and a corpus of reliable radiocarbon and OSL chronologies appearing in several regions in India, it is becoming possible to explore the signature of technological change and microlithic adoption in South Asia more clearly.

The subcontinent occupies an interesting geographical position, sitting on the boundary between Saharo-Arabian flora and fauna and the Oriental species. In many important respects, it marks the junction between the classic technological sequence of the west in Africa, Europe, and SW Asia and the more disparate and highly varied sequences of Asia and Oceania. Eastern India, for instance, marks the boundary between biface and cleaver-based industries typical of the Lower Paleolithic in the west (Europe, Africa, and SW Asia) and the virtual absence of such artifacts in the east. In terms of the Middle Paleolithic, the subcontinent lies at the intersection of the distinctive Middle Paleolithic Levallois-Mousterian of SW Asia and the so-called "chopper," "pebble," or "small tool" industries to the east. Likewise, India marks the point at which Levallois and point dominated technology becomes less common, with only rare cases known east of India. The Upper or Late Paleolithic offers a similar junction, between the classic blade and microlith-based Upper Paleolithic of Europe and SW Asia and the total absence of blade and microlith industries to the east and southeast of India (but not the northeast). India, sitting at the crossroads, comprises elements of both – with microblade technology well represented in some areas, but virtually absent in others (e.g., Sri Lanka). India and the subcontinent in general therefore presents a fascinating boundary region in which to examine key issues in dispersals and technological change and hence the question of microlithic origins.

Resolving the chronology of each of these phases is an ongoing issue in Indian archaeology, one not helped by the lack of well-preserved organic material in a tropical climate, making the retrieval of reliable samples for radiocarbon dating very difficult. Likewise, the hominin skeletal record in India is virtually nonexistent before c. 30 ka (Kennedy and Deraniyagala 1989). These combined problems of dating and skeletal preservation place a heavy burden on lithic assemblages in resolving questions of chronology and hominin association. Fortunately problems of chronology are slowly being resolved, particularly with the help of luminescence dating in sediments that lack datable organics. For instance, the chronology of blade and microlith technology in India is undergoing rapid revision, with reported dates of 44 ka now bringing the Indian Late Paleolithic more into line with the age of Upper Paleolithic in the west. Likewise, the Middle Paleolithic was virtually undated except in rare cases until 10 years ago (Baskaran et al. 1986; Sharma and Clark 1983; Pal et al. 2005; Williams et al. 2006).

This paper examines one issue in particular, the nature of the transition between the Middle Paleolithic and microlithic in India. We aim to test the two models mentioned above by examining whether there is evidence for a clean break in technological traditions with the appearance of microlithic technology or whether the microlithic can be seen to emerge gradually from the Middle Paleolithic. We also consider climatic evidence for the timing of dispersals and the adoption of the microlithic.

The Timing of *Homo sapiens'* Arrival in India

The paucity of human fossils in India has made assigning particular industries or major transitions to hominin species impossible, except to note that microlithic industries dating later than 36 ka are unambiguously associated with *Homo sapiens* in all cases where human remains are found until relatively recent times (Alchin 1963; Mishra et al. 2013). Much older *Homo sapiens* remains are known to the east and west of India, including Southeast Asia and Australia, suggesting that our species was likely in India before 50 ka. The hominin identity of the makers of Middle Paleolithic industries remains completely unknown. This presents major difficulties for unraveling the history of local biological evolution, colonization, replacement, and diffusion and places a heavy burden on the lithic industries to help discern likely hominin associations.

Genetic evidence offers valuable clues to the population history of South Asia. Recent genetic research has revealed that India was the gateway to subsequent colonization of Asia and Australia and saw the first major population expansion of modern human populations anywhere outside of Africa, in the order of a 500% increase (Metspalu et al. 2004; Endicott et al. 2007; Atkinson et al. 2008; Soares et al. 2009; Rasmussen et al. 2011). South Asia therefore provides a crucial stepping-stone in early modern migration to Southeast Asia and Oceania. Genetic evidence also indicates that India was likely home to more than half of the first *Homo sapiens* populations living outside of Africa, as a result of the first and fastest major non-African population expansion (Atkinson et al. 2008). It is thought that the M lineage likely arose or became distinct from other Eurasian lineages in India and that most non-Indian M lineages, including those in Southeast Asia and Oceania, emerged in India (Metspalu et al. 2004). Recent analyses offer growing support for a multiple dispersal model (Rasmussen et al. 2011; Reich et al. 2011). Rasmussen et al. (2011) study of ancient DNA from an Australian Aboriginal hair indicates that Australians were the product of an early dispersal 75–62 ka and that a second distinct wave is identifiable in East Asians 38–25 ka. Since early immigrants to Australia likely passed through India, estimated divergence dates of 75–62 ka also likely apply to the *Homo sapiens* population that colonized South Asia. Recent revision of the human genomic mutation rate provides another line of nuclear DNA evidence that problematizes mtDNA-based out of Africa estimates. Measurements of the nuclear genomic mutation rate in contemporary humans have indicated a value that is approximately half of that previously estimated from fossil calibration (Scally and Durbin 2012). When applied to the analysis of the time of separation of African and non-African populations, the revised rate yields an out of Africa chronology in the range of 130–90 ka (Scally and Durbin 2012: 748). Blinkhorn and Petraglia argue that younger estimates of out of Africa on the basis of mtDNA data may, it is argued, derive from processes of later gene flow and drift/selection and/or complex demographic factors including bottlenecks (Scally and Durbin 2012: 751; Boivin et al. 2013).

The climatic evidence would also support entry into India before 75 ka. Monsoon strength, as reconstructed from several regional datasets, including carbon isotopes, dust-particle spikes, and $d^{18}O$, was high during this period, and humid phases such as this would have allowed entry through the arid Thar Desert corridor, whereas dry phases would have acted as a barrier to human entry with the Thar Desert expanded to meet the coast (Thompson et al. 1997; Achyuthan et al. 2007; Petraglia et al. 2009a, 2010). Furthermore, Arabia experienced growing aridity after 80 ka, and coastal productivity would also have declined with weakening of the monsoon (Parton et al. 2015). Indeed, archaeological evidence indicates hominin populations were present until 75 ka (Petraglia et al. 2012) and for a short period around 55 ka (Delagnes et al. 2012; Parton et al. 2015), in association with Middle Paleolithic tools. The lithic industries of these sites however may show stronger affinities with the Levant than with Africa or India (Petraglia et al. 2010, 2012; Clarkson 2014; Parton et al. 2015).

Models for Microlithic Origins

In the last decade, archaeologists have put forward two contrasting models for the origins of the microlithic in India. Mellars (2006; Mellars et al. 2013) has argued that *Homo sapiens* left Africa sometime after 60 ka with a distinctive microlithic technology, arriving in South Asia by no later than 50 ka. Other cultural components, Mellars suggests, included engraved cross-hatched designs and the manufacture of perforated disc-shaped beads. Mellars contends that this package was lost due to successive founder effects and changes in raw material availability, resulting in a much-simplified technology arriving in Australia without microlithic artifacts. For the purposes of this paper, we will call this the *dispersal model*.

Mishra et al. (2013) have recently offered support for this model, arguing that the Mehtakheri microblade industry of the Narmada Valley dates to 44 ± 2 ka and is therefore in the right timeframe for modern human dispersal into the region with

East/South African-like microlithic technology. Furthermore, she argues the Middle Paleolithic of India has clear affinities with the Late Acheulean, whereas the microlithic boundary is sharp. Finally she argues populations in Southeast Asia and Australia were likely colonized from East Asia, not India, where archaic populations resided until *Homo sapiens* brought microlithic technology c.50 ka (Mishra et al. 2013:Fig. 2:2). The older dates for microliths certainly makes this model more plausible; however, the notion of colonization of India from Asia is clearly at odds with the genetic evidence which points to South Asia as the likely location for expansion of the M lineage.

The second model, proposed by Clarkson and Petraglia and their colleagues (Clarkson et al. 2009; Petraglia et al. 2009a; Clarkson 2014), posits that the microlithic most likely arose in South Asia, in response to deteriorating climatic conditions, at the end of a warm wet phase when populations were already large. It is proposed that modern humans were already in India by at least 80 ka and that the appearance of the microlithic can be seen as a local development out of Middle Paleolithic technologies. We will call this the *adaptive model.*

The adaptive model follows recent literature on microlithic origins around the world as seeing these technologies as providing particular advantages at times of greater foraging risk. This is because microliths are standardized implements suitable for hafting into multipurpose, reusable composite tools that were readily maintained. Rapidly repaired multipurpose tools are desirable in riskier conditions, where activities might be unsuccessful and/or where the consequences of failure were severe (e.g., Hiscock 2002). Elsewhere, archaeological tests of this model typically involve determining whether higher rates of microlith production coincide with periods of higher risk (e.g., Attenbrow et al. 2009).

The Middle Paleolithic to Microlithic Transition

In India, the Middle Paleolithic was once portrayed as "an enigmatic group of stone industries which fall typologically and stratigraphically between the hand-axe (Acheulean) industries on one side and the microlithic on the other" (Allchin 1959:1). This notion is now being replaced with a much clearer view of the Middle Paleolithic. Middle Paleolithic core technology is typically a mix of Levallois, discoidal, single, and multiplatform core reduction systems. There is preferential exploitation of cobbles of fine-grained materials such as limestone, chert and chalcedony, and periods when quartz is common. Middle Paleolithic assemblages are often rich with retouched flake/scraper types of different configurations, such as denticulates, notches, burins, blades, and points with evidence of increased hafting. Middle Paleolithic sites are often found close to sources of high-quality stone and are rarely buried within quaternary sequences in the peninsular region (Korisettar and Rajaguru 2002:332).

The timing of the emergence of Middle Paleolithic traditions within India is now becoming better understood, with early dates for well-defined Middle Paleolithic assemblages dating to around 80–100 ka (Baskaran et al. 1986; Clarkson et al. 2012; Blinkhorn et al. 2013). Middle Paleolithic assemblages are found in highly varied environmental settings, indicating a high level of behavioral flexibility, with sites located in sand dunes (Misra 2001), along streams, on hill slopes (Haslam et al. 2012a), within large valleys (Petraglia et al. 2007; Clarkson et al. 2012), and in rock-shelters. Most sites are however closely linked with water sources and appear in river valleys much like the previous Acheulean mobility patterns (Pappu and Deo 1994; Paddayya 1987).

The microlithic refers to assemblages found with backed artifacts. These industries are marked by the production of small cores and flakes and an increasing preference for cryptocrystalline materials. Not all microlithic industries are made on microblades, as in Sri Lanka, for instance, where microliths are typically made on small flakes, often of quartz. In other parts of South Asia, microliths are clearly associated with systematic microblade production using unidirectional or bidirectional blade cores. The technology evolves through time with indications of the use of punch and pressure blade production by the time of the Neolithic (Clarkson et al. 2009). Microlithic assemblages sometimes contain bone tools, incised ostrich eggshell, and seashell beads, such as at Jwalapuram 9 (Clarkson et al. 2009), Batadombalena and Fa Hien Cave (Perera et al. 2011), and Patne (Sali 1989).

The microlithic should now be differentiated from that of Upper or Late Paleolithic in our opinion (James and Petraglia 2009). The Upper or Late Paleolithic is used in India to refer to a more nebulous group of industries that contain higher proportions of blades, microblades, or "flake blades" with many small retouched tools and a general lack of Levallois or radial core technology. Many of these industries do not contain backed artifacts. In some regions of India, the boundary and technological distinction between Middle Paleolithic and Upper/Late Paleolithic is unclear, due to a lack of well-dated transitional sites (e.g., Sri Lanka) or clear typological markers. We believe that in many cases, these more blade-rich assemblages lacking Levallois are better characterized as late Middle Paleolithic, as demonstrated below.

Having reviewed models for the origins of the microlithic, evidence for modern human settlement, and the general character of Middle Paleolithic and microlithic industries, we now examine the technological sequences for two regions of India where long-term, dated sequences have been reported. These are the Jurreru River Valley in Andhra Pradesh and the Middle Son Valley in Madhya Pradesh. We examine the nature of long-term technological changes and the evidence for sudden or gradual introductions of microlithic technology. We then turn to a discussion of the results.

Regional Case Studies

The Jurreru Valley

The Jurreru River Valley is located in the central part of the state of Andhra Pradesh. It runs in an easterly direction, with steep valley sides and a relatively wide bottom. Archaeological excavations carried out on valley bottom sites (JWP 3, 17, 21, 22, 23), near the valley margins in the south (JWP 20) and in the north, where stratified rock-shelters (e.g., JWP 9) are located, have yielded a long sequence of occupation deposits (Fig. 3.1). Most excavations on the valley floor targeted areas with surface artifacts and underlying Toba ash deposits, exposed during local ash mining activities (Petraglia et al. 2009b). Seven of these open sites fall into the period from c.85 ka to modern times and contain often large and diverse lithic assemblages. Age estimations for Jurreru sites are summarized in Clarkson et al. (2012) and Petraglia et al. (2012), and one site falls into the Late Acheulean period (JWP22).

Occupation surfaces in open valley settings at JWP3 and JWP22 underlie YTT deposits and date to between 85 ± 6 ka and 77 ± 6 ka and 71 ± 8 ka, respectively (Haslam et al. 2012a, b), while those above the ash occur at JWP17, JWP20, JWP21, and JWP23 and date to <55 ka (Petraglia et al. 2009a, b, 2012; Haslam et al. 2012a, and Table 1). Open sites on the valley bottom contain predominantly Middle Paleolithic assemblages with rare backed microliths, microblades, and microblade cores in the topmost sediments of some sites (Fig. 3.2) (Petraglia et al. 2009b). The Jwalapuram 9 rock-shelter sits just above the edge of the northern valley margin and

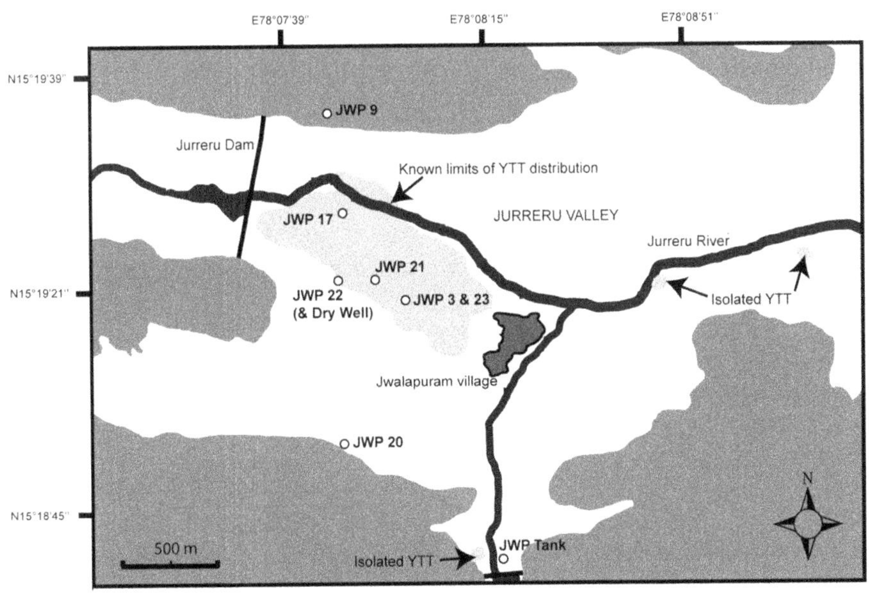

Fig. 3.1 Map of sites in the Jurreru Valley mentioned in the text

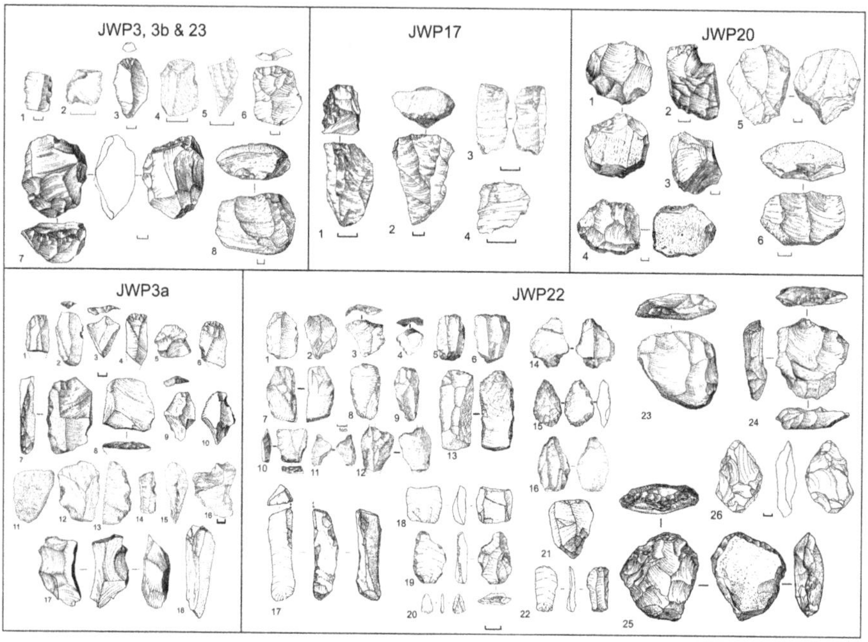

Fig. 3.2 Selected artifacts from the Middle Paleolithic of the Jurreru River Valley. No 26 is a Late Acheulean hand ax from the deep trench JWP22

spans the last 35 ka, dated using AMS radiocarbon, with five strata (A–E) (Clarkson et al. 2009; Petraglia et al. 2009b). JWP9 preserves an exclusively microlithic assemblage with Neolithic and Iron Age occupations in the top two strata (Clarkson et al. 2009; Petraglia et al. 2009b) (Fig. 3.3).

The lithic sequence from the Jurreru Valley is one of the changes with strong underlying elements of continuity in core and scraper technology. The changes to core technology and flake production over time are significant, with reductions in flake size, increases in elongation, decreasing platform faceting, and changes in scar pattern all significant (see Clarkson et al. 2012:170–171, Table 5). The sequence of changes from the oldest Middle Paleolithic (JWP22) through to the youngest Middle Paleolithic (JWP20) of the Jurreru open sites is one of the gradual changes with conservation of specific forms of recurrent Levallois and discoidal core reduction with increasing emphasis placed on single and multiplatform cores, a trend toward higher levels of retouch intensity, and, finally, the appearance of bipolar technology at the end of the Middle Paleolithic sequence (Fig. 3.4). The presence of discoidal and multiplatform cores in the lowest layer at JWP9 may also indicate continuities with the earlier Middle Paleolithic exist even here as seen elsewhere in India, such as at Patne (Sali 1989). Dorsal scar orientations and platform preparation also undergo gradual changes over the sequence, from frequent radial orientations and faceted platforms to unidirectional and proximal orientations, consistent with the

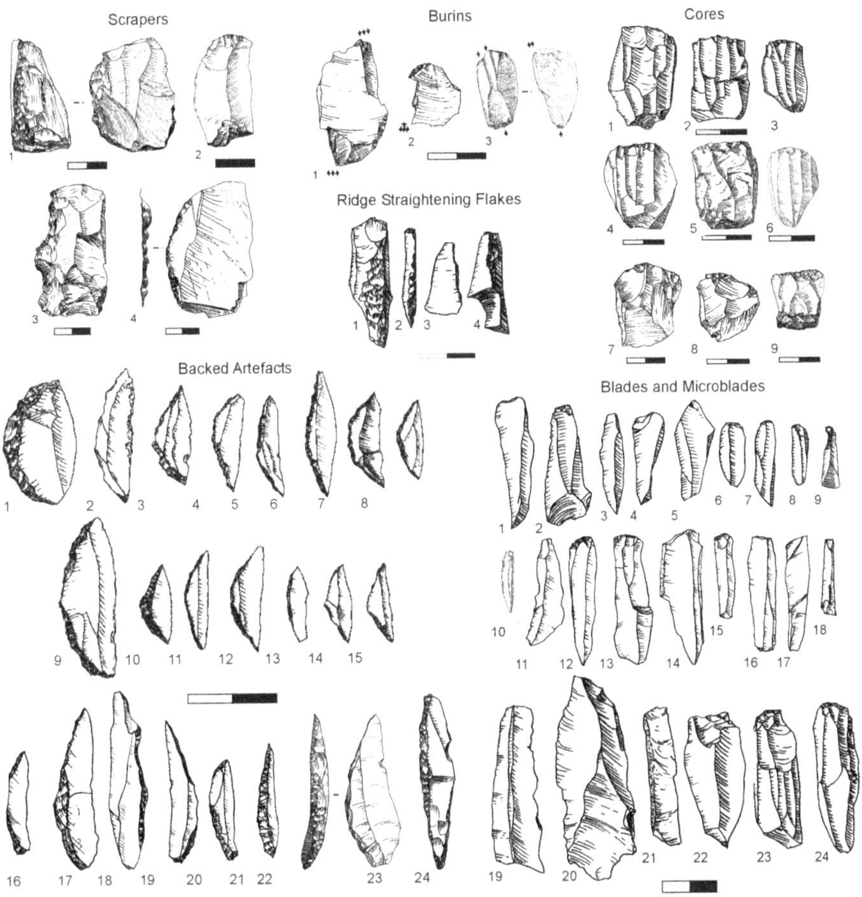

Fig. 3.3 Microliths, microblades, scrapers, and cores from JWP9

change from Levallois to single platform and blade production (Fig. 3.5). Raw materials also undergo marked changes in frequency over the sequence (Fig. 3.6). Local limestone is the dominant raw material in the Middle Paleolithic above and below ash, with chert and chalcedony increasing dramatically at the end of the Middle Paleolithic as seen at JWP17 and in the microlithic period, particularly in the lower layers of JWP9, corresponding to the period from 35 to 16 ka. Local limestone rises in proportion again in the final period. Importantly, microblade production precedes microliths at JWP9, and no microliths are found at the base of the site. Hence the microblades and microliths do not appear suddenly as a package, but appear gradually.

Changes are also evident in the blades produced in the Jurreru Valley over time. Figure 3.7 shows a gradual decrease in blade size (medial width and length) from the Middle Paleolithic through to the microlithic (phase D), a gradual increase in scar number and a gradual increase in elongation (length/width). These changes

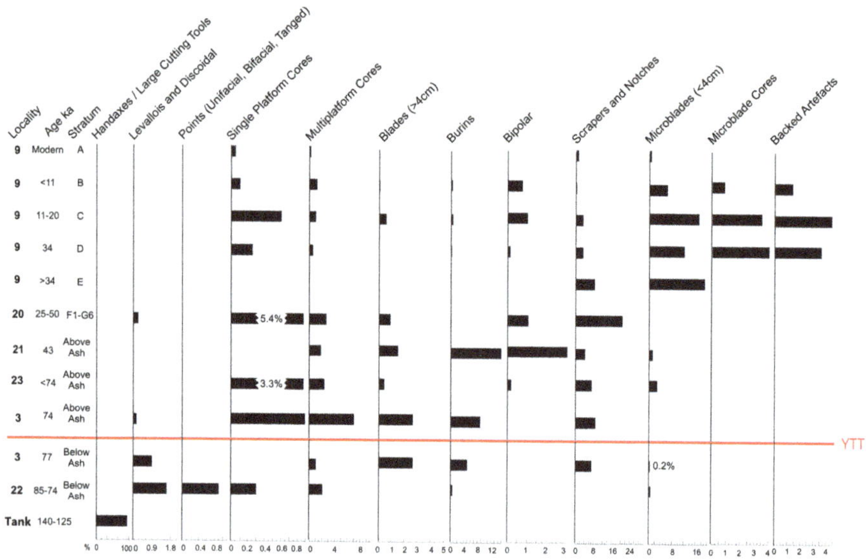

Fig. 3.4 Summary of long-term technological changes in the Jurreru Valley from 140- < 10 ka

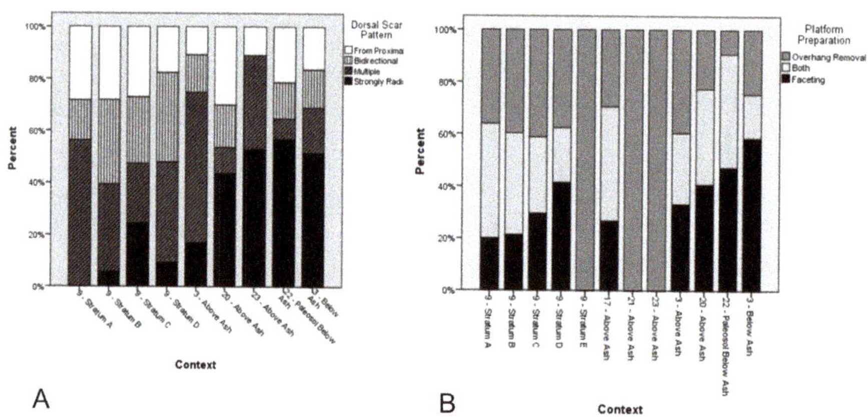

Fig. 3.5 (**a**) Changes in core scar directionality and (**b**) platform preparation in the Jurreru Valley on flakes from pre-YTT (JWP22 and JWP3), to post-YTT (17, 21, 23, 3, and 20), to microlithic (JWP9, strata B–E) over the last 85 ka

suggest that the introduction of microlith production was part of longer-term set of changes to blade production, with high elongation and small size postdating the first introduction microblades and microliths.

In summary, the introduction of the microlithic in the Jurreru Valley appears around the end of a long series of gradual changes in core technology, raw material use, and reduction intensity, reflected in long-term diminution in blade size (i.e., length), increases in elongation (length/width), and dorsal scar number. The intro-

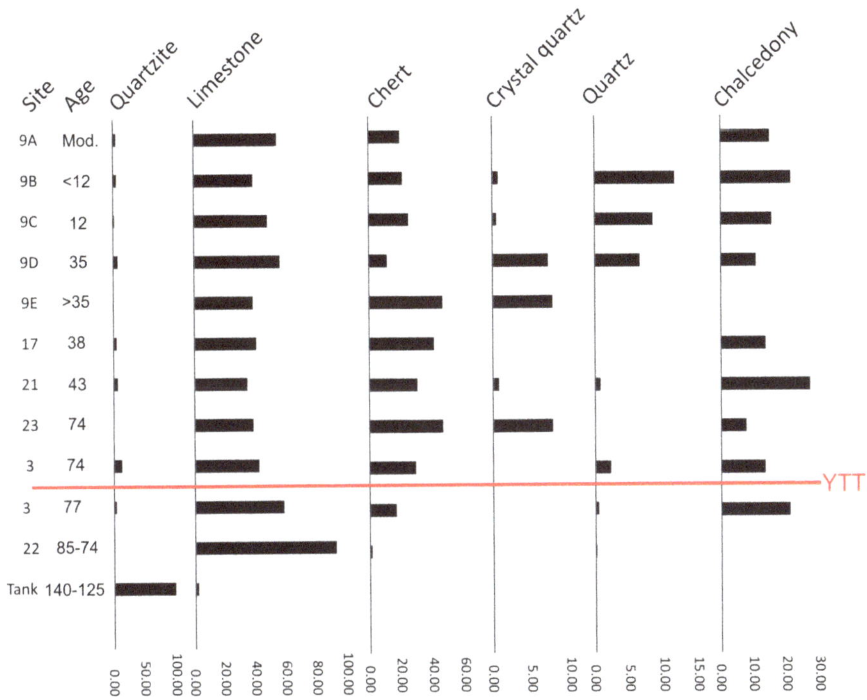

Fig. 3.6 Raw material changes in the Jurreru Valley from 140- < 10 ka

duction of microliths is preceded by the introduction of microblades, indicating these are not introduced as a single package at JWP9. We now examine the sequence from the Middle Son Valley.

The Middle Son Valley

The Middle Son Valley is located in northeastern Madhya Pradesh, south of Allahabad, and southwest of Varanasi. The Middle Son refers to a medial section of the 784 km long Son River that flows into the Ganges near Allahabad (Fig. 3.8). The Middle Son is bordered by the Kaimur Range to the north and the Baghelkhand plateau to the south and shares a confluence with the Gopad River. This region has one of the most comprehensively studied Late Quaternary geological and archaeological records in India, with research dating back to the early 1980s when Sharma and Clark (1983) carried out a systematic and multidisciplinary study of the area. Sharma and Clark developed a sedimentological record for the region between Patpara and the Gopad River that detailed the likely relationship between cultural and geological formations.

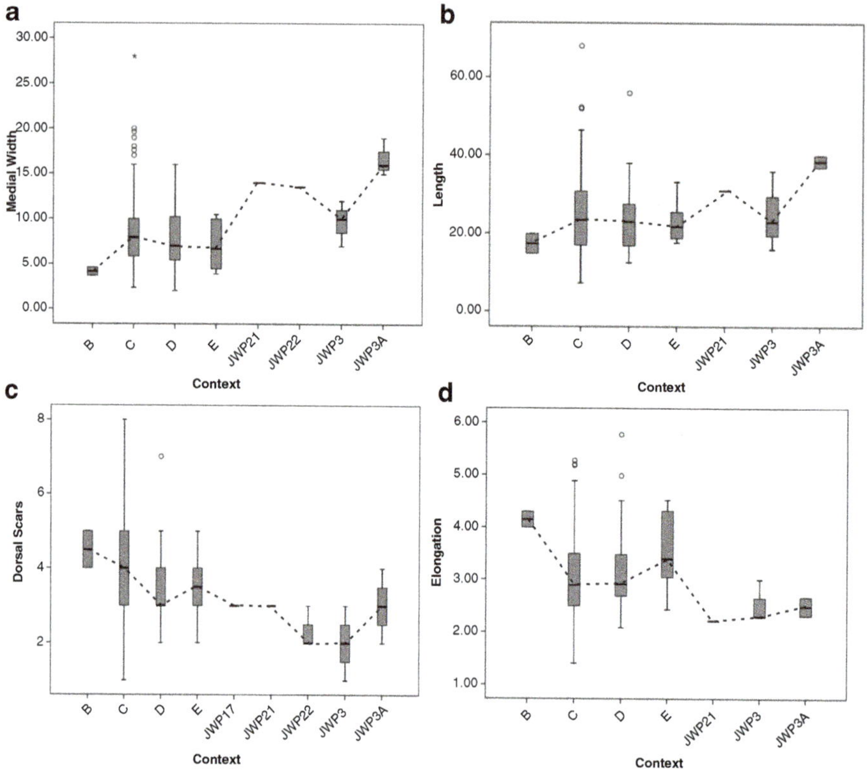

Fig. 3.7 Changes in blade morphology from pre-Toba deposits (JWP3A and JWP22) through post-Toba layers (JWP21 and JWP23) and the microlithic (JWP9, layers B, C, D, and E). (**a**) medial width, (**b**) length, (**c**) number of dorsal scars, (**d**) elongation

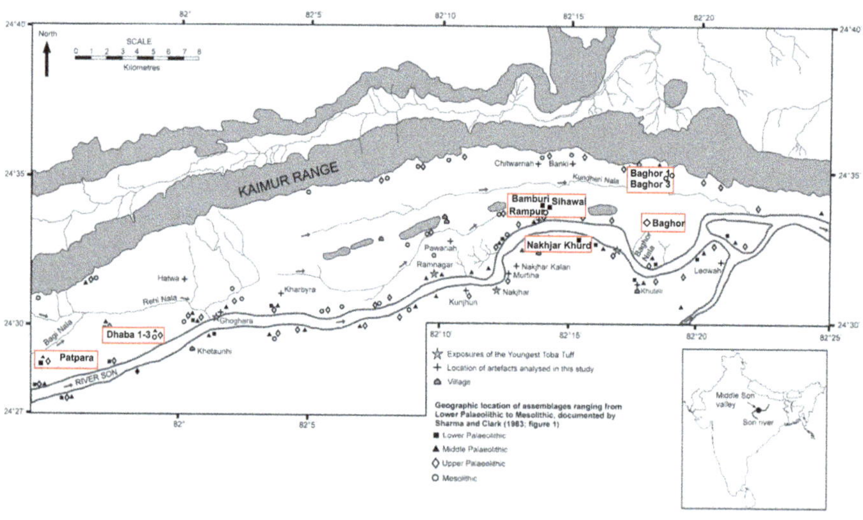

Fig. 3.8 Map of the Middle Son showing the location of sites mentioned in the text (With permission from Sacha Jones)

Williams and Royce (1982) described the four geological formations comprised of alluvial, colluvial, and aeolian deposits that make up the Pleistocene sediments of the Middle Son River region. The colluvial and alluvial deposits overlay Proterozoic sandstone and shale of the Vindhyan Supergroup (Ray et al. 2003) and rise approximately 30–35 m above the lowest level of the river (Williams and Royce 1982; Williams and Clarke 1984). These formations, named Sihawal, Patpara, Khunteli, Baghor, and Khetauni, all contain evidence for different sedimentation regimes, influenced by changes in the river morphology through time.

Studies carried out by Sharma and Clark (1983) document a rich sequence of lithic artifacts recovered from excavated and surface collections, situated on both banks of the Son River. Artifacts span the Acheulean, through the Middle Paleolithic and microlithic, and into the Neolithic periods (Sharma and Clark 1983; Clark and Dreiman 1983; Kenoyer et al. 1983a, b; Kenoyer and Pal 1983; Misra et al. 1983; Sussman et al. 1983). The sites of importance for this study are Patpara, Dhaba, and Baghor, as these span the Middle Paleolithic to microlithic transition.

Artifacts from Patpara in the southwest of the Middle Son study area were assigned to the Lower and Middle Paleolithic based on the morphology of the assemblage of artifacts collected from a surface scatter and excavation. The range of stone artifact types found at Patpara includes scrapers of varying edge modification and retouch intensity, notched artifacts, burins, hand axes, cleavers, backed artifacts, and cores including Levallois, discoidal, and blade cores (Blumenschine et al. 1983: 47–60).

Artifacts from Baghor were located in the northeast of the study area on the north banks of the Son River. The artifacts were found in the Baghor coarse member dated between 39 and 26 ka and are assigned to the Upper Paleolithic and microlithic. Artifacts from Baghor were mostly of chert (92.8%), and artifact types include blade and microblade cores, flakes, retouched flakes, and microblades and blades. Backing was common with some occurrence of double edge backing and some denticulation. Notches are present on microblades (Kenoyer et al. 1983a: 122–123).

Baghor 1 and 3, located in the Baghor fine member, are characterized as microlithic assemblages due to the diminutive size of the artifacts, with none longer than 60 mm in length (Blumenschine et al. 1983: 164). Backed artifacts were common alongside scrapers, notched artifacts, microblade cores, hammerstones, grinding stones, stone discs, and burnishers (Blumenschine et al. 1983: 169). Radiocarbon dates suggest these sites date to between 8 and 6 ka.

The Dhaba locality is in the southwest of the study area and consists of three archaeological sites (Dhaba 1, 2, and 3) on the north banks of the Son River (Haslam et al. 2012b). Each of the three archaeological sites was excavated as a step trench placed into hillslope sediments. Dhaba 1 consists of a dense Middle Paleolithic artifact concentration, with artifacts visibly eroding from sediments at several points up the slope. Very few artifacts were recovered from the upper part of the excavation, with the lower portion of the site preserving chert and silicified limestone flakes and cores, and a single large quartzite flake from near the base of the excavation. Dhaba 2 and 3 had stratified chert and limestone Middle Paleolithic artifacts, with a dense concentration of cryptocrystalline microblade and small flake artifacts higher up in the sequence at Dhaba 3. OSL dates on feldspar have been obtained from the sites and crypotephra studies have revealed the presence of YTT at Dhaba 1. Dhaba 1 has been dated to between 78–64 ka, Dhaba 3 lower to 44–47 ka, and Dhaba 2 and Dhaba 3 upper to 37–26 ka.

Figure 3.9 provides a combined technological sequence for the Middle Son sites, showing likely ages and relative chrono-stratigraphic positioning. A clear technological progression is evident in this figure from Late Acheulean biface dominated industries dating to around 120–140 ka; to radial/Levallois reduction sites with no bifaces dating to around 78–64 ka; to blade and burin industries dating to between around 55–39 ka; and finally a shift to microblade and backed artifact production in the last 43 ka. As in the Jurreru sequence, microblades precede backed artifact production in the Middle Son, in this case by several thousand years. This again suggests the microlithic is not a single package, but appeared in stages. The ages of all sites older than 36 ka are luminescence dates with fairly large error ranges, and hence there is some uncertainty over the exact timing of major transitions. Many of the older sites shown in Fig. 3.10 have no associated radiometric dates and hence rely on presumed relative stratigraphic positioning rather than absolute dates. A representative sample of artifacts from each major phase in the Middle Son is shown in Fig. 3.10.

The changes in core technology over the period are mirrored in changes in dorsal scar orientations for select sites reexamined in 2010 by the authors, as shown in Fig. 3.11. There is a gradual decline in radial scar orientations as Levallois wanes and a rise in mixed, bidirectional, and proximal orientations as core technology shifts toward single and multiplatform and then blade core technology. Changes in platform preparation also show gradual changes over time

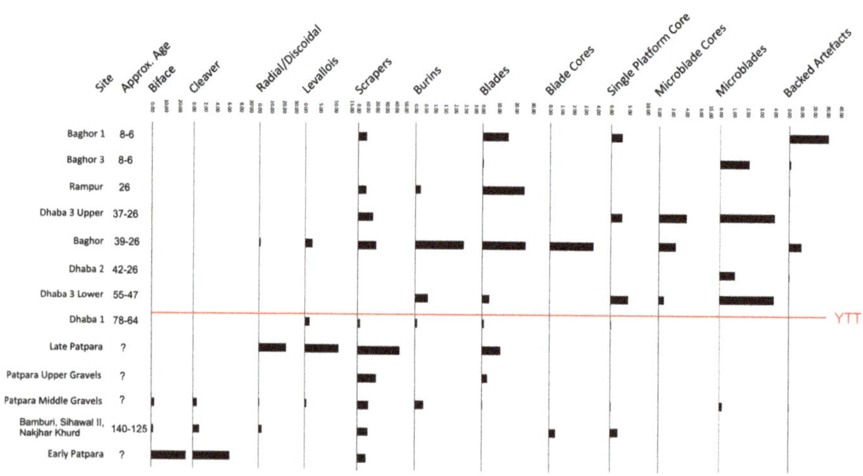

Fig. 3.9 Major technological changes in the Middle Son Valley, summarized by site and period

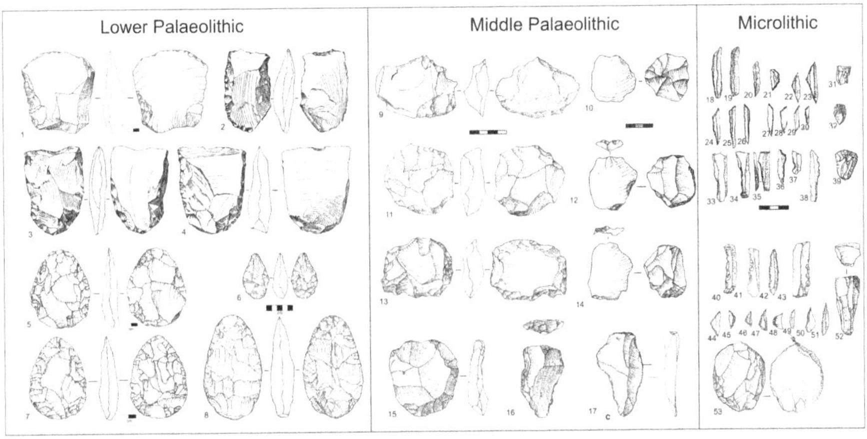

Fig. 3.10 Artifacts from Middle Son sites. *1–4* cleavers, *5–8* hand axes, *9, 11, 13*, and *15* Levallois cores, *10, 12, 14, 16*, and *17* Levallois flakes and points, *18–23, 24–30, 33–38, 40–51* microliths, *31, 32*, and *39* microblade cores, *52* blade core, *53* burin

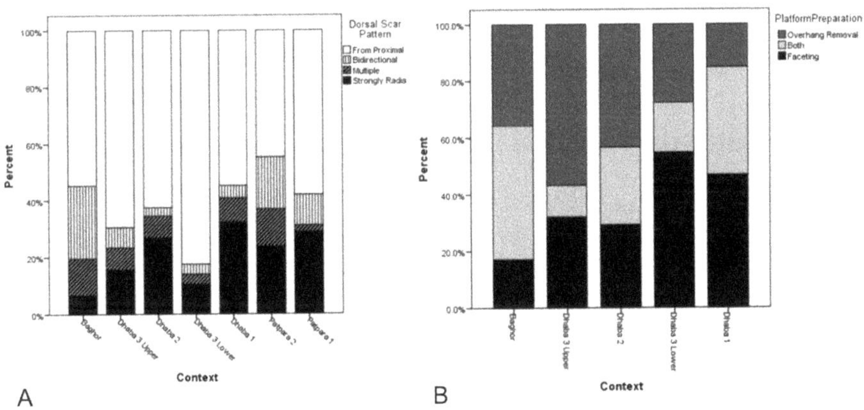

Fig. 3.11 (**a**) Changes in core scar directionality and (**b**) platform preparation, in the Middle Son Valley over the last 120 ka

(Fig. 3.11), with a change from predominantly faceted platforms to increasingly plain platforms reflecting the slow shift from Levallois to single platform to blade cores.

Another clear gradual change is evidenced in raw material selection through time. Figure 3.12 shows a gradual change from quartzite, to limestone, to quartz, and, finally, to cryptocrystalline materials like chert and chalcedony. The pace of change roughly mirrors, or is slower, than that of technological change.

Changes in the characteristics of unretouched blades from Middle Son sites are evident over the sequence from 78–6 ka (Fig. 3.13). Blades are larger and less elongate early on at Dhaba. Blades diminish in medial width and length

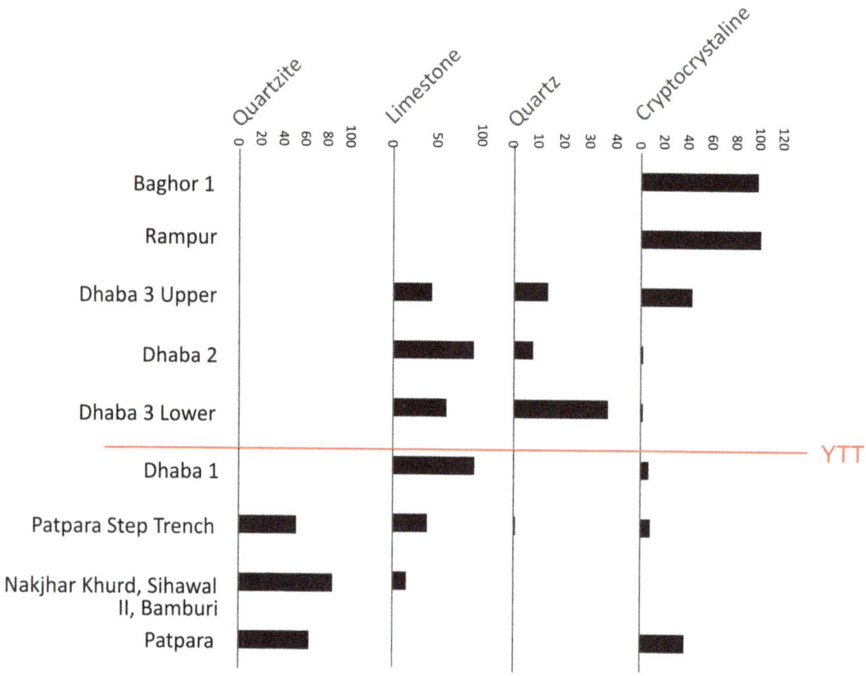

Fig. 3.12 Changes in raw material proportions in the Middle Son Valley by site

gradually in the lead up to the introduction of the Microlithic at c.43 ka at Dhaba 2. The changes in blade form indicate that blade production altered gradually and that increases in elongation and blade length took place after microblade and then microlith introductions. The changes in blade form over the sequence are significant for each variable except elongation (ANOVA, df = 62, medial width = 0.02; length = 0.008; scar count = 0.037; elongation = 0.899).

Discussion

A great deal of parity exists in the timing, sequence, and character of technological changes in these two regions of India. The transition from the Middle Paleolithic to the microlithic manifests as a gradual trend away from Levallois and toward single platform core technology between 70 and 40 ka. These changes can be seen as a gradual shift toward microblade production sometime between 55 and 47 ka in the Son and before 35 ka in the Jurreru, in the lead up to the first appearance of microliths at 43 ka in the

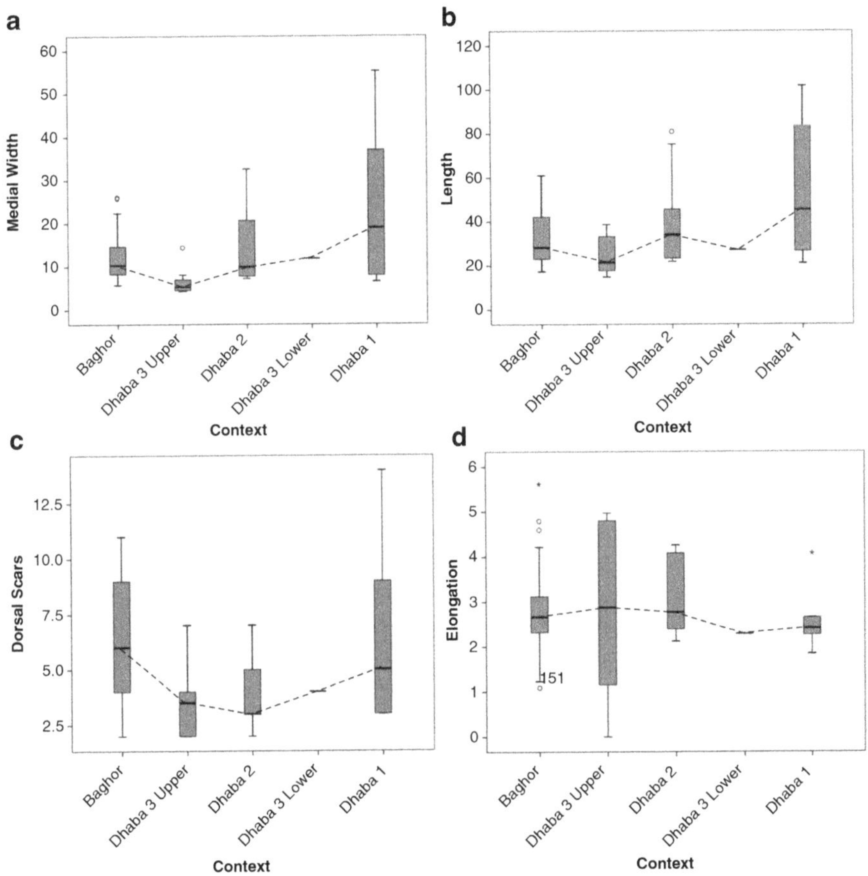

Fig. 3.13 Changes in blade characteristics over the sequence in select Middle Son sites examined by the authors. (**a**) medial width, (**b**) length, (**c**) number of dorsal scars, (**d**) elongation

Middle Son and 35 ka in the Jurreru. This gradual shift is accompanied by an increase in blades, burins, and blade cores and a shift to finer-grained raw materials (chert).

Finally, the transition to the full-blown "microlithic" witnesses further diminution in blade size and increased elongation, with increased use of chalcedony in the Jurreru and cryptocrystalline stone (chalcedony and chert) in the Middle Son. Microliths also become more common later in time, particularly after 35 ka in the Jurreru and in the Holocene in the Middle Son. In both the Jurreru and the Middle Son, microblade production appears to precede large-scale microlith production.

We return now to consideration of the models for the origins of microliths in South Asia. The first important finding is the lack of abrupt change in lithic industries in either of the regions examined where long and relatively well-dated sequences exist. This is contrary to the predictions of the dispersal model. Continuous directional change is evident in the successive introduction and loss of technologies, underwritten by more gradual changes in raw material use. The gradual nature of assemblage change from Middle Paleolithic to microlithic makes a sudden entry of *Homo sapiens* with dramatically different microlithic technology seem unlikely and is more consistent with the adaptive model.

The adaptive model also posits that the appearance of the microlithic in South Asia was likely tied to climatic changes in the lead up to the LGM. Populations, having grown and prospered, would likely have experienced heightened risk as a result of declining resources and increased variability. This transition may be gradual, as also seen in Australia (Hiscock 2002) and East Africa (Ambrose 2002; Brandt et al. 2012; Pleurdeau et al. 2014), with early microliths appearing before widespread production. Figure 3.14 plots monsoon strength over the last 80,000 years (Achyuthan et al. 2007; Petraglia et al. 2010), against dates for microlithic sites in South Asia. The oldest known dates from key sites are also plotted. We see from this graph that as monsoon strength begins to decline, the number of dates associated with microlithic sites increases, suggesting that microlithic sites perhaps became more common as climate began to deteriorate. Furthermore, very old dates associated with microliths from the Ratnapura beds in Sri Lanka suggest the possibility that microliths may also have appeared briefly between 64–70 ka (Deriyanagala 1992: 433) when the monsoon was also weak. Unfortunately, little is known about these microliths or their context, and future work must further explore claims for their great age (see also claims for microliths at 150 ka at Patirajawela (Abeyratne 1996).

Interestingly, the oldest microlithic sites are all located far from the ocean in northern inland river valleys (Fig. 3.15) (Dhaba 2, 43.0 ± 0.9 ka; Mehtakheri, 44 ± 4 ka; Kana, 42 ± 4 ka) (Mishra et al. 2013; Baskak et al. 2014), where the effects of a weakening monsoon would be most keenly felt. The land bridge between Sri Lanka and mainland India was always open during the period of microlithic adoption; yet Sri Lankan microliths appear somewhat later than those in the north. One reason may be that it took time for new technologies to diffuse to the south. Another reason may be that rainforest communities were unaffected at first by deteriorating climatic conditions and that Sri Lankan populations did not experience significant economic risk of the kind that might trigger a technological response until 6 ka or more after the first appearance of the Microlithic in the north.

Conclusion

For now, the question of whether the microlithic evolved independently in India, or diffused into India >50 ka from the west, remains contentious. New chronological and technological data, as well as better paleoclimatic records, are required to resolve this issue. However, current lithic evidence for major transitions in two regions of India leads us to several important conclusions:

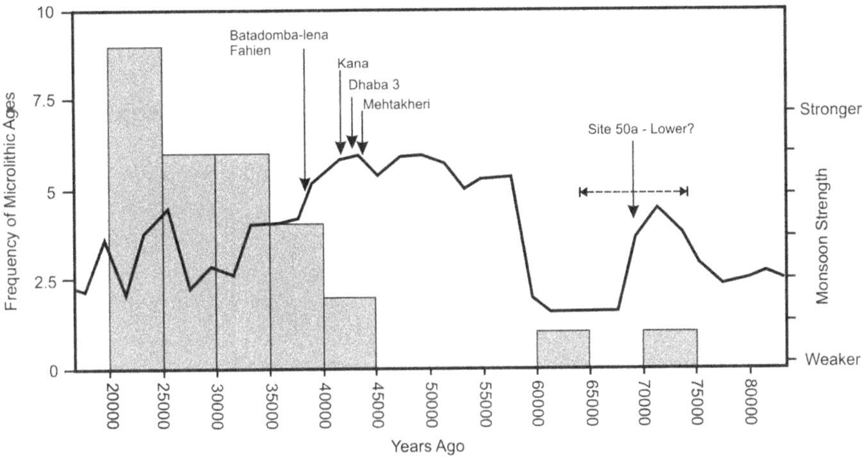

Fig. 3.14 Relationship between the first appearance of microliths and monsoon strength in South Asia. *Gray bars*, number of dates for microlithic sites; *black arrows*, oldest dates for key microlithic sites. Monsoon strength data from Achyuthan et al. (2007). Microlithic dates from Petraglia et al. (2009a, b)

1. Technological change in the lead up to the microlithic is best characterized as gradual.
2. On current evidence microlithic industries were not present in South Asia prior to 44 ka, and the oldest sites are inland where climatic changes would be swiftest to take effect.
3. Lewis et al. (2014) comparison of South Asian and African microliths revealed little similarity.
4. The shift to microblade production comes at the end of a gradual sequence of changes in blade characteristics.
5. The earliest microlithic may have arisen during a time when larger populations allowed higher innovation rates (Petraglia et al. 2009a), but only became common once climatic conditions had worsened, as also seen in Australia (Attenbrow et al. 2009).
6. Microblade technology precedes microlith production in at least two regions.
7. Microlithic technology was adopted later in the south where rainforest vegetation may have buffered local populations from deteriorating climatic conditions experienced in the northern interior. The Sri Lankan microlithic lacks microblades and also contains points, which are likely to be a local development.
8. The gradual adoption of microlithic technology occurs alongside a weakening monsoon with few early microliths appearing in the interior north prior to their spread across South Asia.

We conclude that the current evidence better supports the adaptive microlithic origin model in South Asia. It seems this took place over many millennia, with an early origin in the central north, regions that are today highly drought prone, when

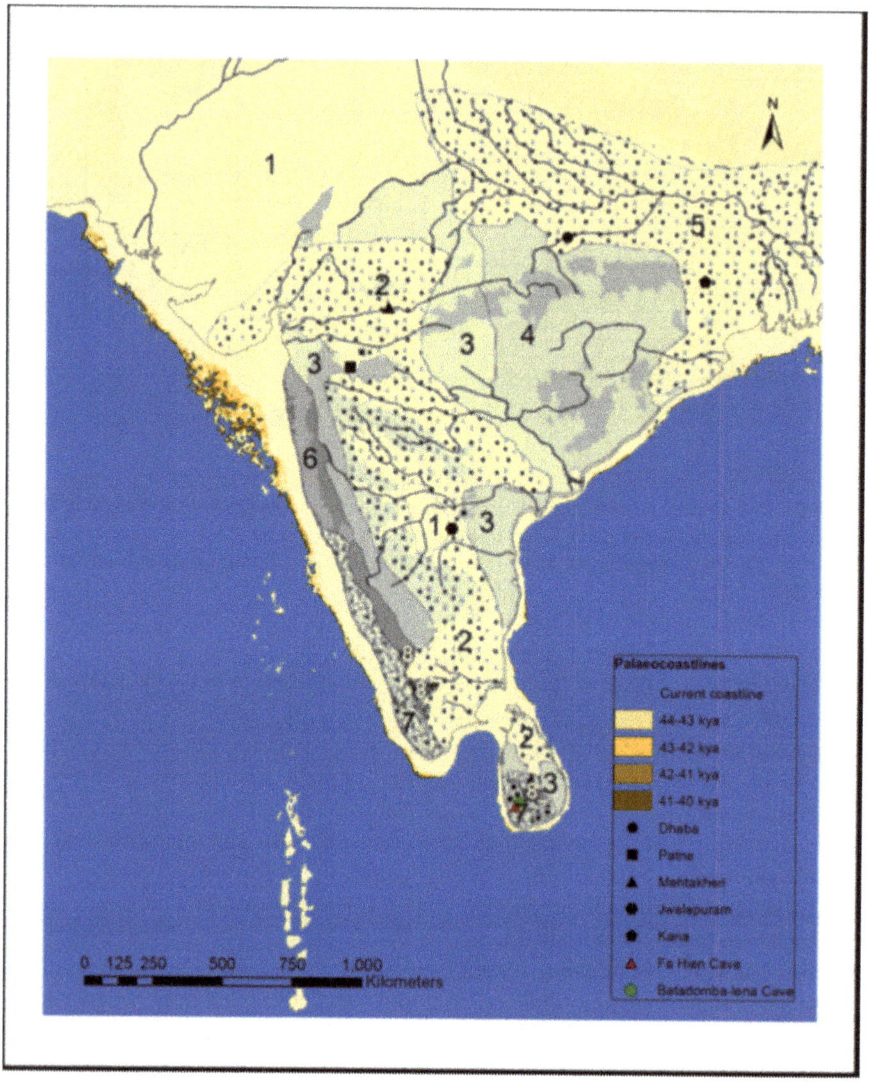

Fig. 3.15 Location of early microlithic sites in South Asia plotted onto sea levels between 40 and 44 ka and reconstructed vegetation zones (excluding now drowned areas) (from Petraglia et al. 2009a, b: Fig. 2). Environmental key: *1* desert, *2* tropical savanna/woodland-grass mosaic (including the tropical evergreen zones of India), *3* dry deciduous woodlands, including teak, *4* dry deciduous woodlands, including *Shorea* and *Hopea*, *5* moist tropical woodland and grassland mosaic (in particular that of eastern India and Ganges plain in which Dipterocarpaceae including *Shorea* are frequent; this zone extends to Southeast Asia), *7* moist tropical woodlands (including both moist deciduous zones like those in India and true tropical rainforests), *8* tropical montane vegetation; map data from GEBCO_2014 Grid, version 20141103, http://www.gebco.net'. Sea level heights taken from Sahul Time (Matthew Coller, http://sahultime.monash.edu.au/)

populations were likely large, and later spreading across South Asia as climatic conditions deteriorated. Widespread technological changes occur in the late Middle Paleolithic, which appear to lead to microblade production, with full-blown microlithic appearing after that. We have discussed the reasons for adopting microlithic industries in the face of worsening climatic conditions at length elsewhere (Clarkson et al. 2009; Petraglia et al. 2009a; Hiscock et al. 2011). The evidence now seems in favor of the adoption of the microlithic as just such a response in South Asia, one that was apparently mirrored in other places and at different times in response to periods of aridity and/or increased climatic variability (Hiscock et al. 2011).

Acknowledgments We thank Erick Robinson and Frédéric Sellet for the invitation to present at the SAAs in Hawaii and to write this paper. We also thank all of our valued Indian colleagues in the various regions where we have worked for the last 15 years. Indian archaeology is always on our minds and we ever delighted by the prospect of new field seasons and collaborations. We wish to thank the ASI for permission to conduct fieldwork in India and the American Institute of Indian Studies for facilitating our research. Our research was primarily supported by grants from the British Academy, the Leverhulme Trust, the European Research Council Grant 295719 (to M.D.P.), and the Australian Research Council-DP0987680 (to C.C and M.P) and DP0770446 (to C.C.).

References

Abeyratne, M. (1996). *Multi-dating studies of archaeological sites*. Unpublished PhD Thesis, The Australian National University, Canberra.

Achyuthan, H., Quade, J., Roe, L., & Placzek, C. (2007). Stable isotopic composition of pedogenic carbonates from the eastern margin of the Thar Desert, Rajasthan, India. *Quaternary International, 162-3*, 50–60.

Alchin, B. (1959). The Indian middle stone age: Some new sites in central and southern India and their implications. *Bulletin of the Institute of Archaeology, 2*, 1–36.

Alchin, B. (1963). The Indian stone age sequence. *The Journal of the Royal Anthropological Institute of Great Britain and Ireland, 92*(2), 210–234.

Ambrose, S. H. (2002). Small things remembered: Origins of early microlithic industries in sub-Saharan Africa. In R. G. Elston & S. L. Kuhn (Eds.), *Thinking small: Global perspectives on microlithic technologies* (pp. 9–29). Arlington: Archaeological Papers of the American Anthropological Association No. 12.

Appenzeller, T. (2012). Eastern odyssey. *Nature, 485*, 24–26.

Atkinson, Q. D., Gray, R., & Drummond, A. (2008). mtDNA variation predicts population size in humans and reveals a major southern Asian chapter in human prehistory. *Molecular Biology and Evolution, 25*(2), 468–474.

Attenbrow, V., Robertson, G., & Hiscock, P. (2009). The changing abundance of backed artefacts in south-eastern Australia: A response to Holocene climate change? *Journal of Archaeological Science, 36*, 2765–2770.

Balter, M. (2010). Of two minds about Toba's impact. *Science, 327*(5970), 1187–1188.

Bar Yosef, O., & Belfer Cohen, A. (2013). Following Pleistocene road signs of human dispersals across Eurasia. *Quaternary International, 285*, 30–43.

Baskak, B., Srivastava, P., Dasgupta, S., Kumar, A., & Rajaguru, S. N. (2014). Earliest dates and implications of microlithic industries of late Pleistocene from Mahadebbera and Kana, Purulia district, West Bengal. *Current Science, 107*(7), 1167–1171.

Baskaran, M., Marathe, A. R., Rajaguru, S. N., & Somayajulu, B. L. K. (1986). Geochronology of Palaeolithic cultures in the Hiran Valley, Saurashtra, India. *Journal of Archaeological Science, 13*, 505–514.

Blinkhorn, J., Achyuthan, H., Petraglia, M., & Ditchfield, P. (2013). Middle Palaeolithic occupation in the Thar Desert during the upper Pleistocene: The signature of a modern human exit out of Africa? *Quaternary Science Reviews, 77*, 233–238.

Blinkhorn, J., & Petraglia, M. D. (2014). Assessing models for the dispersal of modern humans to South Asia. In M. Porr & R. Dennell (Eds.), *Southern Asia, Australia, and the search for human origins* (pp. 64–75). Cambridge: Cambridge University Press.

Blumenschine, R. J., Brandt, S. A., & Clark, J. D. (1983). Excavations and analysis of middle Palaeolithic artifacts from Patpara, Madhya Pradesh. In G. R. Sharma & J. D. Clark (Eds.), *Palaeoenvironments and prehistory in the middle Son Valley* (pp. 39–99). Allahabad: Abinash Prakashan.

Boivin, N., Fuller, D. Q., Dennell, R., Allaby, R., & Petraglia, M. D. (2013). Human dispersal across diverse environments of Asia during the upper Pleistocene. *Quaternary International, 300*, 32–34.

Brandt, S. A., Fisher, E. C., Hildebrand, E. A., Vogelsang, R., Ambrose, S. H., Lesur, J., & Wang, H. (2012). Early MIS 3 occupation of Mochena Borago Rockshelter, southwest Ethiopian highlands: Implications for late Pleistocene archaeology, palaeooenvironments and modern human dispersals. *Quaternary International, 274*, 38–54.

Clark, J. D., & Dreiman, R. (1983). An occurrence with small blade technology in the upper member of the Baghor formation at the Baghor III locality. In G. R. Sharma & J. D. Clark (Eds.), *Palaeoenvironments and prehistory in the middle Son Valley* (pp. 197–208). Allahabad: Abinash Prakashan.

Clarkson, C. (2014). East of Eden: Founder effects and the archaeological signature of modern human dispersal. In M. Porr & R. Dennell (Eds.), *Southern Asia, Australia, and the search for human origins* (pp. 76–89). Cambridge: Cambridge University Press.

Clarkson, C., Petraglia, M., Korisettar, R., Haslam, M., Boivin, N., Crowther, A., Ditchfield, P., Fuller, D., Miracle, P., Harris, C., Connell, K., James, H., & Koshy, J. (2009). The oldest and longest enduring microlithic sequence in India: 35,000 years of modern human occupation and change at the Jwalapuram locality 9 rockshelter. *Antiquity, 83*(2), 1–23.

Clarkson, C., Jones, S., & Harris, C. (2012). Continuity and change in the lithic Industries of the Jurreru Valley, India, before and after the Toba eruption. *Quaternary International, 258*, 165–179.

Delagnes, A., Tribolo, C., Bertran, P., Brenet, M., Crassard, R., Jaubert, J., Khalidi, L., Mercier, N., Nomade, S., Peigne, S., Sitzia, L., Tournepiche, J.-F., Al-Halibi, M., AlMosabi, A., & Macciarelli, R. (2012). Inland human settlement in southern Arabia 55,000 years ago: New evidence from the Wadi Surdud middle Paleoolithic site complex, western Yemen. *Journal of Human Evolution, 63*(3), 452–474.

Deriyanagala, S. U. (1992). *The prehistory of Sri Lanka: An ecological perspective*. Colombo: Department of Archaeological Survey, Government of Sri Lanka.

Endicott, P., Metspalu, M., & Kivisild, T. (2007). Genetic evidence on modern human dispersals in South Asia: Y chromosome and mitochondrial DNA perspectives: The world through the eyes of two haploid genomes. In M. D. Petraglia & B. Allchin (Eds.), *The evolution and history of human populations in South Asia* (pp. 229–244). Amsterdam: Springer.

Haslam, M., Clarkson, C., Roberts, R. G., Bora, J., Korisettar, R., Ditchfield, P., Chivas, A., Harris, C., Smith, V., Oh, A., Eksambekar, S., Boivin, N., & Petraglia, M. (2012a). JWP22: A southern Indian middle Palaeolithic occupation surface sealed by the 74 ka Toba eruption: Further evidence from Jwalapuram locality 22. *Quaternary International, 258*, 148–164.

Haslam, M., Harris, C., Clarkson, C., Pal, J. N., Shipton, C., Crowther, A., Koshy, J., Bora, J., Ditchfield, P., Ram, H. P., Price, K., Dubey, A. K., & Petraglia, M. (2012b). Dhaba: An intital report on an Acheulean, middle Palaeolithich and microlithic locality in the middle Son Valley, north central India. *Quaternary International, 258*, 191–199.

Hiscock, P. (2002). Pattern and context in the Holocene proliferation of backed artefacts in Australia. In R. G. Elston & S. L. Kuhn (Eds.), *Thinking small: Global perspectives on Microlithization* (pp. 163–177). Arlington: American Anthropological Association.

Hiscock, P., Clarkson, C., & Mackay, A. (2011). Big debates over little tools: Ongoing disputes over microliths on three continents. *World Archaeology, 43*(4), 660–671.

James, H., & Petraglia, M. (2009). The lower to middle Palaeolithic transition in South Asia and its implications for hominin cognition and dispersals. In M. Camps & P. Chauhan (Eds.), *Sourcebook of Palaeolithic transitions* (pp. 255–264). Dordrecht: Springer.

Kennedy, K. A. R., & Deraniyagala, S. U. (1989). Fossil remains of 28,000-year-old hominids from Sri Lanka. *Current Anthropology, 30*(3), 394–399.

Kenoyer, M., & Pal, J. N. (1983). Report on the excavation and analysis of an upper Acheulean assemblage from Sihawal II. In G. R. Sharma & J. D. Clark (Eds.), *Palaeoenvironments and prehistory in the middle Son Valley* (pp. 23–38). Allahabad: Abinash Prakashan.

Kenoyer, J. M., Mandal, D., Misra, V. D., & Pal, J.N. (1983a). Preliminary report on excavations at the late Palaeolithic occupation site at Baghor I locality. In G. R. Sharma & J. D. Clark (Eds.), *Palaeoenvironments and prehistory in the middle Son Valley* (pp. 117–142). Allahabad: Abinash Prakashan.

Kenoyer, J. M., Clark, J. D., Pal, J. N., & Sharma, G. R. (1983b). An upper Palaeolithic shrine in India?. *Antiquity, 57*, 88–94.

Korisettar, R., & Rajaguru, S. N. (2002). The monsoon background and the evolution of prehistoric cultures of India. In K. Paddayya (Ed.), *Recent studies in IndiaArchaeology* (pp. 316–348). New Delhi: Munshiram Manoharlal Publishers PVT Ltd..

Lewis, L., Perera, N., & Petraglia, M. (2014). First technological comparison of Southern African Howiesons Poort and South Asian microlithic industries: An exploration of inter-regional variability in microlithic assemblages. *Quaternary International, 350*, 7–25.

Mellars, P. (2006). Going east: New genetic and archaeological perspectives on the modern human colonization of Eurasia. *Science, 313*, 796–800.

Mellars, P., Gori, K. C., Carr, M., Soares, P. A., & Richards, M. B. (2013). Genetic and archaeological perspectives on the initial modern human colonization of southern Asia. *Proceedings of the National Academy of Sciences, 110*(26), 10699–10704.

Metspalu, M., Kivisild, T., Metspalu, E., Parik, J., Hudjashov, G., Kaldma, K., Serk, P., Karmin, M., Behar, D. B., Thomas, M., Gilbert, P., Endicott, P., Mastana, S., Papiha, S., Skorecki, K., Torrini, A., & Villems, R. (2004). Most of the extant mtDNA boundaries in south and southwest Asia were likely shaped during the initial settlement of Eurasia by anatomically modern humans. *BMC Genetics, 5*(1), 1.

Misra, V. D., Rana, R. S., Clark, J. D., & Blumenschine, R. J. (1983). Preliminary excavations at the son river section at Nakjhar Khurd. In G. R. Sharma & J. D. Clark (Eds.), *Palaeoenvironments and prehistory in the middle Son Valley* (pp. 101–116). Allahabad: Abinash Prakashan.

Misra, V. N. (2001). Prehistoric human colonization of India. *Journal of Biosciences, 26*, 491–531.

Mishra, S., Chauhan, N., & Singhvi, A. (2013). Continuity of microblade technology in the Indian subcontinent since 45 ka: Implications for the dispersal of modern humans. *PloS One, 8*(7), 1–14.

Paddayya, K. (1987). The stone age cultural Systems of the Baichbal Valley, Gulburga District, Karnataka: A preliminary report. *Bulletin of the Deccan College Post-Graduate and Research Institute, 46*, 77–100.

Pal, J. N., Williams, M. A. J., Jaiswal, M., & Singhvi, A. K. (2005). Infrared stimulated luminescence ages for prehistoric cultures in the son and Belan valleys, north Central India. *Journal of Interdisciplinary Studies in History and Archaeology, 1*, 51–62.

Pappu, R. S., & Deo, S. G. (1994). *Man-land relationships during Palaeolithic times in the Kaladgi Basin, Karnataka*. Pune: Deccan College.

Parton, A., White, T. S., Parker, A. G., Breeze, P. S., Jennings, R., Groucutt, H. S., & Petraglia, M. D. (2015). Orbital-scale climate variability in Arabia as a potential motor for human dispersals. *Quaternary International, 382*, 82–97.

Perera, N., Kourampas, N., Simpson, I. A., Deraniyagala, S. U., Bulbeck, D., Kamminga, J., Perera, J., Fuller, D., Szabó, K., & Oliveria, N. (2011). People of the ancient rainforest:

Late Pleistocene foragers at the Batadomba-lena rockshelter, Sri Lanka. *Journal of Human Evolution, 61*, 254–261.

Petraglia, M. D., Korisettar, R., Boivin, N., Clarkson, C., Ditchfield, P., Jones, S., Koshy, J., Lahr, M. M., Oppenheimer, C., Pyle, D., Roberts, R., Schwenninger, J. L., Arnold, L., & White, K. (2007). Middle Palaeolithic assemblages from the Indian subcontinent before and after the Toba super-eruption. *Science, 317*, 114–116.

Petraglia, M., Clarkson, C., Boivin, N., Haslam, M., Korisettar, R., Chaubey, G., Ditchfield, P., Fuller, D., James, H., Jones, S., Kivisild, T., Koshy, J., Lahr, M.M., Metspalu, M., Roberts, R., & Arnold, L. (2009a). Population increase and environmental deterioration correspond with microlithic innovations in South Asia ca.35,000 years ago. *Proceedings of the National Academy of Sciences, 106*(3), 12261–12266.

Petraglia, M., Korisettar, R., Bai, M. K., Boivin, N., Clarkson, C., Cunningham, K., Ditchfield, P., Fuller, D. Hampson, J., Jones, S., Koshy, J., Miracle, P., Oppenheimer, C., & White, K. (2009b). Cave and rockshelter records, the Toba super-eruption and forager-farmer interactions in the Kurnool District, India. *Journal of Eurasian Prehistory: A Journal for Primary Archaeological Data, 6*(1–2), 119–166.

Petraglia, M. D., Haslam, M., Fuller, D. Q., Boivin, N., & Clarkson, C. (2010). Out of Africa: New hypotheses for the dispersal of Homo sapiens along the Indian Ocean rim. *Annals of Human Biology, 37*(3), 288–311.

Petraglia, M. D., Ditchfield, P., Jones, S., Korisettar, R., & Pal, J. N. (2012). The Toba volcanic super-eruption, environmental change, and hominin occupation history in India over the last 140,000 years. *Quaternary International, 258*, 119–134.

Pleurdeau, D., Hovers, E., Assefa, Z., Asrat, A., Pearson, O., Bahain, J.-J., & Man Lam, Y. (2014). Cultural change or continuity in the late MSA/early LSA of southeastern Ethiopia? The site of Goda Buticha, Dire Dawa area. *Quaternary International, 343*, 117–135.

Rasmussen, M., Guo, X., Wang, Y., Lohmueller, K. E., Rasmussen, S., Albrechtsen, A., Skotte, L., Lindgreen, S., Metspalu, M., Jombart, T., Kivisild, T., Zhai, W., Eriksson, A., Manica, A., Orlando, L., De La Vega, F. M., Tridico, S., Metspalu, E., Nielsen, K., Ávila-Arcos, M. C., Moreno-Mayar, J. V., Muller, C., Dortch, J., Gilbert, M. T. P., Lund, O., Wesolowska, A., Karmin, M., Weinert, L. A., Wang, B., Li, J., Tai, S., Xiao, F., Hanihara, T., van Driem, G., Jha, A. R., Ricaut, F.-X., de Knijff, P., Migliano, A. B., Gallego Romero, I., Kristiansen, K., Lambert, D. M., Brunak, S., Forster, P., Brinkmann, B., Nehlich, O., Bunce, M., Richards, M., Gupta, R., Bustamante, C. D., Krogh, A., Foley, R. A., Lahr, M. M., Balloux, F., Sicheritz-Pontén, T., Villems, R., Nielsen, R., Wang, J., & Willerslev, E. (2011). An aboriginal Australian genome reveals separate human dispersals into Asia. *Science, 334*, 94–98.

Ray, N., Currat, M., & Excoffier, L. (2003). Intra-deme molecular diversity in spatially expanding populations. *Molecular Biology and Evolution, 20*, 76–86.

Reich, D., Patterson, N., Kircher, M., Delfin, F., Nandineni, M. R., Pugach, I., et al. (2011). Denisova admixture and the first modern human dispersals into Southeast Asia and Oceania. *The American Journal of Human Genetics, 89*, 1–13.

Sali, S. A. (1989). *The upper Palaeolithic and Mesolithic cultures of Maharashtra*. Pune: Deccan College Post Graduate and Research Institute.

Scally, A., & Durbin, R. (2012). Revising the human mutation rate: Implications for understanding human evolution. *Nature Reviews Genetics, 13*, 745–753.

Sharma, G. R., & Clark, J. D. (1983). *Palaeoenvironments and prehistory in the middle Son Valley*. Allahabad: Abinash Prakashan.

Soares, P., Ermini, L., Thomson, N., Mormina, M., Rito, T., Rohl, A., Salas, A., Oppenheimer, S., Macaulay, V., & Richards, M. B. (2009). Correcting for purifying selection: An improved human mitochondrial molecular clock. *American Journal of Human Genetics, 84*, 740–749.

Sussman, C., Blumenschine, R. J., Clark, J. D., & Misra, B. B. (1983). Preliminary report on excavations at the Mesolithic occupation site at Baghor II locality. In G. R. Sharma & J. D. Clark

(Eds.), *Palaeoenvironments and prehistory in the middle Son Valley* (pp. 161–196). Allahabad: Abinash Prakashan.

Thompson, L. G., Yao, T., Davis, M. E., Henderson, K. A., Mosley-Thompson, E., Lin, P. N., Beer, J., Synal, H. A., Cole-Dai, J., & Bolzan, J. F. (1997). Tropical climate instability: The last glacial cycle from a Qinghai-Tibetan ice core. *Science, 276*, 1821–1825.

Williams, M., & Clarke, M. F. (1984). Late quaternary environments in north-central India. *Nature, 308*, 633–635.

Williams, M. A. J., Pal, J. N., Jaiswal, M., & Singhvi, A. K. (2006). River response to quaternary climatic fluctuations: Evidence from the Son and Belan valleys, north-central India. *Quaternary Science Reviews, 25*, 2619–2631.

Williams, M., & Royce, K. (1982). Quaternary geology of the middle Son valley, north central India: Implications for prehistoric archaeology. *Palaeogeography, Palaeoclimatology, Palaeoecology, 38*, 139–162.

Chapter 4
The Hoabinhian of Southeast Asia and its Relationship to Regional Pleistocene Lithic Technologies

Ben Marwick

Introduction

The Hoabinhian represents a certain way of making stone artifacts, especially sumatraliths, during the late Pleistocene and early Holocene in island and mainland Southeast Asia. Although it has been a widely accepted and used concept in the region for several decades, its relationships to key patterns of technological and paleoenvironmental change both at a global scale and in adjacent regions such as East Asia, South Asia, and Australia are currently poorly understood. What makes these relationships especially intriguing is that the geographical locations of Hoabinhian assemblages are at strategic points on the arc of dispersal from Africa to Australia. This highlights the potential of Southeast Asia as a key source of data on the range and diversity of technological behaviors of the colonizers of the southern arc. The archaeology of the regions neighboring the Hoabinhian area has relevance for assessing its industrial affiliations. In this paper I examine select Pleistocene stone artifact assemblages from South Asia, East Asia, and Southeast Asia to investigate patterns of change in stone artifact technology and paleoenvironments. I place the findings within broader debates over the development of the Southeast Asian Paleolithic (Fig. 4.1).

Understanding the Hoabinhian can be challenging due to the use of diverse and abstract concepts such as "culture," "industry," "tradition," and "techno-complex" (Moser 2001). However, if we consider specifically how the stone artifacts that

B. Marwick (✉)
Department of Anthropology, University of Washington, Seattle, WA, USA

Center for Archaeological Science, University of Wollongong, Wollongong, NSW, Australia
e-mail: bmarwick@uw.edu

E. Robinson, F. Sellet (eds.), *Lithic Technological Organization and Paleoenvironmental Change*, Studies in Human Ecology and Adaptation 9,
DOI 10.1007/978-3-319-64407-3_4

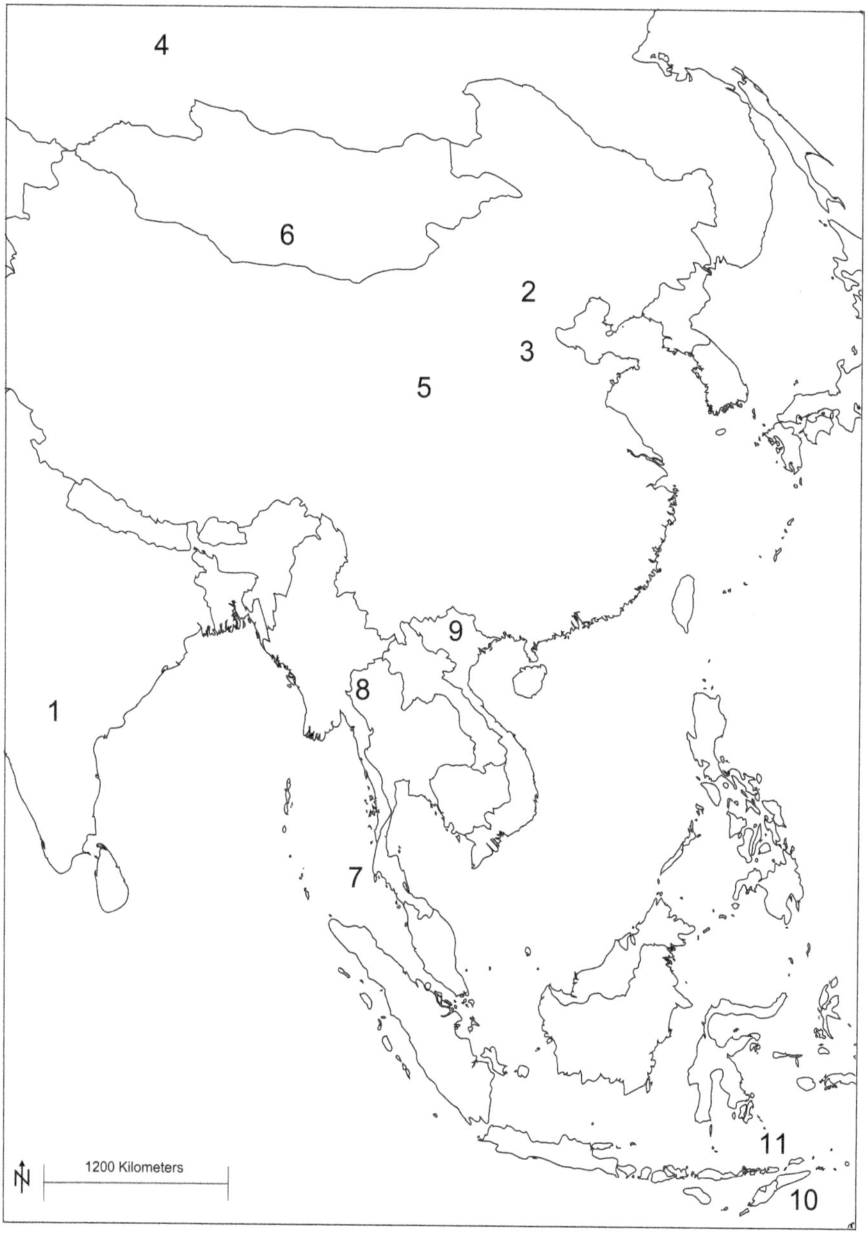

Fig. 4.1 Map of key locations discussed in the text. (1) Jurreru Valley; (2) Donggutuo; (3) Zhoukoudian; (4) Kara Bom; (5) Shuidonggou; (6) Tsagaan Agui; (7) Lang Rongrien; (8) Tham Lod; (9) Tham Khoung and Xom Trai; (10) Jerimalai; (11) Liang Bua and Mata Menge

characterize Hoabinhian sites have been described, we see a clear and consistent theme that unites Hoabinhian sites throughout Southeast Asia. This begins with one of the earliest definitions of the Hoabinhian, drafted in 1932 at the First Congress of Prehistorians of the Far East. At that meeting, French archaeologists working in northern Vietnam proposed that the Hoabinhian "is characterized by tools often worked only on one face, by hammerstones, by implements of sub-triangular section, by discs, short axes and almond shaped artifacts" (Matthews 1966). Forty years later, the lithic elements of the Hoabinhian were similarly defined by Gorman (1972), based on his work on northwest Thailand: "A generally unifacial flaked tool tradition made primarily on water rounded pebbles and large flakes detached from these pebbles, core tools ('sumatraliths') made by complete flaking on one side of a pebble and grinding stones also made on rounded pebbles." Recent literature continues this theme, with Chitkament et al. (2015) noting that sumatraliths are the "the signature of the Hoabinhian technical traditions" and Ji et al. (2016) similarly summarizing as follows: "the larger Hoabinhian tools are consistently shaped on cobbles and are distinguished by a consistent plano-convex cross-section. The common theme for formal tools centers around the unifacial flaking of water-rounded cobbles, with flaking usually around the circumference of a unifacial tool, which has been called the "sumatralith"" (Fig. 4.2).

Although the chronological and geographical ranges of this technology continue to be contested (Moser 2001, 2012), there appears to be a consensus among current usage of the term "Hoabinhian" that the sumatraliths, and closely related artifact forms, are the central defining attributes of Hoabinhian assemblages. In the following sections, I briefly summarize the late Pleistocene stone artifact technologies of

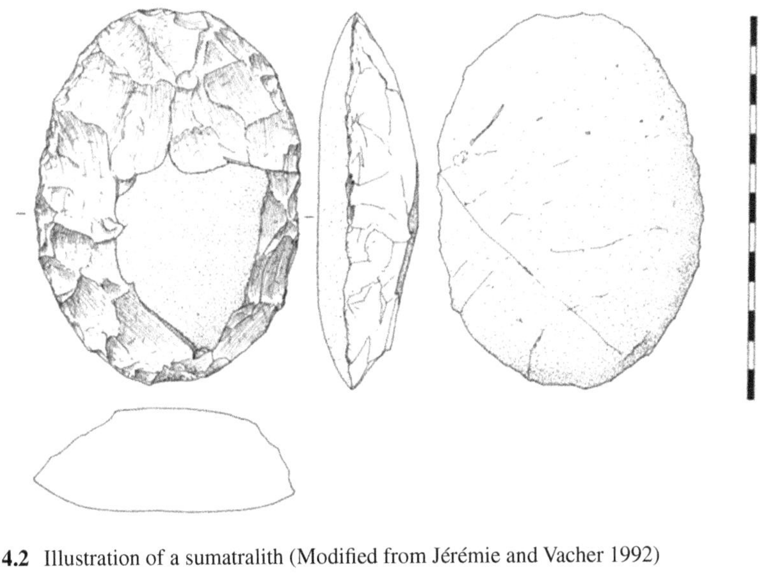

Fig. 4.2 Illustration of a sumatralith (Modified from Jérémie and Vacher 1992)

a selection of regions surrounding the core Hoabinhian area of mainland Southeast Asia. I conclude with a discussion that hypothesizes some potential relationships between the Hoabinhian and surrounding technologies.

South Asia

A richly detailed record of the South Asian Paleolithic comes from several open sites and one rock-shelter in the Jurreru Valley in the Kurnool district of southern India. Much of the work in this area has focused on understanding the impact of the Youngest Toba Tuff (YTT) event at 73.8 ka on human populations (Petraglia et al. 2007; Storey et al. 2012). The two pre-YTT assemblages (n = 1889) have clear Middle Stone Age affinities with Levallois products (~1%), scrapers (~8%), notches (~2%), blades (~1%), and discoidal cores (~0.5%) (Clarkson et al. 2012). There are very few differences between the pre- and post-YTT assemblages. Raw materials are dominated by limestone before and after the YTT event, with a gradual increase in chert and chalcedony much later (Clarkson et al. 2012). Most of these differences occur among flakes, flaked pieces, notches, sidescrapers, and multiplatform cores and are in the order of a 1% change. The post-YTT decline in flakes and increase in flaked pieces, which are numerically dominant in both periods, are probably more indicative of fragmentation than a behavioral shift. The other changes represent a gradually increasing emphasis on single and multiplatform cores and a trend toward higher levels of retouch intensity (Clarkson et al. 2012). The taxonomic affiliation of the makers of these assemblages is uncertain (Appenzeller 2012; Mellars et al. 2013), but the Middle Stone Age affinities suggest a *H. sapiens* population before and after the YTT event (Petraglia et al. 2007). These technological continuities suggest that the YTT event did not have the drastic population differentiation effects that Ambrose (1998) claims would follow from drastically diminished human population sizes.

In contrast to this gradual change, a period of substantial ecological, technological, and demographic transition at 35 ka has been documented in this region. At this time there was a large increase in production rates of microblades coincident with an increasingly fragmented and variable environment and indications of newly separated human populations in India (Petraglia et al. 2009). The microliths are initially elongate asymmetric backed point forms (i.e., trapezes and triangles) followed by greater proportions of symmetric forms (i.e., lunates) and, in the later part of the sequence, artifacts backed along both margins (thought to be drills) (Clarkson et al. 2009, 2012). One influential explanation of this dominance of microliths was that they indicated an eastward dispersal of modern humans from Africa (Mellars 2006). However, anatomically modern humans appear in South Asia before the proliferation of microliths (Perera et al. 2011), and the continuities of the Middle Paleolithic assemblages into the later Pleistocene do not support the idea of population replacement. Instead, Clarkson et al. (2009) propose that the more variable climate motivated people to produce more microliths to support an increased reliance on barbed projectiles as a highly reliable and maintainable hunting technology.

East Asia

In shifting our attention to East Asia, we note that the vocabulary and methods of stone artifact analysis also shift, adding an element of complexity in making regional comparisons. As in South Asia, there is no clear division between the Middle and Upper Paleolithic. At Donggutuo, dated to about 1 mya in the Nihewan Basin of northern China, Schick et al. (1991) describe an assemblage dominated by unmodified flakes and flake fragments (62%) and small numbers of retouched pieces (sidescrapers, end scrapers, notches, points, awls, and burins (~1%) and cores (5%)). The raw material is mostly variable quality chert/fine-grained quartzite (96%) with smaller amounts of limestone and volcanic rocks. They characterize the assemblage as simple, casual, and minimally modified. Hou (2003) has also noted that the assemblage contains sidescrapers, end scrapers, notches, points, awls, and burins and a type of radial discoid core with small elongate flake scars that is similar to the microlithic wedge-shaped cores of northern China during the Upper Paleolithic (Wang et al. 2005). Levallois products have been noted at a few locations, such as Zhoukoudian and Kara Bom in the north (Brantingham et al. 2001; Boëda et al. 2013), but their presence in early assemblages in China is controversial (Derevianko 2011).

Like the South Asian sequence from the Jurreru Valley, the East Asian sequence from the Lower to Upper Pleistocene is characterized by changes in degree rather than kind (Gao and Norton 2002). For example, over time the diversity of retouched flake types gradually increased, direct hard hammer percussion became more consistent with less variation in flake sizes, and the frequency of block-on-block or anvil-rested large flake production dropped to zero. At about 30 ka – slightly earlier in the Mongolian Gobi to the north, suggesting a north-south gradient of change – a more sudden set of changes occur in the East Asian sequence. The appearance of this set marks the start of the Initial Upper Paleolithic and includes microblade tools (microblades, microblade cores, backed artifacts) and in some assemblage blade tools (blade cores, prepared cores, flakes with facetted platforms, triangle flakes, sidescrapers, end scrapers, points, borers, gravers, notches, and notch-denticulates) (Gao and Norton 2002; Hou et al. 2013). At one of the best documented sites for this period, Shuidonggou in northern China, Levallois core reduction strategies also appear to become prominent at this time. Analysis of the rich assemblage from the Shuidonggou localities suggests that Middle Paleolithic technologies, such as high proportions of sidescrapers and notched-denticulate tools, continued after 30 ka with the chief novelty being the addition of an emphasis on the production of blades from Levallois-like prepared cores (Brantingham et al. 2001; Pei et al. 2012). Increasingly preferential raw material use accompanies the proliferation of blades in North and East Asia, for example, at Tsagaan Agui in Mongolia, exotic high-quality cherts and chalcedonies appear in the assemblage for the first time (Brantingham et al. 2001). Current explanations for the emphasis on blades and preference for high-quality raw materials at 30 ka in East Asia center on adaptation and risk minimization. Blade production is a highly standardized process and

Levallois core geometries optimize core productivity, so these two qualities contribute toward a continuous and predictable supply of useful stone tools (Brantingham et al. 2001). However, the motivation for an increased effort to optimize stone tool production and minimize the risk of being without useful stone is currently unclear. There is little evidence of new climatic conditions in the millennia leading up to 30 ka (Wang et al. 2001). One possibility is that there appears to have been an expansion of lakes between ~40 and 35/30 ka in the deserts of western China due to an intensified South Asian summer monsoon circulation (Yang and Scuderi 2010). Without knowing more about the direct effects of this lake expansion (e.g., if stone sources were submerged by the lake expansion), it is difficult to link it to the increased effort in technological risk minimization suggested by microlithic technologies. This raises the possibility that less visible factors, such as demographic dynamics, may be more relevant (cf. Collard et al. 2005).

Southeast Asia

Continuing the arc through to Southeast Asia, we find archaeological sequences that seem to have little in common with South and East Asia. On the mainland there are two especially well-documented lithic sequences spanning the late Pleistocene, Lang Rongrien in Peninsula Thailand and Tham Lod in northwest Thailand. Lang Rongrien has a date from the lower part of the deposit of 43,000 cal BP (Anderson 1997). The Pleistocene artifact assemblage is small, with only 48 stone, bone, and antler artifacts. The stone component is mostly unaltered flakes with some small core tools with unifacially retouched denticulate cutting edges and pieces resembling burins and small blades. Raw materials are mostly locally available geyserite, chert, and silicified limestone. With such a small assemblage, it is not surprising that the assemblage appears relatively simple, but its small size also limits the representativeness of this assemblage for the Southeast Asian Pleistocene.

A larger assemblage comes from Tham Lod in northwest Thailand where the excavated deposit spans 10,000–40,000 cal BP (Marwick and Gagan 2011; Marwick 2013). This is an unusual sequence because it is uninterrupted, unlike many other mainland Southeast Asian sites that tend not to have archaeological material during 25–15 ka (O'Connor and Bulbeck 2014). Although flaked artifacts are more numerous, with 2713 flakes and cores, the flake technology shares many similarities with Lang Rongrien. Raw materials are mostly sandstone (53%) and quartzite (30%) with small amounts of other sedimentary rocks. Of the 2280 flakes at Tham Lod, 62 are blades and only 18 showed unambiguous signs of retouch. When they are present, blades vary between 1 and 7% of the assemblage in excavation units. Retouch was not extensive, with a mean index of invasiveness value of 0.12 and a mean value of retouch as a proportion of a flake's perimeter of 0.28. The distribution of retouch on flakes does appear to be patterned with the lateral margins near the platform showing retouch in more than half of the sample. There are a very small number of more distinctive pieces such as bipolar, end scrapers, bifacial pieces, and platform rejuvenation flakes. As for Lang Rongrien, it is difficult to interpret the behavioral significance of such a small sample of retouched pieces.

A more revealing component of the Tham Lod assemblage is the cores. Twenty-six cores from excavation area one are distinctive as classical sumatralith forms made from metaquartzite or sandstone, the holotype for the Hoabinhian. The earliest sumatralith appears in an excavation unit dated to 23,000 cal BP, and there is a peak in Sumatralith numbers at about 21,000 cal BP ($n = 7$) and then a uniform distribution throughout the sequence. These are not the earliest sumatraliths in the region, with earlier specimens in Vietnam at Tham Khoung (33,150 cal BP) and China at Xiaodong rock-shelter (43,500 cal BP) (Ji et al. 2016). Similarly, the peak in sumatralith discard at Tham Lod appears to be in synchrony with the Hoabinhian elsewhere in mainland Southeast Asia. For example, at Xom Trai the peak discard event of the Hoabinhian is dated to 17–18,000 cal BP (Van Tan 1997). Other distinctive Pleistocene technological elements identified in Vietnam such as the Nguomian, a pre-Hoabinhian assemblage characterized by high frequencies of small flake tools and blades, have not been identified in the Tham Lod assemblage. A pre-Hoabinhian assemblage of nine small flakes made on whitish-yellowish chert has been reported from Ngeubhinh Mouxeu rock-shelter in northern Laos (Zeitoun et al. 2012). Although this component of the assemblage is very small, it may be related to the Pleistocene flake component at Lang Rongrien in indicating a general Pleistocene adaptation with a preference for small artifacts made from cryptocrystalline stone, contrasting with the larger cobble artifacts made from coarser-grained stone typical of the Hoabinhian.

The degree of heterogeneity we currently observe in the few late Pleistocene lithic assemblages available from mainland Southeast Asia suggests that technological stage systems such as that used in Vietnam may have limited applicability beyond the sites where the systems were first applied. However, the similar timing of the first sumatralith at Tham Lod and the peak discard at Xom Trai may point to a connection between technological change and climate fluctuations as they are approximately coincident with a Heinrich event (H1) that has been documented in Chinese speleothem paleoclimate archives (Wang et al. 2001). Heinrich events result from the release of large volumes of ice into the North Atlantic, resulting in global climate fluctuations, often including extremely cold periods. Equivalent speleothem paleoclimate archives have not yet been produced for mainland Southeast Asia, so these paleoenvironmental mechanisms for technological change are currently tenuous. The oxygen isotope paleoclimate record from Tham Lod is currently the most relevant source of insights in the relationship between late Pleistocene climate change and technological change (cf. Marwick and Gagan 2011). But it lacks the chronological resolution required to robustly demonstrate a link between Heinrich events and technological change. If such a link can be shown in future works, it may suggest a process of technological adaptation in Southeast Asia comparable to the proliferation of microliths in South Asia.

While microliths are virtually absent from the Tham Lod assemblage, there are cores that indicate a Levallois-like radial reduction strategy similar to assemblages in South and East Asia. Twenty-eight cores have a radial geometry, the first appearing at 34,000 cal BP, and half of them appear between that date and 24,000 cal BP. These radial cores do not differ significantly in their distributions of mass, number of flake scars, or amount of cortex from other cores, and so the radial cores probably reflect a process that is well integrated into the technological behaviors evident in the assemblage (Marwick 2008). They are present throughout the assemblage, at their most abundant making up 60% of the cores in the excavation unit dated to 33,000 cal BP. Discoidal cores are absent from the assemblage.

The heterogeneity of Pleistocene technology in Southeast Asia can be further illustrated by comparing Tham Lod to a similarly long and rich lithic sequence recovered from Jerimalai rock-shelter in East Timor. Excavations by O'Connor et al. (2011) recovered a large number of stone artifacts (n = 9752), bone points, fishhooks, and shell beads dating to the terminal Pleistocene. The stone artifacts are made from a locally available high-quality chert. In contrast to Tham Lod and Lang Rongrien, the lithics at Jerimalai can be characterized as a faceted radial and rotated core technology and an informal scraper assemblage (O'Connor et al. 2011). Analysis of the depth-age data in Reepmeyer et al. (2011) indicates three discrete depositional episodes at Jerimalai. The earliest episode spans about 42–40 ka, followed by a period of relatively slow sediment accumulation from about 40 to 10 ka. This slow rate of deposition is probably explained by the glacial maximum spanning 30–18 ka, when the site would have been up to 120 m above sea level and thus much further from the sea (O'Connor 2007). Finally, there is a period of relatively fast deposition from 10 to 5 ka which O'Connor (2007) has previously noted that corresponds to the stabilization of sea levels and establishment of a coastal environment near the site at the start of the Holocene. Despite the variations in depositional environments over time, there is no indication of any major changes in the lithic technology (O'Connor et al. 2011; Reepmeyer et al. 2011). Only subtle differences based on small counts of artifacts are evident, and these are not suggestive of major technological reorganization (Marwick et al. 2016). These similarities between depositional episodes indicate a highly stable technological system that appears to be insensitive to major climatic events such as the glacial maximum and the transition from the Pleistocene to the Holocene.

Although the Jerimalai assemblage has no sumatraliths or other Hoabinhian elements, it shares with Tham Lod the presence of radial cores and, with many other Pleistocene assemblages in Southeast Asia, an absence of recurring formal types and a long-term technological conservatism. Jeremalai is also relevant for assessing the taxonomic significance of the lithic assemblages from Liang Bua and Mata Menge, sites that are associated with *Homo floresiensis* (Brumm et al. 2006; Brumm et al. 2010). Moore et al. (2009) identify a group of larger flakes in their assemblage that appear to have been produced off-site and transported to the site where they are discarded as retouched flakes. A similar reduction strategy is identified in the much older Mata Menge assemblage with its small proportions of retouched pieces that tend to be larger than unretouched flakes. This has a parallel in the Jerimalai assemblage with the reduction pathway that resulted in larger retouched flakes. Moore et al. (2009) have noted that the assemblages produced by *H. sapiens* and *H. floresiensis* at Liang Bua are substantially the same. Marwick et al. (2016) have made a detailed study of the similarities in the exploitation, use and discard of stone resources with Liang Bua and Mata Menge. A key implication of their findings is that there is no compelling evidence that separates the makers of the Jerimalai assemblage from the *H. floresiensis* assemblage at Liang Bua and an unknown earlier ancestor that made the assemblage at Mata Menge. While we do not have skeletal remains identifying the hominids responsible for the Jerimalai assemblage, O'Connor et al. (2007, 2011) have argued that they were *H. sapiens* because of the faunal assemblage that includes pelagic species such as tuna which could only have been captured in deep offshore waters with fishhooks (one was found in an early Holocene layer) and probably watercraft.

Discussion

This survey of key late Pleistocene lithic assemblages in three regions of the eastern hemisphere allows us to engage with a few global debates on technological change. The data presented here lend little support to Mellars' (2006) claim of a single hominin dispersal responsible for the appearance of microliths out of Africa. The proliferation of microliths in China can be traced to a set of much earlier local technologies that predate Mellars' dispersal event. Microlithic technology in South Asia is most parsimoniously explained as an in situ development from earlier technologies in response to a reduction in summer rainfall and temperature, producing semi-glacial mosaic environments and desert expansion (Hiscock et al. 2011). The timing of the proliferation of microliths in both South and East Asia around 35,000 cal BP is also not consistent with Mellars' claim for 60–50,000 cal BP for the expansion of microlith-bearing modern humans. Similarly, anatomically modern humans appear in both South and East Asia before the proliferation of microliths. These details point to microliths as autochthonous developments rather than part of a migration package.

The absence of microliths from assemblages in Southeast Asia further challenges the model of microlith-bearing modern humans. The assemblages at Tham Lod and Jerimalai are most parsimoniously explained as products of modern humans, and yet little of the distinctive archaeological evidence that Mellars' claims to be characteristic of modern humans is present. A similar situation has been noted in Australian Pleistocene lithic assemblages by Brumm and Moore (2005). In brief, they argue that the absence of classical indicators of modern humans is not evidence of their absence, but of a need to reconsider how modernity is defined. Drawing further on the Australian Pleistocene, Langley et al. (2011) conclude that the search for singular "packages" of behavioral modernity is misguided and the diversity of specific manifestations of early modern human populations was more likely a result of differences in population size and density, interaction, and historical contingency. The absence of typically modern human artifacts such as microlithic blades and backed artifacts in Southeast Asian assemblages such as Jerimalai and Tham Lod reinforces the importance of these kinds of artifacts as adaptions to specific environmental and demographic conditions (Clarkson et al. 2009; Petraglia et al. 2009).

The data discussed here are also relevant to the question of hominin taxonomic affiliations of lithic assemblages. The similarities between the pre- and post-YTT event have been interpreted by Clarkson et al. (2012) indicating that modern humans produced both assemblages, although this claim has been challenged by (Mellars et al. 2013) using genetic evidence. The many points of similarity between Jerimalai, Liang Bua and Mata Menge suggest that as many as three species of hominin were producing very similar assemblages. The overall picture is that these assemblages give little support to the claims of Foley and Lahr (2003) that different hominid species are associated with different stone artifact technologies. The implications are that either the hominins active during the Pleistocene in Southeast Asia were much more similar in their behavioral and cognitive expressions than their taxonomic affiliations would imply or that stone artifact assemblages such as those found in this region either lack sufficient detail (or current approaches to describing the assemblage sufficient detail) to discriminate between the technological behaviors of the three hominins.

An interesting common element in the assemblages here are radial cores, which are often associated with Levallois flake products. Even though Tham Lod has radial cores overlapping with the Hoabinhian, it is tempting to identify the centripetal flaking that characterizes sumatraliths as a variant of radial core technology. The Hoabinhian may thus be a regional expression, as a divergent descendant, of a very widespread radial method of core reduction. This invites the question of whether radial cores and analogous forms are a result of technological behaviors that are homologous, having persisted from when radial cores first proliferated in Africa around 300 ka (Ambrose 2001). The alternative possibility is that radial cores developed as homoplasy or convergent technological evolution (Brumm et al. 2009) as local adaptations. Answering this question is important for accurately contextualizing the relationships between the three major regions discussed here. Clarkson et al. (2012) have had some success using stone artifact data to differentiate modern human core technology from Neanderthals and late Acheulean populations. Future work with their methods on early Hoabinhian assemblages may help to resolve these questions.

The possibility of an ambiguous trace of Levallois technology in the Tham Lod and other Southeast Asian assembles raises the question of why there is no obvious evidence of Levallois technology in Southeast Asia. The frequent association of bifacial pieces in Acheulean assemblages preceding Levallois technology elsewhere in the world (Schick 2002; Lycett 2007) is also problematic for Southeast Asia, where Pleistocene bifaces are less frequent (Moncel et al. 2017). Two current hypotheses are relevant to these questions. First, the hypothesis is that complex stone artifacts were not produced because of the abundance and ease of using bamboo to make complex tools (Pope 1989). West and Louys (2007) have shown experimentally that it is possible to distinguish cut marks on bone made by bamboo from marks made by stone, but these methods are yet to be applied to an archaeological assemblage. Similarly, Bar-Yosef et al. (2012) have replicated complex forms such as bifacial hand axes and preferential Levallois cores and flakes on raw materials common in Pleistocene Southeast Asian assemblages such as metaquartzite and indurated sandstone. They further experimented with bamboo tools for cutting pig meat and hide and found that while bamboo knives were effective at meat-cutting, they were ineffective for cutting the skin.

These two experimental studies do not directly confirm the bamboo hypothesis, but indicate that an important priority in future research should be the analysis of microscopic cut marks, usewear and residues on archaeological fauna, and stone artifacts to determine the role of bamboo tools during the Pleistocene. Of relevance to the Hoabinhian is the observation by Bar-Yosef et al. (2012) that unifacially flaked choppers were more efficient than bifaces at felling bamboo. This lends support to the long-standing claim that Hoabinhian technology was optimized for bamboo harvesting (Gorman 1972). We might speculate that the technology perhaps became abundant during biogeographical events related to the expansion of the temperate and paleotropical woody bamboo clades (cf. Kelchner 2013) during cooler and dryer episodes in Southeast Asia. Current phylogenetic work on the timing of major biogeographical changes in Southeast Asian bamboo populations during the late Pleistocene will help test this hypothesis.

A second hypothesis for the low frequency of Levallois and bifaces in Southeast Asia is that the Pleistocene hunter-gatherer populations were too small for widespread adoption and persistence of shared complex technologies (Lycett 2007). Simulations by Shennan (2001) and numerical analysis by Henrich (2004) demonstrate that larger populations result in more chances of artifact makers copying from more skilled makers. Higher degrees of social interconnectedness in larger populations similarly result in a higher likelihood of encountering a given craft skill and the greater regularity of such encounters. In smaller populations with lower degrees of interconnectedness, encounters of highly skilled craftspeople are rarer, opportunities from copying are fewer, and the probability of technological loss is higher. Lycett (2007) argues that the overall low density of artifacts and site densities in East Asia indicate lower demographic conditions which were the primary constraint leading to small numbers of bifaces and assemblages with Levallois technologies. While this is an intriguing hypothesis, it currently rests on a problematic "absence of evidence" argument, where the absence of Levallois is taken to indicate the presence of a constraint, rather than simply a smaller sample size. It is also difficult to persuasively test with the currently available archaeological data, given uncertainties involved in inferring demography from distributions of small numbers of sites and dates (i.e., <1000, cf. Buchanan et al. 2008). The most persuasive test of this hypothesis is likely to come from genetic material recovered from ancient human remains. These are an especially rare type of find in any region, but recent developments in high-throughput sequencing of highly fragmented and morphologically indistinct bone may improve recovery of human aDNA from archaeological assemblages (Murray et al. 2013). Genetic evidence currently available from South Asia poses a challenge to this hypothesis, suggesting relatively high populations during the late Pleistocene (Atkinson et al. 2008).

Conclusion

The most parsimonious interpretation of the assemblages described here is that they indicate multiple independent technological innovations indicating different degrees of environmental adaptation. The model of a single migration out of Africa that populated the eastern hemisphere with modern humans characterized by a package of technology is not well supported by the data presented here. While a single population diaspora is not disproven, the characteristic package of modernity is not uniformly evident in these assemblages. The common thread of radial core technology may hint at a shared ancestry, and the Hoabinhian may be a local Southeast Asian expression of this heritage. Future work involving geometric morphometric analysis of cores, pioneered by Clarkson et al. (2012), may be helpful for clarifying the taxonomic affiliations of the late Pleistocene technologies and their phylogenetic relations.

The relationship between lithic technology and paleoenvironments in East and Southeast Asia has many unresolved questions, mostly due to the shortage of well-dated, high-resolution archaeological records that are suitable for correlating with paleoclimate records. In Southeast Asia we still lack high-resolution paleoclimate records. This means we cannot be sure if correlation between environmental change and behavioral adaptation is simply less strong than in other regions or that we do not yet have a large and more detailed enough dataset to make credible claims. A priority for future work should be to identify locations that are amenable to study by interdisciplinary teams that can produce detailed archaeological and paleoenvironmental data spanning the late Pleistocene, and for the data collected to be openly shared (Marwick 2017). In South Asia there is a strong suggestion that the appearance of microliths is related to habitat fragmentation, in East Asia an environmental motive for the appearance of microliths is not readily apparent, and in Southeast Asia, microliths did not appear at all. Instead, in mainland Southeast Asia, the Hoabinhian is a distinctive Pleistocene lithic component, which may have some affinity with South Asian assemblages with its Levallois-like radial reduction strategy. We might tentatively suggest that the first appearance of the Hoabinhian is related to environmental conditions influencing bamboo forests. However the long-term conservatism in the assemblages from Jerimalai in island Southeast Asia indicates that environmental changes need not have any impact on technological choices in Southeast Asia. Finally, the possible influence of demographic dynamics on the near-absence of bifaces and Levallois from Southeast Asia hints at an equally important role for demographic dynamics along with paleoenvironmental changes.

Acknowledgments Thanks to Rasmi Shoocongedej for the opportunity to be involved in the work at Tham Lod, to Sue O'Connor for the access to the Jerimalai lithic assemblage, to Ya-Mei Hou for the advice on the Donggutuo assemblage, and to Scot Kelchner for the advice on bamboo phylogenetics. Any errors remain my own.

References

Ambrose, S. H. (1998). Late Pleistocene human population bottlenecks, volcanic winter, and differentiation of modern humans. *Journal of Human Evolution, 34*(6), 623–651.

Ambrose, S. H. (2001). Paleolithic technology and human evolution. *Science, 291*(5509), 1748–1753.

Anderson, D. D. (1997). Cave archaeology in Southeast Asia. *Geoarchaeology, 12*(6), 607–638.

Appenzeller, T. (2012). Human migrations: Eastern odyssey. *Nature, 485*(7396), 24–26.

Atkinson, Q. D., Gray, R. D., & Drummond, A. J. (2008). mtDNA variation predicts population size in humans and reveals a major Southern Asian chapter in human prehistory. *Molecular Biology and Evolution, 25*, 468–474.

Bar-Yosef, O., Eren, M. I., Yuan, J., Cohen, D. J., & Li, Y. (2012). Were bamboo tools made in prehistoric Southeast Asia? An experimental view from South China. *Quaternary International, 269*, 9–21.

Boëda, E., Hou, Y. M., Forestier, H., Sarel, J., & Wang, H. M. (2013). Levallois and non-Levallois blade production at Shuidonggou in Ningxia, North China. *Quaternary International, 295*, 191–203.

Brantingham, P. J., Krivoshapkin, A. I., Jinzeng, L., & Tserendagva, Y. (2001). The initial upper paleolithic in northeast Asia. *Current Anthropology, 42*(5), 735–747.

Brumm, A., Aziz, F., van den Bergh, G. D., Morwood, M. J., Moore, M. W., Kurniawan, I., Hobbs, D. R., & Fullagar, R. (2006). Early stone technology on Flores and its implications for Homo floresiensis. *Nature, 441*(7093), 624–628.

Brumm, A., Kurniawan, I., Moore, M. W., Suyono, S., Jatmiko, R., Morwood, M. J., & Aziz, F. (2009). Early Pleistocene stone technology at Mata Menge, central Flores, Indonesia. In F. Aziz, M. J. Morwood, & G. D. van den Bergh (Eds.), *Pleistocene geology, palaeontology and archaeology of the Soa Basin, Central Flores, Indonesia* (pp. 119–137). Bandung: Geological Survey Institute Bandung. Special Publication 36.

Brumm, A., & Moore, M. W. (2005). Symbolic revolutions and the Australian archaeological record. *Cambridge Archaeological Journal, 15*(2), 157–175.

Brumm, A., Moore, M. W., van den Bergh, G. D., Kurniawan, I., Morwood, M. J., & Aziz, F. (2010). Stone technology at the Middle Pleistocene site of Mata Menge, Flores, Indonesia. *Journal of Archaeological Science, 37*(3), 451–473.

Buchanan, B., Collard, M., & Edinborough, K. (2008). Paleoindian demography and the extraterrestrial impact hypothesis. *Proceedings of the National Academy of Sciences, 105*(33), 11651–11654.

Chitkament, T., Gaillard, C., & Shoocongdej, R. (2015). Tham Lod rockshelter (Pang Mapha district, north-western Thailand): Evolution of the lithic assemblages during the late Pleistocene. *Quaternary International, 416*, 151–161.

Clarkson, C., Jones, S., & Harris, C. (2012). Continuity and change in the lithic industries of the Jurreru Valley, India, before and after the Toba eruption. *Quaternary International, 258*, 165–179.

Clarkson, C., Petraglia, M., Korisettar, R., Haslam, M., Boivin, N., Crowther, A., Ditchfield, P., Fuller, D., Miracle, P., & Harris, C. (2009). The oldest and longest enduring microlithic sequence in India: 35 000 years of modern human occupation and change at the Jwalapuram Locality 9 rockshelter. *Antiquity, 83*(320), 326–348.

Collard, M., Kemery, M., & Banks, S. (2005). Causes of toolkit variation among hunter-gatherers: a test of four competing hypotheses. *Canadian Journal of Archaeology, 29*, 1–19.

Derevianko, A. P. (2011). Three scenarios of the Middle to Upper Paleolithic transition: Scenario 2: The Middle to Upper Paleolithic transition in continental East Asia. *Archaeology, Ethnology and Anthropology of Eurasia, 39*(1), 2–27.

Foley, R., & Lahr, M. M. (2003). On stony ground: Lithic technology, human evolution, and the emergence of culture. *Evolutionary Anthropology: Issues, News, and Reviews, 12*(3), 109–122.

Gao, X., & Norton, C. J. (2002). A critique of the Chinese 'Middle Palaeolithic'. *Antiquity, 76*(292), 397–412.

Gorman, C. (1972). Excavations at Spirit Cave, North Thailand: Some interim interpretations. *Asian Perspectives, 13*, 79–107.
Henrich, J. (2004). Demography and cultural evolution: How adaptive cultural processes can produce maladaptive losses-The Tasmanian case. *American Antiquity, 69*(2), 197–214.
Hiscock, P., Clarkson, C., & Mackay, A. (2011). Big debates over little tools: ongoing disputes over microliths on three continents. *World Archaeology, 43*(4), 653–664.
Hou, Y.-M., Yang, S.-X., Dong, W., Zhang, J.-F., & Liu, Y. (2013). Late Pleistocene representative sites in North China and their indication of evolutionary human behavior. *Quaternary International, 295*, 183–190.
Hou, Y. M. (2003). Naming and preliminary study on the category of the "Donggutuo core". *Acta Anthropologica Sinica, 22*(4), 79–292.
Jérémie, S., & Vacher, S. (1992). Le Hoabinhien en Thaïlande: un exemple d'approche expérimentale. *Bulletin de l'Ecole française d'Extrême-Orient, 79*(1), 173–209.
Ji, X., Kuman, K., Clarke, R. J., Forestier, H., Li, Y., Ma, J., Qiu, K., Li, H., & Wu, Y. (2016). The oldest Hoabinhian technocomplex in Asia (43.5 ka) at Xiaodong rockshelter, Yunnan Province, southwest China. *Quaternary International, 400*, 166–174.
Kelchner, S. A. (2013). Higher level phylogenetic relationships within the bamboos (Poaceae: Bambusoideae) based on five plastid markers. *Molecular Phylogenetics and Evolution, 67*(2), 404–413.
Langley, M. C., Clarkson, C., & Ulm, S. (2011). From small holes to grand narratives: the impact of taphonomy and sample size on the modernity debate in Australia and New Guinea. *Journal of Human Evolution, 61*(2), 197–208.
Lycett, S. J. (2007). Why is there a lack of Mode 3 Levallois technologies in East Asia? A phylogenetic test of the Movius-Schick hypothesis. *Journal of Anthropological Archaeology, 26*(4), 541–575.
Marwick, B. (2008). *Stone artefacts and human ecology at two Rockshelters in NorthwestThailand.* Unpublished PhD, The Australian National University.
Marwick, B. (2013). Multiple optima in Hoabinhian flaked stone artefact palaeoeconomics and palaeoecology at two archaeological sites in Northwest Thailand. *Journal of Anthropological Archaeology, 32*(4), 553–564.
Marwick, B. (2017). Computational reproducibility in archaeological research: Basic principles and a case study of their implementation. *Journal of Archaeological Method and Theory, 24*(2), 424–450.
Marwick, B., Clarkson, C., O'Connor, S., & Collins, S. (2016). Early modern human lithic technology from Jerimalai, East Timor. *Journal of Human Evolution, 101*, 45–64.
Marwick, B., & Gagan, M. K. (2011). Late Pleistocene monsoon variability in northwest Thailand: An oxygen isotope sequence from the bivalve Margaritanopsis laosensis excavated in Mae Hong Son province. *Quaternary Science Reviews, 30*(21–22), 3088–3098.
Matthews, J. M. (1966). A review of the 'Hoabinhian' in Indo-China. *Asian Perspectives, 9*, 86–95.
Mellars, P. (2006). Going east: New genetic and archaeological perspectives on the modern human colonization of Eurasia. *Science, 313*(5788), 796.
Mellars, P., Gori, K. C., Carr, M., Soares, P. A., & Richards, M. B. (2013). Genetic and archaeological perspectives on the initial modern human colonization of southern Asia. *Proceedings of the National Academy of Sciences, 110*(26), 10699–10704.
Moncel, M.-H., Arzarello, M., Boëda, E., Bonilauri, T., Chevrier, B., Gaillard, C., Forestier, H., Yinghua, L., Sémah, F., & Zeitoun, V. (2017). Assemblages with bifacial tools in Eurasia (second part). What is going on in the East? Data from India, Eastern Asia and Southeast Asia. *Comptes Rendus Palevol*. doi:10.1016/j.crpv.2015.09.010.
Moore, M. W., Sutikna, T., Morwood, M. J., & Brumm, A. (2009). Continuities in stone flaking technology at Liang Bua, Flores, Indonesia. *Journal of Human Evolution, 57*(5), 503–526.
Moser, J. (2001). *Hoabinhian: Geographie und Chronologie eines steinzeitlichen Technocomplexes in Südostasien*. Lindensoft: Köln.

Moser, J. (2012). The Hoabinhian definition—In the past and today: a short historical review of defining the Hoabinhian. In *Crossing Borders: Selected Papers from the 13th International Conference of the European Association of Southeast Asian Archaeologists* (Vol. 1). Singapore. NUS Press.

Murray, D. C., Haile, J., Dortch, J., White, N. E., Haouchar, D., Bellgard, M. I., Allcock, R. J., Prideaux, G. J., & Bunce, M. (2013). Scrapheap challenge: A novel bulk-bone metabarcoding method to investigate ancient DNA in faunal assemblages. *Scientific Reports, 3*(3371), 1–8.

O'Connor, S. (2007). New evidence from East Timor contributes to our understanding of earliest modern human colonisation east of the Sunda Shelf. *Antiquity, 81*(313), 523–535.

O'Connor, S., & Bulbeck, D. (2014). Homo sapiens societies in Indoneasia and South-Eastern Asia. In V. Cummings, P. Jordan, & M. Zvelebil (Eds.), *The Oxford handbook of the archaeology and anthropology of hunter-gatherers* (pp. 346–367). Oxford: Oxford University Press.

O'Connor, S., Ono, R., & Clarkson, C. (2011). Pelagic fishing at 42,000 years before the present and the maritime skills of modern humans. *Science, 334*(6059), 1117–1121.

Pei, S., Gao, X., Wang, H., Kuman, K., Bae, C. J., Chen, F., Guan, Y., Zhang, Y., Zhang, X., Peng, F., & Li, X. (2012). The Shuidonggou site complex: new excavations and implications for the earliest Late Paleolithic in North China. *Journal of Archaeological Science, 39*(12), 3610–3626.

Perera, N., Kourampas, N., Simpson, I. A., Deraniyagala, S. U., Bulbeck, D., Kamminga, J., Perera, J., Fuller, D. Q., Szabó, K., & Oliveira, N. V. (2011). People of the ancient rainforest: Late Pleistocene foragers at the Batadomba-lena rockshelter, Sri Lanka. *Journal of Human Evolution, 61*(3), 254–269.

Petraglia, M., Clarkson, C., Boivin, N., Haslam, M., Korisettar, R., Chaubey, G., Ditchfield, P., Fuller, D. Q., James, H., Jones, S., Kivisild, T., Koshy, J., Lahr, M. M., Metspalu, M., Roberts, R., & Arnold, L. (2009). Population increase and environmental deterioration correspond with microlithic innovations in South Asia ca. 35,000 years ago. *Proceedings of the National Academy of Sciences, 106*(30), 12261–12266.

Petraglia, M., Korisettar, R., Boivin, N., Clarkson, C., Ditchfield, P., Jones, S., Koshy, J., Lahr, M. M., Oppenheimer, C., Pyle, D., Roberts, R., Schwenninger, J.-L., Arnold, L., & White, K. (2007). Middle Paleolithic assemblages from the Indian subcontinent before and after the Toba Super-Eruption. *Science, 317*(5834), 114–116.

Pope, G. G. (1989). Bamboo and human evolution. *Natural History, 98*(10), 48–56.

Reepmeyer, C., O'Connor, S., & Brockwell, S. (2011). Long-term obsidian use at the Jerimalai rock shelter in East Timor. *Archaeology in Oceania, 46*(2), 85–90.

Schick, K., Toth, N., Qi, W., Clark, J. D., & Etler, D. (1991). Archaeological perspectives in the Nihewan Basin, China. *Journal of Human Evolution, 21*(1), 13–26.

Schick, K. D. (2002). A comparative perspective on Paleolithic cultural patterns. In T. Akazawa, K. Aoki, & O. Bar-Yosef (Eds.), *Neandertals and modern humans in Western Asia* (pp. 449–460). Dordrecht: Springer.

Shennan, S. (2001). Demography and cultural innovation: A model and its implications for the emergence of modern human culture. *Cambridge Archaeological Journal, 11*(1), 5–16.

Storey, M., Roberts, R. G., & Saidin, M. (2012). Astronomically calibrated 40Ar/39Ar age for the Toba supereruption and global synchronization of late Quaternary records. *Proceedings of the National Academy of Sciences, 109*(46), 18684–18688.

Van Tan, H. (1997). The Hoabinhian and before. *Bulletin of the Indo-Pacific Prehistory Association (Chiang Mai Papers, Volume 3), 16*, 35–41.

Wang, H., Deng, C., Zhu, R., Wei, Q., Hou, Y., & Boëda, E. (2005). Magnetostratigraphic dating of the Donggutuo and Maliang Paleolithic sites in the Nihewan Basin, North China. *Quaternary Research, 64*(1), 1–11.

Wang, Y. J., Cheng, H., Edwards, R. L., An, Z. S., Wu, J. Y., Shen, C. C., & Dorale, J. A. (2001). A high-resolution absolute-dated Late Pleistocene monsoon record from Hulu Cave, China. *Science, 294*(5550), 2345–2348.

West, J. A., & Louys, J. (2007). Differentiating bamboo from stone tool cut marks in the zooarchaeological record, with a discussion on the use of bamboo knives. *Journal of Archaeological Science, 34*(4), 512–518.

Yang, X., & Scuderi, L. A. (2010). Hydrological and climatic changes in deserts of China since the Late Pleistocene. *Quaternary Research, 73*(1), 1–9.

Zeitoun, V., Forestier, H., Pierret, A., Chiemsisouraj, C., Lorvankham, M., Latthagnot, A., Chanthamoungkhoun, T., & Norkhamsomphou, S. (2012). Multi-millennial occupation in northwestern Laos: Preliminary results of excavations at the Ngeubhinh Mouxeu rock-shelter. *Comptes Rendus Palevol, 11*(4), 305–313.

Chapter 5
Horizons of Change: Entanglement of Paleoenvironment and Cultural Dynamics in Australian Lithic Technology

Peter Hiscock

Introduction

Australian research into ancient lithic technology has often been polarized between models that depict technological change as a response to paleoenvironmental change and models that presume paleoenvironmental change was not the principle stimulus for social change. Both propositions offer only partial representations of cultural change, especially when large-scale evolution in technology has occurred. A powerful framework for defining the articulation of technological, social, and environmental changes is niche construction theory (NCT), a perspective focused on the capacity of organisms, in this case humans, to modify natural selection in their environment and in that way to contribute to the direction of their own evolution (Odling-Smee et al. 2003; Laland and O'Brien 2010). Niche construction therefore concentrates not only on the solutions that organisms find to survive in and exploit their environment but also how those solutions modify conditions in ways that change conditions confronting the organism. Application of a niche construction framework offers substantial value in understanding technological change and will form part of the analysis in this chapter.

Australia provides a vista of considerable technological change during the terminal Pleistocene and Holocene. One widespread instance is the development of microliths and subsequently an intensification of microlith production and geographic expansion of microlith assemblages. Understanding these complex temporal and spatial patterns requires an exploration of both environmental and social contexts. Furthermore the character of microliths themselves has been puzzling. They are small, sometimes very small, and yet often standardized and precisely made. We know that many of the specimens were used as tools, and yet it is also

P. Hiscock (✉)
Department of Archaeology, University of Sydney, Sydney, NSW, Australia
e-mail: peter.hiscock@sydney.edu.au

E. Robinson, F. Sellet (eds.), *Lithic Technological Organization and Paleoenvironmental Change*, Studies in Human Ecology and Adaptation 9,
DOI 10.1007/978-3-319-64407-3_5

clear that the objects carried social meaning. In this chapter, I examine these patterns in three steps: first, describing the spatial and temporal patterns of microlith production to depict the evolutionary change; second, characterizing the climatic and landscape contexts of those evolutionary changes; and third to offer speculations of the niche construction dynamics that might have been in operation.

Australian Backed Artifacts

In Australia, archaeologists have adopted the term "backed artifacts" to redefine the category often labeled as "microlith" elsewhere in the world. These backed artifacts are flakes with steep retouch along one or more margins. Their distinguishing feature is the near 90-degree retouch that was usually accomplished with the use of bipolar techniques on an anvil. Multivariate analyses have shown that backed artifacts are morphologically distinct from other retouched flakes and that this category contains specimens that have been selected and treated distinctively (Hiscock and Attenbrow 2005). The size of backed artifacts varies considerably between regions, and in some places, they are large; hence the reference to small size, in the term microlith, is not appropriate in the Australian context. Redefining the phenomena in terms of backing that blunted an edge has created a challenging new image of this technological system in Australia.

The new image stems in part from the more technological identification of specimens. By requiring backing retouch, my analyses exclude specimens that were unretouched but previously included in analyses from typological perspectives. There are many reasons that unretouched flakes in particular were misclassified as backed artifacts, including the confusion of dorsal platforms, platform faceting, overhang removal, and heat shattering, with backing retouch. Since error rates in identifying backed artifacts were commonly in the range on 10–40% of analyzed specimens, those previous typological characterizations were often inaccurate. By including only retouched flakes with backing in this research, it is possible to describe production, tool use, and spatial and temporal variations.

Backed artifacts display clear geographical variation across the Australian mainland. They are not found in the northern portions of central and Western Australia. Across the rest of the mainland, they are found in varying densities, with regionally different sizes, shapes, and production systems. One depiction of the spatial differences in form has been expressed metrically with an index of symmetry (Hiscock 2014a). As shown in Fig. 5.1, higher values are more symmetrical, and lower values are less symmetrical. The continental pattern is that asymmetry is more typical of marginal regions. This geographical variation in backed artifact morphology is probably related to the dispersal history of this kind of artifact, as modification of the morphological variation in backed artifact arose from transmission error, technological adjustments of retouching to different blank production strategies, and divergence in the backed forms to facilitate social signaling in learning/production contexts (see Hiscock 2014b).

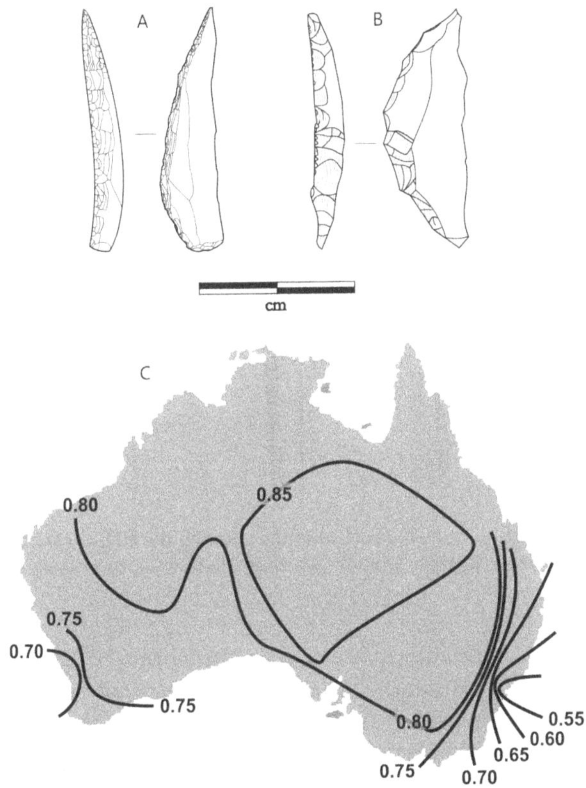

Fig. 5.1 Illustration of the geographical variation in backed artifacts. (**a**) Asymmetrical specimen from the east coast; (**b**) symmetrical specimen from the interior; (**c**) isopleths of the backed artifact symmetry index (Hiscock 2014a) in which higher values are more symmetrical

At any one location, backed artifacts are often extremely uniform in size and shape. The process of creating standardized forms was sometimes assisted by the production of flakes with regular shapes, but it did not depend on standardized blank creation. Specimens were made on a variety of flakes that were selected because they had an appropriate cross section, width, and one straight or gently undulating margin of sufficient length (Hiscock and Attenbrow 1996; Hiscock 2006). This means that prehistoric manufacturing did not involve, typically, mechanically applying a procedure to a uniform blank but required the artisan to select a suitable section of each flake to retouch. Standardization was then achieved by careful and extended retouching of flakes on an anvil.

Backed artifact production typically involved considerable investment. In some regions we know that high levels of backed artifact manufacture were associated with switches to the procurement of higher-quality lithic materials, reflecting a restructuring of economic activities. In some locations, knappers also improved material responsiveness by subjecting the rock to controlled thermal alteration. Heat treatment was often undertaken late in the production process, when the specimens

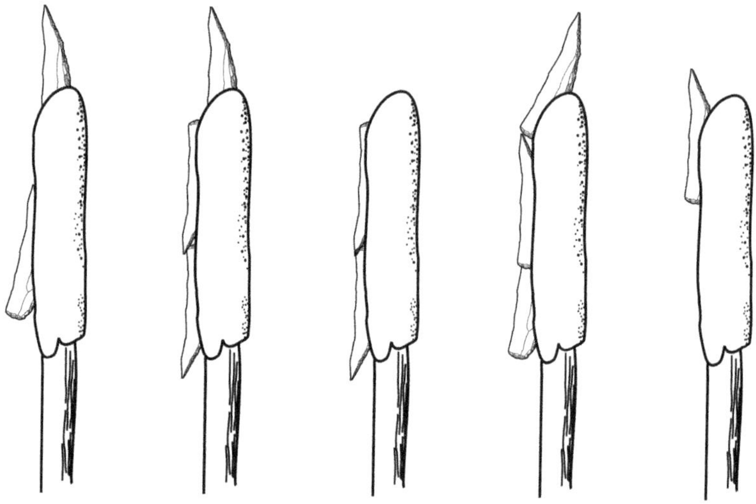

Fig. 5.2 Idealized images of possible backed artifact hafting positions, inspired by McCarthy's (1976) illustration but excluding those options which are unlikely given existing use-wear and residue evidence

were smaller, so that application of heat had a higher chance of successfully improving fracture quality of specimens. Heat-treated pieces of rock were typically knapped with care, and in many instances flake production in preparation for backed artifact manufacture involved regular and precise knapping of small pieces of stone, typically involving careful platform preparation and structured core reduction. The combination of high investment in materials and regular skillful knapping conserved material by extracting a large number of specimens per stone unit, and there may have been some degree of craft specialization involved in the organization of production. If the geographically localized patterns of standardized backed artifacts reflect localized craft traditions and practices, then we can hypothesize that those manufacturing systems reflected social structures involved with learning, such as recognized craft specialists or craft contexts involving apprenticeship learning.

The way backed artifacts were used as tools suggests similar mechanisms. Backed artifacts may have sometimes been used as tools held individually in the hand, but there is abundant evidence that many were hafted as part of composite tools. Resin residues and stains from hafting adhesives have been widely found and suggest that specimens were most commonly positioned with backed edge toward a shaft and with the sharp chord edge parallel or subparallel to the shaft. Residues and stains are typically concentrated along the backed edge, indicating that specimens were held in place by resinous compounds packed around them and encircling or partly encircling the shaft. Figure 5.2 illustrates some of the possible hafting patterns. The length of the shaft remains unknown, and it has often been presumed to be long, reflecting the widely held proposition that backed artifacts were armatures on spears, either hafted along the shaft as barbs and/or at the tip as projectile points. This use of backed artifacts cannot be entirely ruled out, and one recent review finds projectile armatures are still a plausible hypothesis (Fullagar 2016), but the weight of evidence indicates that at least in the southeast of the continent backed artifacts were commonly used on composite craft tools rather than on projectiles.

In southeastern Australia, at places such as the Mangrove Creek catchment, use-residue studies reveal that backed artifacts were primarily used for many tasks involving multipurpose cutting and slicing of both plant and animal materials but rarely as thrown spears (Robertson 2002; Attenbrow et al. 2009; Robertson et al. 2009). Many specimens have residue or wear patterns documenting that they were used for two or more purposes, indicating that the composite tool on which they were hafted was capable of being employed for diverse craft activities. This evidence compels us to abandon visions of Australian microliths as always being spear armatures and eliminates explanations that rely on that proposition, such as suggestions that they had been used as spears for large game hunting (McBryde 1974) or in individual pursuit strategies (Morwood 1986, 1987) or during periods of intensive warfare/violence (Flood 1995; McDonald et al. 2007). The early version of Hiscock's (1994) risk reduction model, which hypothesized that backed artifacts were made in regular shapes to provide easily maintainable spears that would reduce foraging risk in unpredictable environments, is also refuted by this evidence. The functional evidence we have instead reveals that backed artifacts were used to create intricate, sometimes delicate, crafts made from hide, bone, wood, feathers, and other organic materials. These items were scraped, carved, incised, and sliced. None have been preserved or recovered, and so we cannot at this point describe the kinds of craft items being constructed. However there are two possibilities to consider.

It may be that backed artifacts were being employed as a craft tool to make diverse material culture of directly utilitarian kinds: clothing, bags, hunting gear such as nets/traps, shelters, bedding, boats, and so on. If these were the goods being made, then Hiscock's (1994) risk reduction model should be reformulated to state that regular-shaped backed artifacts were produced as the edges of critical multipurpose craft tools capable of manufacturing a diverse range of utilitarian tools that provided economic advantage and acting as a buffer against foraging risk. In this model artisans maintain standardized backed artifact sizes and shapes because the regularity of artifact form assists in the maintenance of the composite tool which can be kept in a functional state.

A more likely proposition is that a proportion of the craft goods being produced were also employed as social signals, perhaps in the context of creating valued items for exchange and/or for use in performances associated with cult practices. Given the residue evidence of red ochre and feathers being used in making the composite tools, I argue that they were not only used for cutting, scraping, and sawing; they were simultaneously employed to send signals about social phenomena. The key questions for this model are why people invested in production of paraphernalia, and what mechanism was in operation that produced uniform specimens? These questions arise because it might have been possible to send social signals with a composite tool, or to use the tool to make paraphernalia that sent the signal, simply by hafting suitable unretouched flakes. Small stone artifacts hafted into a composite tool will sometimes be barely visible and, at a distance, it could be hard to distinguish a backed artifact from a simple and less costly flake. So, why invest in the production of backed artifacts?

It must have been important for members of any forager group to know that these items were produced to specified norms. This implies that there were circumstances in which there was scrutiny of the truthfulness of the signal—truthfulness being the accurate production relative to socially constructed standards. There are two mechanisms I will discuss here that are capable of maintaining production regularity across time and space. The first is the power of master-apprentice relationships to construct and maintain normative traditions of artifact production. In apprenticeship learning, there is a need for group approval: usually, the establishment of "authenticity" through public performance. Even informal guilds of craft people might apply significant stabilizing pressure on performance if reputation and exchange opportunities were dependent on meeting normative judgments (see Hiscock 2014b). A second mechanism for scrutinizing truthfulness of the signal is a physical exchange between maker and user, in which the receiver can examine the specimens in detail and comment on any departure from the expected form. This idea has been proposed as an explanation of backed artifact standardization in the Howieson's Poort industries of southern Africa, but not explored in detail, and not proposed in the Australian context (Hiscock et al. 2011). In the African context, discussions typically focus on the idea that public signaling coordinated relationships between neighboring groups, and therefore the proposed models expect that exchanges were across group boundaries, a proposition that would predict backed artifacts were often made from material exotic to the local area. Within Australia, this expectation is not generally met: backed artifacts are not typically made on materials from distant areas; instead they are often made from good-quality local rock. And so, if this process was in play, I would predict exchanges of backed artifacts were normally within group, probably between social units such as subsections.

My purpose here is not to test which of these or other stabilizing mechanisms were operating but rather to examine the paleoenvironmental contexts of the production behaviors. Thus far, I have shown that Australian backed artifacts are best characterized as being carefully made, relatively energy expensive, and locally regular in form. They were usually inserted into multipurpose composite craft tools, and in those portions of southeastern Australia where detailed functional studies have been carried out, they were used to scrape, carve, incise, and slice hide, bone, wood, flesh, and feathers; they were rarely if ever employed as attachments to thrown spears. Their major role was, therefore, in craft production of organic artifacts, which probably had both utilitarian and signaling purposes. The maintenance of a standard and distinctive form for a hard to see element in elaborate craft tools is taken to indicate that backed artifacts were socially scrutinized; their form was judged against agreed norms, perhaps in apprenticeship/guild contexts or in contexts of exchange. Investigations of why this craft production and social signaling operated can be examined by discussing the context of chronological changes in production rates.

Every region in which they are known archaeological sequences shows a single significant proliferation event, in which the production rates of backed artifacts increased substantially for a short period and then decreased to very low levels before backed artifact manufacture ceased altogether in the last millennium. Low chronological resolution makes the length of the event difficult to measure, but it is probably between a few hundred years and slightly more than a thousand years. The antiquity of the proliferation varies regionally and may generally be later in the arid zone than in better-watered continental margins. In the well-studied Sydney Basin, the proliferation occurred between 3500 and 2500 years ago, and this timing appears to apply broadly along the eastern seaboard. Archaeological examinations of why this proliferation occurred have focused on identifying the social and paleoenvironmental conditions at the time when the proliferation began. The rationale for these investigations is that context, at the time when a dramatically increased investment in construction of craft tools and paraphernalia began, is likely to provide an insight into the selective pressures operating.

Environmental Change and Its Articulation to Technology

There is evidence for climate change in the Holocene, commencing or intensifying around 4000 years ago, or slightly before that time. At this time, there was enhanced variability in rainfall and reduced precipitation related to the strengthening of the El Niño-Southern Oscillation (ENSO) system (e.g., Shulmeister and Lees 1995; Tudhope et al. 2001; Andrus et al. 2002; Koutavas et al. 2002; Lynch et al. 2007). Reductions of effective precipitation were probably mirrored in a reduction in the gross resource availability in many landscapes, diminishing carrying capacity, and increased the patchiness of resource distribution. Archaeologists have suggested that decreased productivity and heightened resource variability acted as powerful selective pressures to modify landscape use, foraging strategies, and associated technological systems (e.g., Rowland 1999; Hiscock 2006; Asmussen and McGuiness 2013). One indication of human responses to resource reductions during the Holocene may be the increased foraging for lower-ranked foods such as moths, toxic nuts, and grass seeds, a shift in foraging strategies that can be characterized as an expansion of diet breadth (see David and Lourandos 1998: 212). Reduction of procurement costs and risks through an emphasis on different technological strategies would constitute a companion response. Hiscock (1994, 2002, 2006) proposed that in the circumstance increased standardization of lithic tools would have enhanced the readiness of tools, effectively creating a partial buffer against reduced foraging uncertainty. Key tests offered for the backed artifact proliferation as a response to heightened foraging risk were (1) the multifunctionality of backed specimens (Attenbrow et al. 2009) and (2) the covariation of backed artifact production rates and climatic conditions (Hiscock 2002, 2006).

A clear but imprecise coincidence exists between the start of the period of intensive backed artifact production and the onset of an ENSO-dominated climate that involved reduced effective precipitation and increased climatic variability. An increasingly high-resolution record of climate proxies is being created by several

Table 5.1 Summary of trends in effective precipitation and backed artifact abundance for the southeastern region of Australia

Years BP	Effective precipitation	Backed artifact production
0–2000	Increasing but still variable	Low
2–4000	Low and variable	High
4–5000	Declining	Increasing
>5000	High	Very low

From Hiscock (2002)

lines of paleoenvironmental research (e.g., Black et al. 2008; Conroy et al. 2008; Petherick et al. 2013), but the cultural timeline is currently of far lower resolution, limiting analysis of the covariation in temporal phases. An example is Table 5.1, taken from Hiscock (2002). The relationship as presented shows an inverse relationship between effective precipitation and backed artifact production rates in southeastern Australia. Hiscock's interpretation was that the onset of more variable climatic conditions created a context in which resource distribution and availability were less easily and reliably mapped or predicted. Adjustments to foraging practices and social interactions were made to reduce the increased risk attached to these circumstances, and a component of the response was the increased use of craft tools containing a specific insert to produce organic craft items. When effective precipitation increased in the last 2000 years, backed artifact production rates declined as the technological system responded to other pressures affecting foragers.

The environmental changes that occurred 3000–4000 years ago were not limited to climate change. Human introduction of the dingo triggered altered faunal structures with serious implications for foragers. The effectiveness of this new high-order predator is revealed by its likely role in exterminating the thylacine from mainland Australia (Fillios et al. 2012; Letnic et al. 2012), but the really dramatic change was wrought by its impact on large and small fauna. Comparisons between modern regions, with and without dingos, indicate that the dingo would have suppressed kangaroos to a small fraction of their numbers (Fillios et al. 2010). The onset of this direct competition with human foragers created resource reductions at almost the same time as the El Niño amplification magnified the climatically driven resource decrease. While long-term climatic fluctuations of greater size had occurred in the Pleistocene, this conjunction of both climate-derived and competition-derived depression of terrestrial resources has few parallels in Australian prehistory.

Reduction of terrestrial fauna required altered foraging strategies. A shift in hunting emphasis toward smaller game may have occurred, although elements of the small mammal fauna may also have been depleted. The economic balance between hunting and gathering probably shifted, as we have evidence of increased exploitation of lower-ranked plant resources, as ENSO intensification increased subsistence risks and lowered productivity (Asmussen and McInnes 2013). Additionally, as terrestrial game availability reduced and coastal ecosystems stabilized following sea-level stabilization marine resources probably took on a significant role. In concert these multiple resource shifts would have required new economic systems that pursued the different resource structures and available biomass that were in place by around three and a half thousand years ago.

There are also claims that population increases occurred at this time. Recently such claims have been based on summed probabilities of radiometric dates (e.g., Turney and Hobbs 2006; Williams 2013), with interpretations of measureable population increase in the late Holocene. If such demographic changes were true, they occurred during a time of terrestrial resource depletion, and the economic and social reorientation already mentioned must have been capable of supplying food for significantly greater numbers of people. However, it is not clear that demographic interpretations of these data are sound, and it may be that the new economic systems yielded a new archaeological pattern without radically larger populations (see Attenbrow and Hiscock 2015). A reorganization of landscape use might translate into altered site abundance, and Attenbrow (2003) has documented shifts in the dispersal or clustering of activities and therefore in the potential number of sites/dates preserved. If there were higher densities of people in the late Holocene landscape and/or higher mobility or smaller residential group size, then burning may have become more widespread with consequences for resource availability.

The outcome of these cultural and environmental interactions was the transformation of the ecological space occupied by human foragers, with substantial changes initiated approximately 3000–4000 years ago. It was this transformed niche that provided the context for the backed artifact proliferation.

Technology and Environmental Context

Backed artifact production was emphasized during the proliferation event, in new economic and social contexts. However the backed artifacts themselves were not normally used directly as extractive tools; their role typically was for processing resources. While some of the processing may have been involved in food preparation, the primary use of hafted backed artifacts was in the construction of craft objects: things made from hide, wood, bone, feathers, etc. The objects manufactured may have been extractive tools such as wooden spears, boomerangs, throwing and digging sticks, and so on. Craft objects may have additionally been made for storage and transport of things including food, or for shelter, clothing, and other practical purposes. Such objects may all have had value in negotiating economic actions during the onset and intensification of conditions of lower and less predictable resource availability, around 3000–4000 years ago. It is additionally likely that the craft items being manufactured were providing social signals that acted to mediate human actions. David and Lourandos (1998) suggested that the stone tools such as backed artifacts acted as social mechanisms in the negotiation of territory and group composition. It is now clear that while backed artifacts may themselves have had social meaning, their low visibility allowed them to be viewed only in limited transactions between individuals. Those transactions may have been important in validating subsequent public signals and social actions, and it was the craft items (perhaps elaborately decorated) being manufactured with the backed artifacts that were being used as part of social negotiations in processes of cultural changes. Those changes were situated in a context of resource scarcity and unpredictability and may therefore have involved patterns of migration, altered territorial boundedness, and new social arrangements for resource access. The public signaling represented by this craft production, like the signaling represented by rock art, is often thought to have been used in expressing identity and in constructing and/or managing rules about access to resources (e.g., David and Lourandos 1998). Such connections of public signaling to resource use are plausible but require independent testing.

Conclusion

This niche construction framing of the relationship between ancient technology and paleoenvironment reveals how unhelpful were previous debates about whether the backed artifact proliferation was driven by "social" or "environmental" factors. No such simple dichotomy can be extracted from the evidence for the cultural and natural processes that interacted to create niches in the mid- to late Holocene. Similarly, lithic technologies were entangled in the varied processes by which people occupied those niches, technologies being widely emphasized/adopted simultaneously in response to environmental contexts and social contexts. At least some elements of technological change in Australia over the last few millennia are clearly synchronized to paleoenvironmental transitions, and yet the mechanisms which sync technology and environment were not direct or simple. The intensified use of microliths, backed artifacts, in the Holocene of southeastern Australia reflects an emphasis on the production of craft objects, for acquisition/processing of resources and for conveying social information in changing times.

References

Andrus, C. F. T., Crowe, D. E., Sandweiss, D. H., Reitz, E. J., & Romanek, C. S. (2002). Otolith $\delta^{18}O$ record of mid-Holocene sea surface temperatures in Peru. *Science, 295*, 1508–1511.

Asmussen, B., & McInnes, P. (2013). Assessing the impact of mid-to-late Holocene ENSO-driven climate change on toxic Macrozamia seed use: A 5000 year record from Eastern Australia. *Journal of Archaeological Science, 40*, 471–480.

Attenbrow, V. (2003). Habitation and land use patterns in the Upper Mangrove Creek catchment, New South Wales central coast, Australia. *Australian Archaeology, 57*, 20–32.

Attenbrow, V., & Hiscock, P. (2015). Dates and demography. Are radiometric dates a robust proxy for long-term prehistoric demographic change? *Archaeology in Oceania, 50*, 29–35.

Attenbrow, V., Robertson, G., & Hiscock, P. (2009). The changing abundance of backed artefacts in south-eastern Australia: A response to Holocene climate change? *Journal of Archaeological Science, 36*, 2765–2770.

Black, M. P., Mooney, S. D., & Attenbrow, V. (2008). Implications of a 14,200 year contiguous fire record for understanding human-climate relationships at Goochs Swamp, New South Wales, Australia. *The Holocene, 18*, 437–447.

Conroy, J., Overpeck, J., Cole, J., Shanahan, T., & Steinitz-Kannan, M. (2008). Holocene changes in eastern tropical Pacific climate inferred from a Galápagos lake sediment record. *Quaternary Science Reviews, 27*, 1166–1180.

David, B., & Lourandos, H. (1998). Rock art and socio-demography in northeast Australian prehistory. *World Archaeology, 30*, 193–219.

Fillios, M., Crowther, M., & Letnic, M. (2012). The impact of the dingo on the thylacine in Holocene Australia. *World Archaeology, 44*, 118–134.

Fillios, M., Gordon, C., Koch, F., & Letnic, M. (2010). The effect of a top predator on kangaroo abundance in arid Australia and its implications for archaeological faunal assemblages. *Journal of Archaeological Science, 37*, 986–993.

Flood, J. (1995). *Archaeology of the dreamtime*. Sydney: Collins.

Fullagar, R. (2016). Uncertain evidence for weapons and craft tools: Functional investigations of Australian microliths. In R. Iovita & H. Kastuhiro (Eds.), *Multidisciplinary approaches to the study of stone age weaponry* (pp. 159–166). Dordrecht: Springer.

Hiscock, P. (1994). Technological responses to risk in Holocene Australia. *Journal of World Prehistory, 8*, 267–292.

Hiscock, P. (2002). Pattern and context in the Holocene proliferation of backed artefacts in Australia. In R. G. Elston & S. L. Kuhn (Eds.), *Thinking small: Global perspectives on microlithization* (pp. 163–.177). Arlington: American Anthropological Association. Archaeological Papers of the American Anthropological Association (AP3A) number 12.

Hiscock, P. (2006). Blunt and to the point: Changing technological strategies in Holocene Australia. In I. Lilley (Ed.), *Archaeology of Oceania: Australia and the Pacific Islands* (pp. 69–95). Oxford: Blackwell.

Hiscock, P. (2014a). Geographical variation in Australian backed artefacts; trialling a new index of symmetry. *Australian Archaeology, 79*, 124–130.

Hiscock, P. (2014b). Learning in lithic landscapes: A reconsideration of the hominid "tool making" niche. *Biological Theory, 9*, 27–41.

Hiscock, P., & Attenbrow, V. (1996). Backed into a corner. *Australian Archaeology, 42*, 64–65.

Hiscock, P., & Attenbrow, V. (2005). *Australia's eastern regional sequence revisited: Technology and change at Capertee 3*. Oxford: Archaeopress.

Hiscock, P., Clarkson, C., & Mackay, A. (2011). Big debates over little tools: Ongoing disputes over microliths on three continents. *World Archaeology, 43*, 653–664.

Koutavas, A., Lynch-Stieglitz, J., Marchitto, T. M., Jr., & Sachs, J. P. (2002). El Niño-like pattern in ice age tropical Pacific Sea surface temperature. *Science, 297*, 226–230.

Laland, K., & O'Brien, M. J. (2010). Niche construction theory and archaeology. *Journal of Archaeological Method and Theory, 17*, 303–322.

Letnic, M., Fillios, M., & Crowther, M. (2012). Could direct killing by larger dingoes have caused the extinction of the thylacine from mainland Australia? *PloS One, 7*, 1–5.

Lynch, A. H., Beringer, J., Kershaw, A. P., Marshall, A., Mooney, S., Tapper, N., Turney, C., & van der Kaars, S. (2007). Using the palaeorecord to evaluate climate and fire interactions in Australia. *Annual Review of Earth and Planetary Sciences, 35*, 215–239.

McBryde, I. (1974). *Aboriginal prehistory in New England*. Sydney: Sydney University Press.

McCarthy, F. D. (1976). *Australian aboriginal stone implements* (2nd ed.). Australian Museum Trust.

McDonald, J., Donlon, D., Field, J., Fullagar, R., Brenner Coltrain, J., Mitchell, P., & Rawson, M. (2007). The first archaeological evidence for death by spearing in Australia. *Antiquity, 81*, 877–885.

Morwood, M. J. (1986). The archaeology of art: Excavations at Maidenwell and Gatton Shelters, S.E. Queensland. *Queensland Archaeological Research, 3*, 88–132.

Morwood, M. J. (1987). The archaeology of social complexity in South-east Queensland. *Proceeding of the Prehistoric Society, 53*, 337–350.
Odling-Smee, F. J., Laland, K. N., & Feldman, M. W. (2003). *Niche construction: The neglected process in evolution*. Princeton, NJ: Princeton University Press.
Petherick, L., Bostock, H., Cohen, T. J., Fitzsimmons, K., Tibby, J., Fletcher, M.-S., Moss, P., Reeves, J., Mooney, S., & Barrows, T. (2013). Climatic records over the past 30 ka from temperate Australia ? A synthesis from the Oz-INTIMATE workgroup. *Quaternary Science Reviews, 74*, 58–77.
Robertson, G. (2002). Birds of a feather stick: Microscopic feather residues on stone artifacts from Deep Creek Shelter, New South Wales. In S. Ulm, C. Westcott, J. Reid, A. Ross, I. Lilley, J. Prangnell, & L. Kirkwood (Eds.), *Barrier, borders, boundaries* (pp. 175–182). Brisbane: University of Queensland.
Robertson, G., Attenbrow, V., & Hiscock, P. (2009). Multiple uses for Australian backed artifacts. *Antiquity, 83*, 296–308.
Rowland, M. J. (1999). Holocene environmental variability: Have its impacts been underestimated in Australian pre-history? *The Artifact, 22*, 11–48.
Shulmeister, J., & Lees, B. G. (1995). Pollen evidence from tropical Australia for the onset of an ENSO-dominated climate at c. 4000 BP. *The Holocene, 5*, 10–18.
Tudhope, A. W., Chilcott, C. P., McCulloch, M. T., Cook, E. R., Chappell, J., Ellam, R. M., Lea, D. W., Lough, J. M., & Shimmield, G. B. (2001). Variability in the El Niño southern oscillation through a glacial-interglacial cycle. *Science, 291*, 1511–1517.
Turney, C. S. M., & Hobbs, D. (2006). ENSO influence on Holocene aboriginal populations in Queensland, Australia. *Journal of Archaeological Science, 33*, 1744–1748.
Williams, A. N. (2013). A new population curve for prehistoric Australia. *Proceedings of the Royal Society B-Biological Sciences, 280*, 1–10.

Chapter 6
Human Adaptive Responses to Environmental Change During the Pleistocene-Holocene Transition in the Japanese Archipelago

Kazuki Morisaki, Masami Izuho, and Hiroyuki Sato

Introduction

In this chapter, we discuss human adaptive responses to environmental change during the Pleistocene-Holocene transition in the Japanese archipelago, focusing on correlations between lithic technological/human behavioral strategies and paleoenvironmental change. Our time period of focus is 19,000 ~ 10,000 cal BP (15,000 ~ 9000 ^{14}C BP), from the Final Paleolithic to the Initial Jomon period.

In Japan, past chronological studies established by Jomon pottery typology, lithic typology, and radiometric dates have isolated three stages that occurred equally over all regions of Japan: from (1) the initial stage characterized by a microblade industry; to (2) the Mikoshiba industrial stage composed by large foliate and lanceolate bifacial points, large ground axes, and small quantities of pottery; and finally, to (3) the Incipient Jomon stage characterized by stemmed points (Okamoto 1979; Inada 1986; Kurishima 1991; Okamura 1997). After the 1990s, however, accumulation of new archaeological data and radiocarbon dates has suggested a more complicated spatiotemporal mosaic of cultural complexes during this transitional period (Inada 1993; Imamura 1999; Odai-yamamoto I site excavation team 1999; Kodama 2001; Taniguchi and Kawaguchi 2001; Anzai 2002; Kudo 2005; Taniguchi 2011; Mitsuishi 2013), though lithic studies from this period are still few and only at a regional scale of analysis. Recent research has proposed a correlation

K. Morisaki (✉)
Monuments and Sites Division, Agency for Cultural Affairs-Japan, Tokyo, Japan
e-mail: morisaki@mext.go.jp

M. Izuho
Archaeology Laboratory, Faculty of Social Sciences and Humanities, Tokyo Metropolitan University, Tokyo, Japan

H. Sato
Department of Archaeology, Faculty of Letters, The University of Tokyo, Tokyo, Japan

E. Robinson, F. Sellet (eds.), *Lithic Technological Organization and Paleoenvironmental Change*, Studies in Human Ecology and Adaptation 9,
DOI 10.1007/978-3-319-64407-3_6

between environmental change and lithic tools or assemblages (Kanomata 2007; Miyoshi 2013). These studies have made an important contribution, but there has been a lack of focus on questions of technological organizational and human behavioral change.

This paper embraces a technological organization perspective for analyzing lithic tools, in which technology, including lithic technology, is seen as a strategy for the manufacture, transportation, use, and discard of subsistence tools (Binford 1979). As Binford (1979, 1980) pointed out, since this strategy is systematically organized according to environmental conditions, a technological organizational perspective allows us to view lithic technology as a human behavioral strategy and better understand the dynamics of human adaptation. Therefore, the economic aspects of technology when dealing with environmental or ecological conditions are emphasized in studies of technological organization that sometimes reference to optimal foraging theory or risk management (Nelson 1991; Bamforth and Bleed 1997; Morisaki et al. 2015).

Accordingly, environmental change should be primarily considered as one of the most significant contexts for studying lithic technological and human behavioral change and diversity. Obviously, it is not the sole determinant of human behavior. Based on archaeological data, adaptive behavioral strategies must have been diverse, even in similar environments. From a technological organizational perspective, however, environmental conditions should be viewed as a constraint on lithic technology and human behavior.

It is well known that environmental conditions from the Late Pleniglacial to the Preboreal fluctuated abruptly. To explain lithic technological and human behavioral change for this period, the influence of environmental changes driven by climatic fluctuation should be addressed.

This paper first discusses the issue of chronology by compiling radiocarbon ages that have recently accumulated throughout the Japanese archipelago. We then analyze diachronic and interregional variability of lithic technology, its organizational characteristics, and its role in reflecting human responses to environmental change. Lastly, we consider human behavioral variation within the context of paleoenvironmental changes during this transitional period.

Paleoenvironments from Late MIS3 to the Beginning of MIS1

Climatic Fluctuation and Geological Setting

The Northern Europe chronozone of the Late Glacial (LG) has traditionally been divided into five stages: Oldest Dryas, Bølling, Older Dryas, Allerød, and Younger Dryas (Fig. 6.1a) (Stuiver et al. 1995). Although there are some studies which adapt these stages to Japanese archaeology, the stages are not always synchronized with the results of high-resolution pollen analyses in the Japanese archipelago (e.g., Lake Suigetsu). Recently, paleoclimate studies of cave stalagmites in China (e.g., Hulu

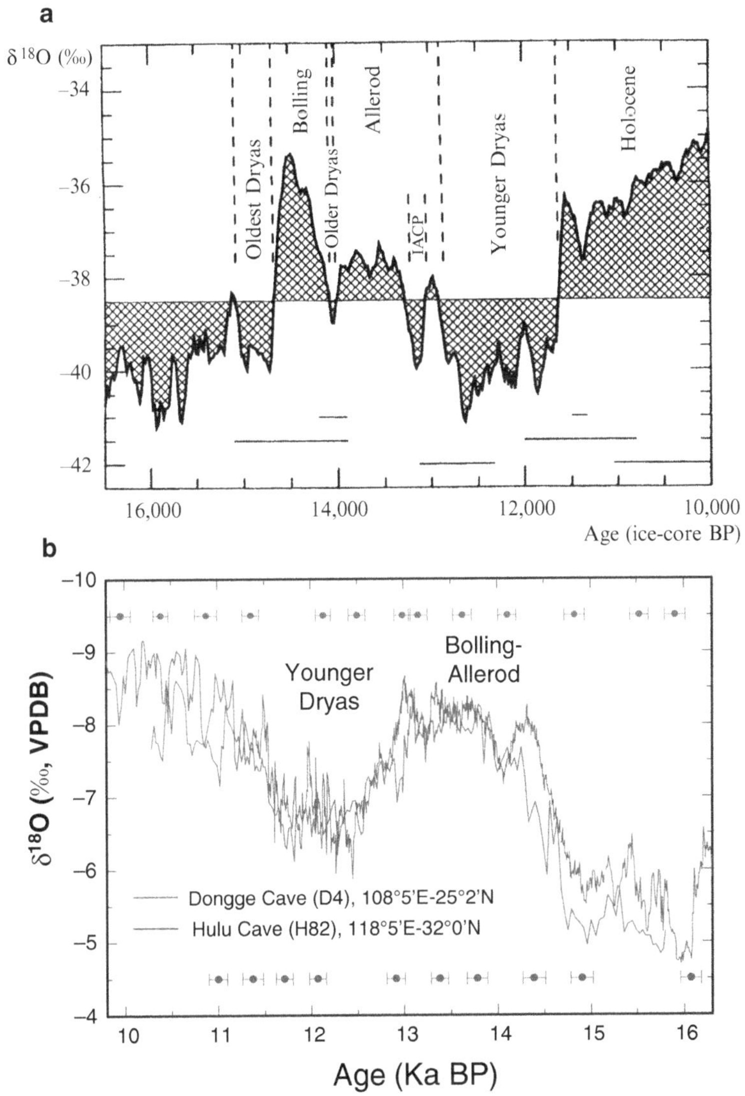

Fig. 6.1 Oxygen isotope record and chronozones in Northern Europe (**a**) (Stuiver et al. 1995) and in East Asia (China) (**b**) (Yuan et al. 2004)

Cave) have revealed millennium-scale fluctuations of Asian monsoon intensity, which correspond to the climatic fluctuations recorded in the Greenland ice cores (Fig. 6.1b) (Wang et al. 2001; Yuan et al. 2004). There are some differences between the oxygen isotope records of Northern Europe and China, but it should be noted that there seem to be at least three relatively distinct synchronic changes of oxygen isotope signatures between the two regions (Wang et al. 2001): the onset of the LG warm period, the Younger Dryas cold event, and the onset of the Holocene.

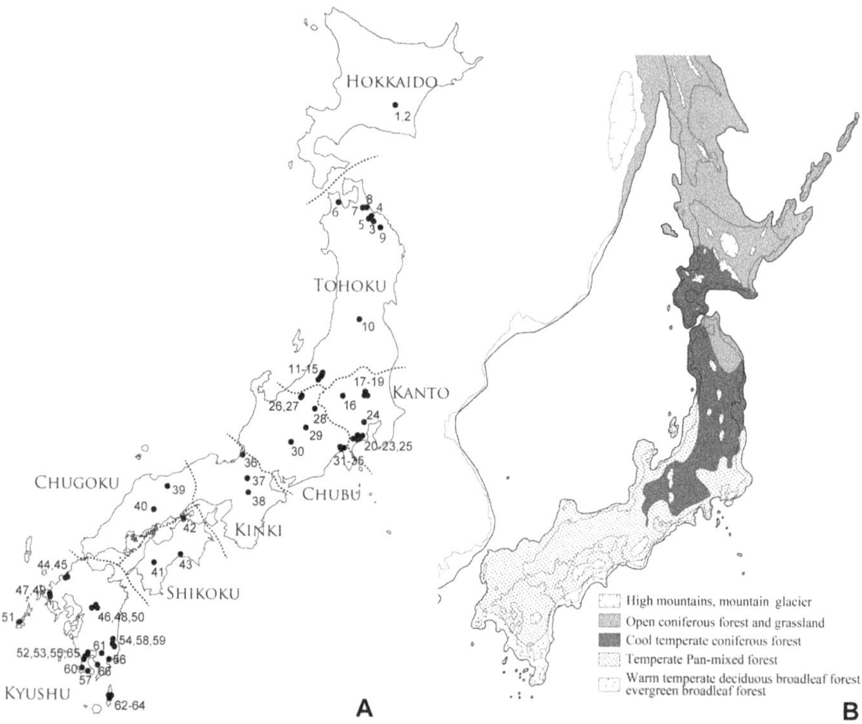

Fig. 6.2 Distribution of sites mentioned in this paper (**a**) and reconstructed paleogeography and vegetation of the Japanese archipelago and surrounding region during the Last Glacial Maximum (**b**) (After Sato et al. 2011a). Site numbers correspond to those in Table 6.2

At present, four reliable chronozones can be recognized within the time period considered in this chapter, based on Nakazawa et al. (2011), Kudo (2012), and Kudo et al. (2011): the Late Pleniglacial (LPG, GS-2: ca. 19000 ~ 14,700 cal BP), the Bølling/Allerød (B/A, GI-1), the Younger Dryas (YD, GS-1), and the Preboreal. Duration of the chronozones is ca. 14,700 ~ 12,800 cal BP for the B/A and ca. 12,800 ~ 11,500 cal BP for YD (Wang et al. 2001). The LG here means the time period from the Bølling/Allerød to the Younger Dryas.

Figure 6.2 shows the reconstructed paleogeography of the Japanese archipelago and surrounding region during the Last Glacial Maximum (LGM). Landmasses of this region during the LGM mainly consisted of two distinct parts: the Paleo-Sakhalin-Hokkaido-Kuril Peninsula and the Paleo-Honshu Island. Hokkaido was the southern part of the Paleo-Sakhalin-Hokkaido-Kuril (SHK) Peninsula, connected by a land bridge between the Sakhalin and the Kuril Islands (Kunashiri and Shikotan Islands). Honshu was attached to Shikoku and Kyushu, forming Paleo-Honshu Island. This island was not connected to the Paleo-SHK Peninsula during the Last Glacial, although distances across the straits were shortened from only a few up to a dozen kilometers (Matsui et al. 1998; Sato et al. 2011b). Hokkaido had long been under cold and dry continental-like climate, until inflow of warm current started and caused a precipitation increase around 15,000 cal BP. After the end of the Younger Dryas (ca. 11,500 cal BP), stable warm and wet climate dominated and changed the Japanese archipelago to the present form.

Flora

Table 6.1 is based on the latest data on vegetation history from the LGM to the Holocene (Takahara 2011). During the LGM, Hokkaido had long been covered by open *Larix* forest and grassland (Igarashi 2008), which never existed in Honshu. Honshu was divided into two vegetation zones: northeastern Honshu was covered by evergreen coniferous forest, and southwestern Honshu was covered by temperate coniferous forest. During the Late Glacial, flora changed as a consequence of the inflow of warm current into the Japan Sea and precipitation increase, which caused the vegetation of Honshu to change to temperate broadleaf forests. Hokkaido was gradually separated from the continent and started to be covered with forests similar to northeastern Honshu as well, though it occurred later than in Honshu.

This description is a little different from that of Fig. 6.2b, because the data source is not completely the same. The vegetation map in Fig. 6.2b tells us it is also possible to estimate higher proportion of broadleaf species in the southwestern Paleo-Honshu forest vegetation during the LGM than Table 6.1. Also, temperate broadleaf forests and broadleaf evergreen forests extended along the Pacific Ocean coastal area during the Terminal Pleistocene (Tsuji 2004). Here we confirm that the

Table 6.1 Vegetation history from the LGM to the Holocene

	Glacial period		Post-Glacial period	
ka BP	Stadial (30–15)	Late Glacial (15–10)	Early (10–7)	Middle (7–4)
Hokkaido	*Larix*	*Evergreen conifer* (Pinaceae)	*Pan mixed*	*Pan mixed*
Tohoku	*Evergreen conifer* (Pinaceae)	*Evergreen conifer* (Pinaceae)	*Temperate broadleaf*	*Temperate broadleaf*
Chubu	*Evergreen conifer* (Pinaceae)	*Temperate broadleaf*	*Temperate broadleaf*	*Temperate broadleaf*
Kanto	*Temperate conifer* (Pinaceae)	*Temperate broadleaf*	*Temperate broadleaf*	*Broadleaf evergreen*
Western Japan (Pacific Ocean side)	*Temperate conifer* (Pinaceae)	*Temperate broadleaf*	*Temperate broadleaf*	*Broadleaf evergreen*
Western Japan (Japan Sea side)	*Temperate conifer* (Pinaceae)	*Temperate broadleaf*	*Temperate broadleaf*	*Temperate conifer* (Pinaceae)

After Takahara (2011)

forest vegetation of southern Paleo-Honshu Island indicates a warmer climate than that of the northern part of the island and that only the Pacific Ocean coastal region of southern Honshu was covered with broadleaf evergreen forests.

Fauna

Recent studies of the formative history of terrestrial fauna on the Japanese archipelago during the terminal Pleistocene can be summarized as follows (Takahashi 2008; Takahashi and Izuho 2012). Before and during the LGM, several kinds of large mammals inhabited the Japanese archipelago. There were two faunal complexes: the Paleoloxodon-Sinomegaceroides complex with Nauman's elephant (*Palaeoloxodon naumanni*), which mainly inhabited Paleo-Honshu Island, and the Mammoth fauna complex with mammoths (*Mammuthus primigenius*), which mainly inhabited southern Paleo-SHK Peninsula, namely, Hokkaido, and areas further north. As a consequence of climatic fluctuations, some animals in each group are thought to have mixed complexes, migrating to the north or south according to species-specific habitat and temperature preferences.

Large mammals of the Paleoloxodon-Sinomegaceroides complex became almost extinct at the onset of LGM due to the climatic deterioration. Terrestrial fauna in Paleo-Honshu Island seemed to be composed of middle and small species like the present time since this period. On the other hand, the mammoth complex lived through the LGM in Hokkaido. However, large mammals of this complex seem to have become extinct or moved to northern regions gradually, having faced the climatic amelioration of the Late Glacial (Kuzmin and Orlova 2004). Terrestrial fauna in Hokkaido seems to have approximated to the present complexes since this period.

Materials and Methods

Data

The study includes a total of 74 assemblages from 66 archaeological sites across the Japanese archipelago, except the Ryukyu Islands which are located in the southernmost Japanese archipelago. We selected the materials which have both radiocarbon ages or firm tephrochronology or pottery typology and a lithic assemblage to establish a reliable archaeological chronology (about the distribution of the sites, see Fig. 6.2a). All data were gleaned from published excavation reports of Paleolithic sites (in Japanese). Most published excavation reports we consulted contained information on lithic tool kit assemblage structure and reduction strategies reconstructed through refit analyses. Unfortunately, organic remains are usually absent at the sites discussed here. However, rich lithic materials are preserved. All these data are summarized in Table 6.2.

Table 6.2 List of archaeological sites from 19,000 to 10,000 cal BP in the Japanese archipelago, with region, pottery association, stone weaponry, reduction technique, occupation intensity, radiocarbon dates of related archaeological assemblages, and references

No.	Region	Site	Pottery	Stone hunting weapon[a]	Primary reduction	Score of occupation intensity	Lab. no.	Method	Material	^{14}C Age (BP)	±1σ	Calendar age (cal BP; 68.2%)[b]		Chronozone	References
1.	Hokkaido	Taisho 3	Yes	Bf	Biface, flake	0	Beta-194631	AMS	Charred residues on pottery	12100	40	14050	13850	B/A	Obihiro City Board of Education (2006)
							Beta-194629	AMS	Charred residues on pottery	12420	40	14680	14320		
2.	Hokkaido	Taisho 6	Yes	Ah, Bf	Flake, blade, biface?	0	Beta-194635	AMS	Charred residues on pottery	9250	40	10510	10300	PB	Obihiro City Board of Education (2005, 2006)
							Beta-194636	AMS	Charred residues on pottery	9480	40	11050	10600		
3.	Tohoku	Takihata	Yes	(GAx)	——	2	Beta-138898	AMS	Charcoal	10260	40	12110	11840	YD	Hashikami Town Board of Education (1999)
4.	Tohoku	Kushibiki	Yes	Ah	Flake	2	Beta-113349	AMS	Charcoal	10030	50	11700	11390	YD~PB	Aomori Prefecture Buried Cultural Property Center (1999)
5.	Tohoku	Kiwada	Yes	n/a	n/a	0?	Unreported	AMS	Charcoal	12360	50	14520	14170	B/A	Aomori Prefecture Nango Village Board of Education (2001)
6.	Tohoku	Odai-yamamoto I	Yes	Ah, Bf	Blade, biface?	0	NUTA-6510	AMS	Charred residues on pottery	12680	140	15310	14760	LPG	Odai-yamamoto I site excavation team (1999)
							NUTA-6506	AMS	Charred residues on pottery	~13780	170	16940	16390		
7.	Tohoku	Akahira I	Yes	Bf	Blade, biface	0	IAAA-61926	AMS	Charcoal	13740	60	16740	16450	LPG	Aomori Prefecture Buried Cultural Property Center (2008)
							IAAA-61927	AMS	Charcoal	13800	70	16850	16540		

(continued)

Table 6.2 (continued)

No.	Region	Site	Pottery	Stone hunting weapon[a]	Primary reduction	Score of occupation intensity	Lab. no.	Method	Material	^{14}C Age (BP)	±1σ	Calendar age (cal BP; 68.2%)[b]		Chronozone	References
8.	Tohoku	Itsukawame	No	Mc (prism)	Microblade	0	IAAA-92228	AMS	Charcoal	13600	30	16470	16280	LPG	Aomori Prefecture Buried Cultural Property Center (2011)
							IAAA-92231	AMS	Charcoal	~15930	40	19300	19100		
9.	Tohoku	Hayasakatai, CL2	No	Mc (wedge)	Microblade	0	Beta-176021	AMS	Charcoal	13450	100	16330	16030	LPG	Iwate Prefecture Buried Cultural Property Center (2004)
10.	Tohoku	Hinata Cave western terrace	Yes	Bf, Ah	Biface, flake	0	n/a	n/a						(B/A[c])	Sagawa and Suzuki (2006)
11.	Tohoku	Kurohime cave	Yes	Ah	Flake	2	IAAA-40495	AMS	Charred residues on pottery	9050	50	10240	10190	PB	Irihirose Village Board of Education, Cave Excavaton Team of Uonuma Region (2004)
							Beta-194820	AMS	Charred residues on pottery	~9850	40	11270	11220		
12.	Tohoku	Unoki-minami	Yes	Bf, Ah	Flake, biface	0	Tka-14593	AMS	Charred residues on pottery	10660	170	12750	12380	B/A~YD	Yoshida et al. (2008)
							Tka-14583	AMS	Charred residues on pottery	~11670	130	13700	13350		
13.	Tohoku	Jin	Yes	Bf, Ah	Biface, flake	2		AMS	Charred residues on pottery	11700	90	13700	13420	B/A	Yoshida et al. (2008), Kobayashi (1981)
							Tka-14552	AMS	Charred residues on pottery	11800	60	13720	13560		
14.	Tohoku	Kubodera-minami	Yes	Bf	Biface, blade	0	Tka-14598	AMS	Charred residues on pottery	12460	90	14850	14320	B/A	Nakazato Village Board of Education (2001)
							Tka-14586	AMS	Charred residues on pottery	~12690	110	15290	14880		

15.	Tohoku	Araya	No	Mc (wedge, boat)	Microblade, biface	1	GrA-5715	AMS	Charcoal	13690	80	16660	16340	LPG	Department of Archaeology Graduate School of Arts and Letters Tohoku University (1990, 2003)
							GrA-5713	AMS	Charcoal	~14250	110	17520	17180		
16.	Kanto	Saishikada Nakajima	Yes	Ah	Flake	1	Beta-128025	AMS	Charcoal	10070	70	11770	11400	YD~PB	Kasakake Town Board of Education (2003)
17.	Kanto	Tako-minamihara	No	n/a	n/a	1	GaK-17981	β	Charcoal	10650	150	12720	12420	YD	Tochigi Archaeological Research Center (1999)
							GaK-17982	β	Charcoal	10240	130	12380	11710		
18.	Kanto	Yakushiji-inaridai	Yes	Ah	Flake	0	Unreported	AMS	Charred residues on pottery	11170	50	13100	13000	B/A~YD	Kobayashi et al. (2009)
							Unreported	AMS	Charred residues on pottery	10750	50	12730	12660		
19.	Kanto	Nozawa	Yes	Ah	Flake	1	IAAA-10051	AMS	Charred residues on pottery	11390	50	13290	13160	B/A	Tochigi Archaeological Research Center (2003)
							IAAA-10050	AMS	Charcoal	~11860	50	13740	13610		
20.	Kanto	Shonan Fujisawa Campus	Yes	Sp, Bf, Ah	Flake	1	Gak-15904	AMS	Charcoal	11350	160	13370	13060	B/A	Keio-gijuku University Department of Archaeology (1993)
21.	Kanto	Manpukuji	Yes	Sp, Bf, Ah	Flake	0	Beta-191840	AMS	Charred residues on pottery	12330	40	14400	14140	B/A	Ariake Cultural Property Research Institute, Manpukuji Sites Excavation Team (2005)
22.	Kanto	Tsukimino-kamino, loc.2	Yes	Sp, Ah	Flake	0	Beta-158196	AMS	Charred residues on pottery	12480	50	14900	14460	B/A	Yamato City Board of Education (1986)
23.	Kanto	Miyagase-kitahara	Yes	Sp? Bf	Flake, biface?	1	Beta-105402	AMS	Charcoal	13020	80	15740	15420	LPG	Kanagawa Archaeology Foundation (1998)
							Beta-105398	AMS	Charcoal	~13060	80	15810	15480		
24.	Kanto	Gotenyama 2N	Yes	Bf	Flake, biface?	1	Beta-196087	AMS	Charred residues on pottery	13560	40	16420	16230	LPG	Kato Construction Company Research Department of Buried Property (2004)
							MTC-05108	AMS	Charcoal	13200	70	15990	15740		

(continued)

Table 6.2 (continued)

No.	Region	Site	Pottery	Stone hunting weapon[a]	Primary reduction	Score of occupation intensity	Lab. no.	Method	Material	^{14}C Age (BP)	±1σ	Calendar age (cal BP; 68.2%)[b]		Chronozone	References
25.	Kanto	Yoshioka B	No	Mc (prism), Bf?	Microblade, flake	0	Tka-11613	AMS	Charcoal	16490	250	20200	19580	LPG	Kanagawa Archaeology Foundation (1999)
							Tka-11599	AMS	Charcoal	16860	160	20540	20130		
26.	Chubu	Nakamachi, loc. BP5a, SQ03	Yes	Sp, Bf, Ah	Flake, biface	0	NUTA2-7388	AMS	Charred residues on pottery	11420	45	13320	13200	B/A	Archaeological Research Center of Nagano Prefecture (2004)
							PLD-1843	AMS	Charred residues on pottery	~12280	110	14520	14030		
27.	Chubu	Seiko-sanso B	Yes	Sp, Bf, Ah	Flake, biface	1	Beta-133847	AMS	Charred residues on pottery	12000	40	13940	13760	B/A	Archaeological Research Center of Nagano Prefecture (2000)
							Beta-133848	AMS	Charred residues on pottery	~12340	50	14470	14150		
28.	Chubu	Tenjin-one	No	Mc (boat)	Microblade, blade	0	Beta-150648	AMS	Charcoal	13290	80	16110	15840	LPG	Saku City Board of Education (2006)
							Beta-150647	AMS	Charcoal	14780	80	18100	17870		
29.	Chubu	Mikoshiba	No	Bf, blade tool	Blade, biface	0	n/a							(LPG)[d]	Hayashi et al. (2008).
30.	Chubu	Hananoko	Yes	Ah	Flake	0	MTC-09201	AMS	Charred residues on pottery	9775	50	11240	11180	PB	Hara et al. (2010)
31.	Chubu	Ikeda B	Yes	Ah	Flake	2	Beta-127648	AMS	Charcoal	9480	50	11060	10600	PB	Shizuoka Prefecture Archaeological Center (2000)
							Beta-127647	AMS	Charcoal	~9590	50	11100	10780		
32.	Chubu	Maruokita	Yes	Ah	Flake	0	Beta-127648	AMS	Charcoal	9480	50	11060	10600	YD~PB	Shizuoka Prefecture Archaeological Center (2009)
							IAAA-80894	AMS	Charred residues on pottery	~10090	40	11810	11410		
33.	Chubu	Kuzuharazawa IV	Yes	Ah	Flake	2	IAAA-71618	AMS	Charcoal	10860	60	12780	12700	YD	Kobayashi (2008)
							IAAA-71620	AMS	Charcoal	10960	60	12890	12730		
34.	Chubu	Yasumiba	No	Mc (boat)	Microblade	1	Gak-604	Beta	Charcoal	14300	700	18260	16450	LPG	Sugihara and Ono (1965)

35.	Chubu	Oshikakubo loc.3	Yes	Ah, Bf?	Flake	3	Beta-167428	AMS	Charcoal	10850	40	12750	12700	YD	Shibakawacho Board of Education (2006)
36.	Kinki	Torihama, 84 trench layer 52–61	Yes	Ah	Flake	0	KSU-1016	Beta	Wood	10070	45	11760	11400	YD~PB	Torihama shell midden research group (1987), Keally et al. (2003), Murakami and Onbe (2008)
							KSU-404	Beta	Wood	10320	45	12380	12010		
		Torihama, 83 trench layer 85	Yes	Ah, Bf?	Flake	0	KSU-1017	Beta	Wood	10290	45	12160	11960	YD	
							KSU-1027	Beta	Wood	10770	160	12850	12530		
37.	Kinki	Aitani-kumahara	Yes	Ah	Flake	1	IAAA-100028	AMS	Charcoal	10870	50	12780	12700	B/A~YD	Matsumuro and Shigeta (2010)
							IAAA-100022	AMS	Charcoal	~11210	50	13130	13040		
38.	Kinki	Kiriyama-wada	Yes	Ah, Sp, Bf	Flake	0	n/a							(B/A)[c]	Archaeological Institute of Kashihara (2002)
39.	Chugoku	Higashi	No	Mc, Bf	Microblade, flake	1	n/a							(LPG)[d]	Hiruzen Educational Association Board of Education (2003)
40.	Shikoku	Taishakukyo-mawatari, layer 4	Yes	Ah, Sp, Bf	Flake	0	HR-330	Beta	Shell	12080	100	14050	13790	B/A	Kawagoe (1995), Takehiro (2008)
41.	Shikoku	Kamikuroiwa, layer 9	Yes	Sp, Ah	Flake	0	Beta-201260	AMS	Charcoal	12530	40	15010	14700	LPG~B/A	Onbe and Kobayashi (2009), Watanabe (1966)
42.	Shikoku	Wasajima	No	Mc (wedge,boat)	Microblade, flake?	0	n/a							(LPG)[d]	Kagawa Prefecture Board of Education (1984)
43.	Shikoku	Okutani-minami	No	Mc (boat), Bf?	Microblade, flake?	0	n/a							(LPG)[d]	Kochi Prefecture Cultural Foundation Buried Cultural Property Center (2001)
44.	N.Kyushu	Obaru D, grid15-3	Yes	Ah	Flake	0	Beta-139873	AMS	Carbonized wood	9170	110	10490	10230	PB	Fukuoka City Board of Education (2002)
							Beta-139874	AMS	Carbonized wood	9210	80	10400	10210		
		Obaru D, grid14	Yes	Ah, Gah	Flake	2	Gak-20568	AMS	Charcoal	10480	30	12530	12400	YD	
							PLD-6288	AMS	Charred residues on pottery	~10880	110	12890	12690		

(continued)

Table 6.2 (continued)

No.	Region	Site	Pottery	Stone hunting weapon[a]	Primary reduction	Score of occupation intensity	Lab. no.	Method	Material	^{14}C Age (BP)	±1σ	Calendar age (cal BP; 68.2%)[b]		Chronozone	References
45.	N.Kyushu	Matsukida	Yes	Gah,Ah	Flake	2	PLD-6290	AMS	Charred residues on pottery	9400	30	10670	10580	PB	Fukuoka City Board of Education (1998), Nishimoto (2009)
							PLD-6289	AMS	Charred residues on pottery	~9630	25	11140	10870		
46.	N.Kyushu	Kawayo F	Yes	Mc (wedge), Ah	Microblade, flake	0	Beta-154841	AMS	Charcoal	12140	50	14120	13940	B/A	Kumamoto Prefecture Board of Education (2003)
							Beta-154931	AMS	Charcoal	~12360	50	14520	14170		
47.	N.Kyushu	Sempukuji, layer 8	Yes	Mc (wedge)	Microblade (bifacial blank)	3	MTC-11296	AMS	Charred residues on pottery	12220	80	14250	13980	B/A	Aso (1985)
48.	N.Kyushu	Takahata-otonohara	Yes	Mc (wedge), Ah	Microblade, flake	0	Beta-213635	AMS	Charred residues on pottery	12470	50	14860	14430	LPG~B/A	Yamato Town Board of Education (2007)
							Beta-213636	AMS	Charred residues on pottery	12570	60	15080	14760		
49.	N.Kyushu	Fukui cave, layer 2–3	Yes	Mc (wedge)	Microblade	0	Unreported	AMS	Charcoal	13180	50	15940	15740	LPG	Sasebo City Board of Education (2013)
							Unreported	AMS	Charcoal	~13410	50	16240	16050		
		Fukui cave, layer 4	No	Mc (boat)	Microblade	0	Unreported	AMS	Charcoal	13580	40	16450	16250		
		Fukui cave, layer 7–9	No	Small flake tool	Flake	1	Unreported	AMS	Charcoal	14230	50	17450	17220		
		Fukui cave, layer 12	No	Mc (prism)	Microblade	1	Unreported	AMS	Charcoal	14670	50	17960	17770		
		Fukui cave, layer 13	No	Flake tool	Flake	2	Unreported	AMS	Charcoal	14600	50	17880	17680		
							Unreported	AMS	Charcoal	~15290	60	18650	18480		
50.	N.Kyushu	Kawahara 3, CL6	No	Mc (prism)	Microblade	0	Beta-135259	AMS	Charcoal	14660	70	17950	17730	LPG	Shiba and Obata (2007)

51.	N.Kyushu	Chaen, layer 5	No	Mc (prism)	Microblade	1	Beta-107730	AMS	Charcoal	15450	190	18910	18510	LPG	Kishiku Town Board of Education (1998)
52.	S.Kyushu	Kakuriyama	Yes	Ah	Flake	2	N-3928	Beta	Carbonized wood	8630	125	9890	9480	PB	Kagoshima Prefecture Board of Education (1981), Taniguchi (2002)
							N-3927	Beta	Carbonized wood	~9110	125	10500	10170		
53.	S.Kyushu	Nagasakobira	Yes	Ah	Flake	2	IAAA-40272	AMS	Charcoal	9280	50	10570	10400	PB	Kagoshima Prefectural Buried Cultural Property Center (2005b)
							IAAA-40273	AMS	Charcoal	~9400	50	10700	10570	PB	
54.	S.Kyushu	Kiwaki	Yes	Ah	Flake	1	MTC-10292	AMS	Charred residues on pottery	9505	25	11060	10700	PB	Miyazaki Prefecture Archaeological Center (2001a), Onbe (2009)
							MTC-10293	AMS	Charred residues on pottery	9430	55	10720	10580		
55.	S.Kyushu	Kenshojo	Yes	Ah, Bf, Mc (prism)	Flake, microblade	4	Beta-163810	β	Charcoal	11220	120	13240	12970	B/A	Kagoshima Prefecture Aira Town Board of Education (2005)
							Beta-163812	AMS	Charcoal	~10920	50	12810	12720		
56.	S.Kyushu	Higashi-kurotsuchida	Yes	n/a	n/a	2	PLD-15892	AMS	Carbonized cotyledon	11530	35	13420	13320	B/A	Setoguchi (1981), Kudo (2012)
							PLD-15893	AMS	Carbonized cotyledon	11555	35	13440	13340		
57.	S.Kyushu	Mizusako, layer 7	Yes	Ah	Flake	3	n/a (layer 7 containing Sz-S <ca. 12800 cal BP>)							(B/A~YD)	Ibusuki Board of Education (2002)
		Mizusako, layer 9	No	Bp, Tr, Mc?	Flake, microblade?	0	n/a (layer 9 containing Ata-Iw <ca. 19000–15000 cal BP>)							(LPG)	
58.	S.Kyushu	Kiyotake Kamiinoharu, loc.5	Yes	Ah, Bf	Flake	3	Unreported	AMS	Charcoal	11380	60	13280	13150	B/A	Kiyotake Town Board of Education (2009)
							Unreported	AMS	Charcoal	~11720	40	13570	13480		

(continued)

Table 6.2 (continued)

No.	Region	Site	Pottery	Stone hunting weapon[a]	Primary reduction	Score of occupation intensity	Lab. no.	Method	Material	^{14}C Age (BP)	±1σ	Calendar age (cal BP; 68.2%)[b]		Chronozone	References
59.	S.Kyushu	Tsukabaru C	Yes	Ah, Mc	Flake, microblade	1	MTC-10288	AMS	Charred residues on pottery	11850	60	13740	13590	B/A	Miyazaki Prefecture Archaeological Center (2001b)
							MTC-10289	AMS	Charred residues on pottery	11750	60	13700	13470		
60.	S.Kyushu	Shikazegashira	Yes	Ah, Mc (prism)	Flake, microblade	3	Beta-118964	AMS	Charred residues on pottery	11780	50	13720	13550	B/A	Kagoshima Prefecture Kaseda City Board of Education (1999)
							Beta-118963	AMS	Charred residues on pottery	11860	50	13740	13610		
61.	S.Kyushu	Kiriki-mimitori, CL3	Yes	Ah, Mc?	Flake, microblade?	2	Beta-139159	AMS	Charcoal	11800	110	13750	13490	B/A	Kagoshima Prefectural Buried Cultural Property Center (2005a)
							Beta-139160	AMS	Charcoal	11690	110	13710	13400		
		Kiriki-mimitori, CL2	No	Bp, Tr	Flake	0	n/a (CL2 containing Tkn-bs <ca. 19100 cal BP>)							(LPG)	
62.	S.Kyushu	Okunonita	Yes	Ah, GAh	Flake	1	MTC-09141	AMS	Charred residue on pottery	11740	60	13700	13460	B/A	Kagoshima Prefecture Nishino-omote City Board of Education (1995)
63.	S.Kyushu	Sankakuyama I	Yes	Ah, GAh	Flake	3	MTC-05834	AMS	Charred residues on pottery	12080	70	14030	13810	B/A	Kagoshima Prefecture Buried Cultural Property Center (2006)
							IAAA-31697	AMS	Charred residues on pottery	~11050	70	13010	12820		

64.	S.Kyushu	Onigano	Yes	Ah, GAh	Flake	3	Beta-177289	AMS	Charred residues on pottery	11880	60	13760	13600	B/A	Kagoshima Prefecture Nishino-omote City Board of Education (2004)
							Beta-177290	AMS	Charred residues on pottery	~12180	40	14140	14000		
65.	S.Kyushu	Yokoi-takenoyama	Yes	Ah, Mc (boat, prism)	Microblade, flake	0	n/a (archaeological assemblage contained below Sz-S<ca. 12800 cal BP>)							LPG~B/A	Kagoshima City Board of Education (1990)
66.	S.Kyushu	Nishimaruo, layer 7b	No	Mc (boat, prism)	Microblade	1	n/a (archaeological assemblage contained below Sz-S<ca. 12801 cal BP>)							LPG~B/A	Kagoshima Prefecture Board of Education (1992)

Ah arrowhead, *GAh* ground arrowhead, *Bf* bifacial point, *Sp* bifacial stemmed point, *Mc* microblade, *Bp* backed point, *Tr* trapezoid, *PD* pit dwelling, *PC* pebble cluster, *H* hearth, *CH* hearth with chimney, *SH* stone-lined hearth, *SP* storage pit, *TP* trap pit, *LPG* Late Pleniglacial, *B/A* Bølling/Allerød, *YD* Younger Dryas, *PB* Preboreal

[a]Microblade core is classified intro three types: wedge-shaped, boat-shaped, and prism

[b]The IntCal 13 calibration curve is used

[c]Chronologically positioned by pottery typology

[d]Chronologically positioned by lithic typology

As past studies reported, there are at least five regions where different pottery types and lithic assemblages developed (Okamura 1997). These regional differences seem to have derived from those formed during the LGM which reflect lithic technological and human behavioral differences (Morisaki 2010). Accordingly, we divided the Japanese archipelago into five regions, namely, Hokkaido, Tohoku, Kanto/Chubu, Kinki/Chugoku/Shikoku, and Kyushu, in this paper (Fig. 6.2a). Only the Kyushu region has been subdivided into northern and southern areas.

Chronology

Most of the assemblages are dated by radiocarbon dates. In collecting the dates, the latest data sources (Nakazawa et al. 2011; Kudo 2012) are referenced. According to the sample evaluation criteria of Graf (2009), almost all samples were collected from clear contexts, and, therefore, the dates are also reliable because samples from bad contexts were already excluded from the data source in advance. All dates were calibrated using the OxCal v.4.2 (Bronk Ramsey 2009, 2013), adopting the IntCal13 radiocarbon age calibration curve (Reimer et al. 2013). When a site has multiple ^{14}C dates, the oldest and youngest dates were listed in Table 6.2. All sites were assigned to the possible chronozone, indicated by the calibrated dates.

Detailed lithic technological and human behavioral study in Hokkaido was recently published (Yamada 2006, 2008), so we therefore relied on this data and only compiled data for sites that were not focused on in that study.

Lithic Technological Analysis

The main focus of this paper is not to establish archaeological chronology but to investigate human behavioral responses to environmental change. Therefore, we mainly discuss lithic technology that reflects human behavioral strategies, focusing on the composition of stone hunting weapons and primary reduction sequences. Four main stone weaponry systems of the time period of focus can be identified: chipped or ground arrowheads, bifacial points, bifacial stemmed points, and microblades which are slotted in organic shafts (Fig. 6.3). Besides them, backed points and trapezoids are seen in only a few sites. Primary reduction is divided into four types: flake, blade, biface, and microblade.

Intensity of Occupation

Various archaeological features such as pit dwellings, pebble clusters, hearths (earth oven), hearths with chimney, stone-lined hearths, storage pits, and trap pits are known within the time period of focus in this paper (Fig. 6.3g–j). Since their

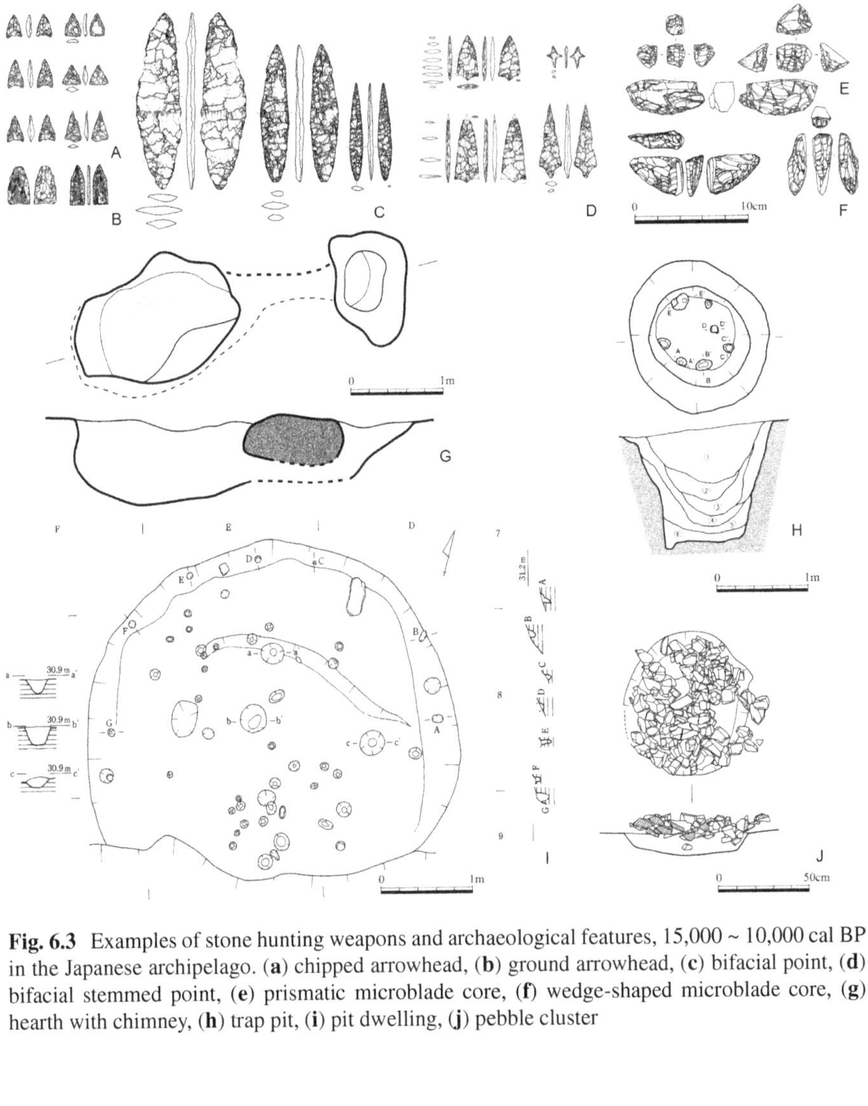

Fig. 6.3 Examples of stone hunting weapons and archaeological features, 15,000 ~ 10,000 cal BP in the Japanese archipelago. (**a**) chipped arrowhead, (**b**) ground arrowhead, (**c**) bifacial point, (**d**) bifacial stemmed point, (**e**) prismatic microblade core, (**f**) wedge-shaped microblade core, (**g**) hearth with chimney, (**h**) trap pit, (**i**) pit dwelling, (**j**) pebble cluster

construction requires much labor, these different features can serve as a proxy of intensified occupation or more sedentary lifeways. To estimate the degree of occupation intensity objectively, we scored the total number of different structures. There are seven types of archaeological features from the sites considered here, with scores ranked from 0 to 7 (with 7 being the highest). High score means labor-intensive occupation, while low scores indicate small investment in occupation activities. Apart from this, we also considered the timing of the appearance of pit dwellings, the most time-consuming structure providing evidence for occupation intensity.

Pottery

A variety of Jomon pottery types are known from the time period considered. Although it is of course important to chronological studies and studies of social interaction, the presence or absence of pottery is only briefly discussed here to describe its appearance within the context of the research questions.

Results

Kyushu Region

From 23 archaeological sites, 30 assemblages were analyzed (Fig. 6.2a, Table 6.2: 44–66). Lithic assemblages can largely be divided into three stages: the Late Pleniglacial, the Bølling/Allerød, and the Preboreal.

Before the Bølling/Allerød, lithic assemblages in southern Kyushu have backed points and trapezoids produced by expedient flake reduction, while those in northern Kyushu already have a microblade industry originally with prismatic microblades and later with wedge-shaped microblade cores on small bifaces (Shiba 2011). In the southern Kyushu, expedient microblade technique, which uses low-quality small lithic raw materials, was adopted belatedly after 17/16,000 cal BP.

At the same time as, or a little earlier than the beginning of the Bølling/Allerød warm period, the microblade industry with wedge-shaped microblade cores was associated with pottery in northern Kyushu. In contrast, the arrowhead industry, concomitant with the expedient boat-shaped microblade industry, started to be used between 16,000 and 14,000 cal BP in southern Kyushu (Morisaki 2015; Morisaki and Sato 2014). Moreover, ground arrowheads (Fig. 6.3b) are also found on Tanegashima Island (Sankakuyama I site, etc.), in the southernmost Kyushu region. Frequent use of ground arrowheads seems to have started earlier than previously mentioned (Miyata 2003). At the end of the Bølling/Allerød, the microblade industry disappeared, and stone hunting weaponry was completely replaced by chipped arrowheads made on blanks produced by expedient flake reduction.

Several types of archaeological features such as pit dwellings, pebble clusters, several kinds of hearths, and trap pits have been reported from sites after the Bølling/Allerød. The average of the occupation intensity score of whole Kyushu during the time period considered is 1.43 and that of northern Kyushu is 0.92, while that of southern Kyushu is 1.82; the score of assemblages during the Bølling/Allerød is 1.25 for northern Kyushu, and 2.50 for southern Kyushu which is the highest score among the regions focused here.

As seen in the example from Obaru D site in Fukuoka Prefecture in northern Kyushu, the cultural complex composed of arrowhead production on flake blanks, Jomon pottery production, and pit dwellings (Fig. 6.3i), the so-called Jomon cultural complex, continued during the Younger Dryas chronozone.

Kinki, Chugoku, and Shikoku Regions

Nine assemblages from eight archaeological sites from Kinki and Shikoku region were investigated (Fig. 6.2a; Table 6.2: 36–43). There is just one site whose archaeological context is evident in Chugoku region. Although samples are few, lithic assemblages can be largely divided into three stages: the Late Pleniglacial, the Bølling/Allerød, and the Preboreal.

Pre-Bølling/Allerød lithic assemblages such as those from the Wasajima and Okutani-minami sites contain microblade industries with mainly boat-shaped microblade cores and bifacial points made on flakes. Although these sites have no radiocarbon dates, their chronological position should be assigned to this time period because the previous period was comprised of different lithic assemblages such as the backed point assemblages (~20,000 cal BP; Morikawa 2010; Morisaki 2010).

Archaeological sites during the Bølling/Allerød are characterized by stemmed points, bifacial points, arrowheads, and pottery. Blanks for these bifacial tools were supplied by simple flake reduction. The stemmed point style (Fig. 6.3d, left) differs from those found in the Kanto region (Fig. 6.3d, right). Bifacial stemmed points and other bifacial points seem to have been phased out of use by the Younger Dryas (YD) at the latest. Arrowhead and simple flake reduction dominated after the onset of the Younger Dryas, as evidenced by assemblage from trench No. 83 of Torihama site and Aidani-kumahara site.

Pit dwellings first appear around the end of the Bølling/Allerød to the YD at the Aidani-kumahara site. A site that is not mentioned in this paper also has four pit dwellings (Kayumi-ijiri site) and which may be positioned to the Bølling/Allerød chronozone on the basis of pottery typology. The average of the occupation intensity score is very low (0.22), but it should be noted that three sites (Okutani-minami site, Taishakukyo-mawatari site, and Kamikuroiwa site) are rock-shelter or cave sites which might have supplemented for housing.

Kanto/Chubu Region

Twenty assemblages from twenty archaeological sites were analyzed (Fig. 6.2a; Table 6.2: 16–35). In this region, lithic assemblages can be largely divided into three stages; the Late Pleniglacial, the Bølling/Allerød, and the Preboreal.

Pre-Bølling/Allerød lithic assemblages comprise bifacial points on flakes and a microblade industry, first with prismatic and later with boat-shaped cores. Some sites of this period in northern Kanto, which lack radiocarbon ages and are not included in this paper, have wedge-shaped bifacial microblade cores. Biface reduction strategies are rare, except for some sites in the northern Chubu region. Only a few sites, such as the Gotenyama 2 N site and Miyagase-kitahara site, have possible pottery fragments. If the dates given to this pottery assemblage (15,420 ~ 16,420 cal

BP) are correct, then they are almost as old as the oldest plain pottery assemblage from the Odai-yamamoto I site in Aomori prefecture. Judging from the large elaborated lanceolate points, large blade tools, and lack of pottery, the Mikoshiba sites should be placed in this period.

Coinciding with the onset of the Bølling/Allerød, distinct Jomon pottery came into use throughout the region. Lithic assemblages contain bifacial stemmed points, other bifacial points, and arrowheads. With the exception of the blanks of some bifacial points in the northern Chubu region that were produced by bifacial reduction, all stone tools (including hunting weapons) were produced by expedient flake reduction. Although rare, microblade reduction is also recognized at a few sites (e.g., Tsukimino-kamino site loc. 2).

There were only a few bifacial tools which predated the start of the Younger Dryas, at which point stone hunting weapons had different kinds of arrowheads, with blanks prepared by simple flake reduction (Table 6.2).

Some types of archaeological features such as pit dwellings and pebble clusters have been reported from some sites, mostly after the Bølling/Allerød. As seen at the Oshikakubo site, which has 11 pit dwellings from the Younger Dryas chronozone (Shibakawacho Board of Education 2006), pit dwellings increased during this period all around the coastal area of the Pacific Ocean. The average of the occupation intensity score is 0.63 and that of assemblages from the Bølling/Allerød is 0.67.

Tohoku Region

Thirteen assemblages from thirteen archaeological sites were investigated (Fig. 6.2a; Table 6.2:3–15). The lithic assemblages can also be largely divided into three stages but in a little different way from the aforementioned regions because lithic technology during the Younger Dryas chronozone seems similar to those in the Bølling/Allerød in the Tohoku region; the Late Pleniglacial, the Bølling/Allerød to the Younger Dryas, and the Preboreal.

Before the Bølling/Allerød, lithic assemblages were characterized by microblade industries, first with prismatic microblade cores, and later with wedge-shaped microblade cores on bifaces, and boat-shaped microblade cores that were elaborately shaped. In addition, another type of lithic assemblage (e.g., the Odai-yamamoto I site and Akahira I site), which consists of fine blade tools and/or large bifaces, is found during this stage. Plain pottery from Odai-yamamoto I site was dated to 14,760 ~ 16,940 cal BP and is accepted as one of the oldest pottery assemblages in the world (Keally et al. 2003). The relationship between the two types of assemblages is still unknown, but it is noteworthy that the bifacial reduction technique characterizes both.

Lithic assemblages during the Bølling/Allerød witnessed the abandonment of the microblade industry. Bifacial tools (Fig. 6.3c) and arrowheads produced by bifacial reduction and flake reduction replaced it. These tools are highly standardized, and small tools such as arrowheads are made on flakes produced through a curated-biface reduction process (Sagawa and Suzuki 2006). Jomon pottery with a variety of decoration types appeared at this stage.

If the Unoki-minami site could be placed chronologically in the Younger Dryas, it would then imply the continuation of bifacial tools and other tools produced by bifacial reduction (in addition to other flake reduction schemes) throughout the onset of the Post-Glacial (with bifacial points continuing until the Early Jomon period). Lithic technology in this region during the Pleistocene-Holocene transition was therefore characterized by bifacial reduction techniques.

A small number of archaeological features such as pebble clusters and hearths have been reported from a few sites after the Bølling/Allerød. Clear evidence of pit dwellings first appeared during the YD (Kushibiki site), though some rock-shelter sites and one uncertain pit dwelling are known before that. The average of the occupation intensity score is 0.75, and that of assemblages from the Bølling/Allerød is 0.80.

Hokkaido Region

Paleolithic assemblages in Hokkaido (the southern Paleo-SHK peninsula), after 19,000 cal BP, contain a microblade industry which can be primarily characterized by the presence of a wide variety of microblade core types and reduction techniques (Nakazawa et al. 2005; Sato and Tsutsumi 2007).

The assemblages with microblade industries in Hokkaido can be divided into three stages on the basis of the presence of distinct microblade core types, radiocarbon data, and geochronology (Yamada 2006). These stages are an initial early stage (ca. 21,500–18,500 ^{14}C BP; 26,000 ~ 22,000 cal BP), a late early stage (ca. 15,500–13,500 ^{14}C BP; 19,000 ~ 16,000 cal BP), and a late stage (ca. 13,500–11,000 ^{14}C BP; 16,000 ~ 13,000 cal BP) (Morisaki et al. 2015).

Of these, the late early and the late stage are the focus of this paper. The late early stage (before the Bølling/Allerød) lithic assemblage has two types of microblade cores (Sakkotsu and Togeshita), burins, end scrapers, and sidescrapers. Blanks of these tools were bifaces, bifacial thinning flakes, and blades. Togeshita-type cores are made mainly on flakes. Consequently, reduction techniques are composed of microblade, biface, blade, and flake reduction processes. Therefore, tools and cores are highly portable and suit a broad ranging foraging subsistence. The late stage (the Bølling/Allerød) lithic assemblages contain at least three types of microblade cores (Shirataki, Oshorokko, Hirosato), bifacial stemmed points, adzes, axes, burins, end scrapers, and awls. Blank production techniques are composed of microblade, biface, and blade reduction processes.

In southeastern Hokkaido, one Jomon assemblage from the Taisho 3 site that has been dated to the Bølling/Allerød warm period is known. The Jomon pottery was well dated, and many lithic samples were collected from this site, but the lithics and pottery are supposed to be different from the Hokkaido cultural tradition and possibly left by migrants or by culturally related groups from the Tohoku region (Yamahara 2008).

To date, there has been no archaeological site firmly dated to the Younger Dryas in Hokkaido, so lithic technology and archaeological features for this period are undefined. There is a possibility that hunter-gatherer population density fluctuated. At least, they did not adopt a sedentary lifeway until the onset of the Holocene, when lithic assemblages containing arrowheads, flake reduction techniques, and Jomon pottery (e.g., Taisho 6 site) appear.

Microblade assemblages in Hokkaido contain highly standardized tools including microblades, burins, drills, end scrapers, and sidescrapers. Bifacial stemmed points and axes appeared during the late stage. Tool kit diversity became higher in the late stage than in the preceding stages. These tools were produced by several reduction techniques in combination. In the late early stage, variability of microblade production technology and tool types was relatively low, while high tool kit diversity and various microblade core types emerged in the later stage.

Hearths (earth oven) are the only features recorded for this region. Features constructed by digging down into the ground seem quite rare until the onset of Holocene, with the first appearance of pit dwellings.

Summary

Changes in lithic technology as well as the earliest occurrence of pottery and pit dwellings are summarized in Fig. 6.4. Three main points can be inferred from our analyses.

The first is that lithic technologies in Hokkaido and Honshu are different throughout the Pleistocene, until the onset of Holocene. Lithic technology in Hokkaido was highly elaborate and similar to its continental northeastern Asia counterpart (Sato 2003; Izuho and Sato 2008; Izuho 2013). They clearly differ from that of the Paleo-Honshu Island (Morisaki 2010; Morisaki et al. 2015).

The second is that Jomon pottery appears widely around 15,000 cal BP, except in Hokkaido (Kudo 2005) and a few other areas. The appearance of pottery coincides with changes in lithic technology toward flake tool industries and with the production of arrowheads by invasive thinning (using local lithic raw materials). This technological change characterizes the specific “Jomon lithic technology” for hunting weaponry systems.

The third is that there were interregional and temporal differences in lithic technology and in the occurrence of archaeological features. Although the timing is synchronous in the southwestern Paleo-Honshu Island, there are also some notable differences in the specific components of lithic assemblages and technology.

Bifacial reduction was the main blank producing technique in the Tohoku region. Chipped arrowheads seem to dominate much later in lithic assemblages of the northeastern Paleo-Honshu Island (Sagawa and Suzuki 2006; Sato et al. 2011b). By contrast, ground arrowheads are used initially in the southernmost Kyushu since the

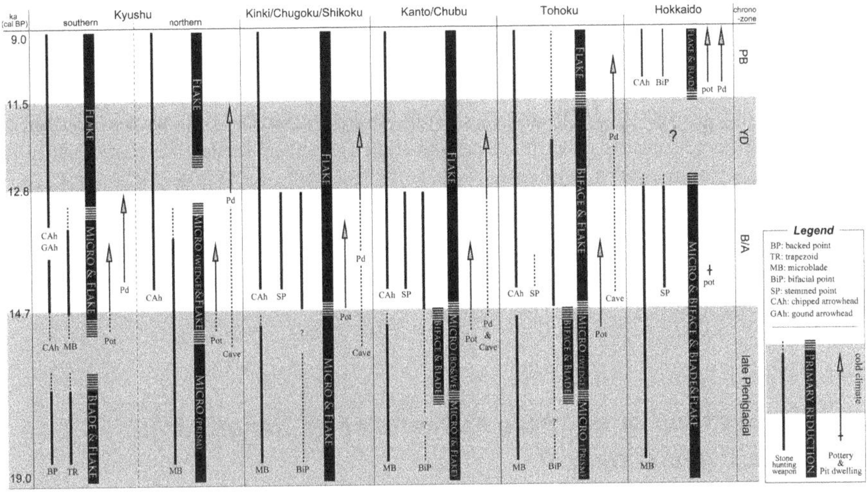

Fig. 6.4 Spatiotemporal diversity of lithic technology and timing of the appearance of pit dwellings and pottery in the Japanese archipelago from 19,000 to 10,000 cal BP

Bølling/Allerød. Meanwhile, bifacial stemmed points are often used in the Kinki, Shikoku, Kanto, and Chubu regions, except in the Kyushu region. As mentioned below, these differences are thought to reflect human behavioral variability. As for archaeological features, during the Bølling/Allerød, a number of pit dwellings and other features were more prevalent in the southern Kyushu region than in other regions (the highest occupation intensity score: 2.50).

Discussion

Lithic Technological Difference Between Hokkaido (the Southern Paleo-SHK Peninsula) and the Paleo-Honshu Island

It is possible to answer why the lithic technologies in Hokkaido differ from those in Honshu. As we mentioned earlier, environmental differences between Hokkaido and Honshu through the Late Pleniglacial must have caused humans to develop different lithic technology and behavioral strategies. Although continental dry and cold climates in Hokkaido were gradually changing to an island climate during the Late Glacial, stable warm and wet climates did not dominate until the end of Younger Dryas (11,500 cal BP). Except for the Taisho 3 site, which is supposed to have been occupied by migrants from the Paleo-Honshu Island during the Bølling/Allerød chronozone, the appearance of Jomon-type lithic assemblages and technology in Hokkaido occurred after the onset of the Holocene. Therefore, the uniqueness of lithic technology in Hokkaido throughout the terminal Pleistocene is plausibly explained by environmental differences with the Paleo-Honshu Island (Okamura 1997; Sato 2008a).

The highly portable, curated tool kit and relatively low tool assemblage richness in the late early stage of Hokkaido microblade assemblages indicate that foragers organized their technology to the dispersed distribution of lithic raw materials in a cold grassland landscape (even during the LG). But high tool kit diversity and various microblade core types in the late stage (the Bølling/Allerød) suggest the possibility that foraging territories were gradually becoming smaller, in keeping with the climatic amelioration and development of a forest environment (Morisaki et al. 2010; Yamada 2006).

The Background of the Wide Appearance of Jomon Lithic Technology and Pottery

The next question is: why did pottery and Jomon lithic technology appear at the same time across Paleo-Honshu Island? It should be noted that the timing clearly coincided with the onset of the Bølling/Allerød. This climatic amelioration was responsible for an increase in precipitation, for the development of a forest landscape similar to the Holocene on Paleo-Honshu Island (Table 6.1), and for the disappearance of large mammals from Pleistocene faunal complex. These abrupt environmental changes must have had a strong impact that caused human populations to shift to a new Holocene type of hunting behavior in the forest environment. This is consistent with the characterization of Jomon lithic technology as supported by flake reduction as primary reduction, suitable to a more sedentary way of life in relatively small foraging areas, in a newly forested landscape. Moreover, it is also of importance that lithic technology changed at the onset of the Bølling/Allerød and did not return to the technology in the previous stage. This is counter to the idea of the Younger Dryas having a strong impact on lithic technology in Japan (Kanomata 2007; Miyoshi 2013). The amount of pottery, however, clearly decreased during the Younger Dryas cooling event, as some researcher already pointed out (Taniguchi 2004; Sato 2008; Nakazawa et al. 2011).

Regional Differences in Human Behavior During the Late Glacial

After 15,000 cal BP, the Jomon lithic technology makes an appearance, with regional differences. If we consider the evidence for the adoption of ground stone technology and the construction of many archaeological features, which are indicators of occupation intensity, then foragers in the southern Kyushu region will transitioned quickly to a sedentary way of life during the Bølling/Allerød chronozone, earlier than other regions on the Paleo-Honshu Island (Amemiya 1993; Okamura 1997). This is supported also by the fact that they adopted relatively expedient primary reduction techniques.

On the other hand, the above data suggests that instead of pit dwellings, rock overhangs and caves were occasionally utilized for shelter in all other regions except southern Kyushu (Suzuki 2009). As opposed to southern Kyushu, foragers in the Tohoku region habitually practiced curated bifacial reduction as a major tool and tool blank producing technique during the Late Glacial, alongside flake reduction (with all tools being elaborately produced). They did not use pit dwellings until the late Younger Dryas. These facts suggest that relatively high mobility continued in the Tohoku region during this period (Sato et al. 2011b).

Meanwhile, judging from lithic technology and archaeological features, foragers in the Kinki, Shikoku, Chubu, and Kanto regions seem to have been more mobile than those in the southern Kyushu region but less mobile than those in the Tohoku region. This interpretation is based on lower numbers of pit dwellings and the lack of evidence for sophisticated reduction techniques (as in the Tohoku region). Additionally, in these regions, the number of pit dwellings increased sharply since the onset of Holocene.

This geographic cline of lithic technology and occupation intensity during the Late Glacial (which includes Hokkaido) seems to coincide with the natural environment (e.g., Fig. 6.2b, Table 6.1). In short, these data imply that microblade industries are mostly related to high mobility and to less favorable environmental conditions, whereas flake industries are related to low mobility and to more favorable environmental conditions. Bifacial industries seem to be positioned between these spectrums. Future research should investigate how this diversity in technologies is related to foragers' adaptations to changing fauna, flora, and landscapes and whether there are additional transformations after the period considered here.

Conclusion

Our reassessment of Late Glacial and early Holocene archaeological chronology and lithic technological changes has revealed regional and diachronic differences. Lithic technologies in Hokkaido and Honshu were clearly different until the onset of Holocene, due to environmental variability between these regions. Most typical attributes of Jomon lithic technology and archaeological features did not appear in Hokkaido until after the onset of the Holocene due to delayed climatic amelioration, yet they appeared at the onset of the Late Glacial in Honshu. As early as this period, there are regional differences in the Jomon lithic technology of Honshu. It is likely that most of this regional variation is a reflection of adaptation to spatiotemporally variable environments.

Recent studies have revealed that exploitation of marine resource and processing in pots dates back to the Incipient Jomon period (Kunikita et al. 2013). This is also supported by other researchers (Craig et al. 2013). Although this paper doesn't have the space to mention these studies, it is likely that when future technological and behavioral studies of lithics integrate the important new results from isotope analyses, a more detailed and dynamic trajectory across time and space of subsistence changes from the Paleolithic to the Jomon period can be expected.

Acknowledgments I am grateful to Drs. Erick Robinson (University of Wyoming) and Frédéric Sellet (University of Kansas) for their invitation to participate in the SAA symposium in Honolulu and to this exciting book. We deeply thank Drs. Dai Kunikita (The University of Tokyo) and Yuichiro Kudo (National Museum of Japanese History) for advising on radiocarbon dating and Drs. Daigo Natsuki (The University of Tokyo) and Shogo Osawa (Nara National Research Institute for Cultural Properties) for helping us prepare this paper.

References

Amemiya, M. (1993). The beginning of sedentary life way in temperate forest. *Kobunkadanso, 30*, 988–1023. (In Japanese).

Anzai, M. (2002). Critique of the migration theory of 'the Mikoshiba-Chnojakubo culture' from the continent. *Busshitsu Bunka, 72*, 1–20. (In Japanese).

Aomori Prefecture Buried Cultural Property Center. (1999). *Aomori prefecture buried cultural property Center excavation report (no.263): Kushibiki site*. Aomori: Aomori Prefecture Board of Education. (In Japanese).

Aomori Prefecture Nango Village Board of Education. (2001). *Nango Village Board of Education Excavation Report (no.3): Kiwada site*. Aomori: Aomori Prefecture Nango Village Board of Education. (In Japanese).

Aomori Prefecture Buried Cultural Property Center. (2008). *Tsubokezawa (I) iseki, Shibayama (I) iseki, Daibougasira iseki, Akahira (I) iseki, Akahira (II) iseki*. Aomori: Aomori prefecture Board of Education. (In Japanese).

Aomori Prefecture Buried Cultural Property Center. (2011). *Aomori prefecture buried cultural property Center excavation report (no.502): Itsukawame (6) site*. Aomori: Aomori Prefecture Board of Education. (In Japanese).

Archaeological Institute of Kashihara. (2002). *Archaeological Institute of Kashihara Excavation Report (no.91): Kiriyama-wada site*. Nara: Nara Prefecture Kashihara Archaeological Institute. (In Japanese).

Archaeological Research Center of Nagano Prefecture. (2000). *Archaeological research Center of Nagano Prefecture Excavation Report (no.49): Joetsu-Jidoshado buried cultural property excavation report vol. 16*. Nagano: Archaeological Research Center of Nagano Prefecture. (In Japanese).

Archaeological Research Center of Nagano Prefecture. (2004). *Archaeological Research Center of Nagano Prefecture excavation report (no.63): An report on excavation with construction of route 18 vol.3*. Nagano: Archaeological Research Center of Nagano Prefecture. (In Japanese).

Ariake Cultural Property Research Institute, Manpukuji Sites Excavation Team. (2005). *Manpukuji sites*. Kanagawa: Ariake Buried Property Research Institute, Manpukuji Sites Excavation Team. (In Japanese).

Aso, M. (1985). *Excavation report of the Senpukuji cave*. Tokyo: Tsukiji shokan. (In Japanese).

Bamforth, D., & Bleed, P. (1997). Technology, flaked stone technology, and risk. In C. Barton & G. Clark (Eds.), *Rediscovering Darwin: Evolutionary theory in archaeological explanation, Archaeological papers of the American Anthropological Association 7* (pp. 267–290). Arlington: American Anthropological Association.

Binford, L. R. (1979). Organization and formation processes: Looking at curated technologies. *Journal of Anthropological Research, 35*(3), 255–273.
Binford, L. R. (1980). Willow smoke and dog's tails: Hunter-gatherer settlement systems and archaeological site formation. *American Antiquity, 45*(1), 4–20.
Bronk Ramsey, C. (2009). Bayesian analysis of radiocarbon dates. *Radiocarbon, 51*(1), 337–360.
Bronk Ramsey, C. (2013). OxCal v4.2.3. https://c14.arch.ox.ac.uk/login/login.php?Location=/oxcal/OxCal.html
Craig, O. E., Saul, H., Lucquin, A., Nishida, Y., Tache, K., Clarke, L., Thompson, A., Altoft, D. T., Uchiyama, J., Ajimoto, A., Gibbs, K., Isaksson, S., Heron, C. P., & Jordan, P. (2013). Earliest evidence for the use of pottery. *Nature, 496*(7445), 351–354.
Department of Archaeology Faculties of Arts and Letters Tohoku University. (1990). *The Araya site*. Miyagi: Department of Archaeology Graduate School of Arts and Letters Tohoku University & Kawaguchi town Board of Education, Sendai. (In Japanese with English abstract).
Department of Archaeology Graduate School of Arts and Letters Tohoku University. (2003). *The Araya site Report of the second and third term excavations*. Miyagi: Department of Archaeology Graduate School of Arts and Letters Tohoku University, Kawaguchi town Board of Education, Sendai. (In Japanese with English abstract).
Fukuoka City Board of Education. (1998). *Fukuoka City Board of Education Excavation Report (No.578): Matsukida Site Cluster : 2nd · 3rd Excavation*. Fukuoka: Fukuoka City Board of Education, Fukuoka. (In Japanese).
Fukuoka City Board of Education. (2002). *Fukuoka City Board of Education Excavation Report (No.741): Obaru D Site Cluster (4)*. Fukuoka: Fukuoka City Board of Education, Fukuoka. (In Japanese).
Graf, K. (2009). 'The good, the bad, and the ugly': Evaluating the radiocarbon chronology of the middle and late upper Paleolithic in the Enisei River valley, south-central Siberia. *Journal of Archaeological Science, 36*, 694–707.
Hashikami Town Board of Education. (1999). *Takihata site excavation report*. Aomori: Hashikami Town Board of Education. (In Japanese).
Hara, H., Onbe, S., Miyata, Y., & Murakami, N. (2010). The radiocarbon dating of pottery from Hananoko site. *Kodai Bunka, 62*(1), 90–98. (In Japanese).
Hayashi, S., & Kami-Ina Archaeological Society. (2008). *Mikoshiba site*. Okayama: Shinmai Shoseki Publication Centre. (In Japanese).
Hiruzen Educational Association Board of Education. (2003). *Higashi Site 1*. Yatsukamura: Hiruzen Educational Association Board of Education. (In Japanese).
Ibusuki Board of Education. (2002). *Ibusuki Board of Education Excavation Report (no.35): Report of the Mizusako archaeological site Vol. 2*. Kagoshia: Ibusuki Board of Education. (In Japanese).
Igarashi, Y. (2008). Climate and vegetation changes since 40,000 years BP in Hokkaido and Sakhalin. In H. Sato (Ed.), *Human ecosystem changes in the northern Circum Japan Sea area (NCJSA) in the late Pleistocene* (pp. 27–41). Kyoto: Research Institute for Humanity and Nature.
Imamura, K. (1999). *In pursuit of the reality of Jomon culture*. Tokyo: Yoshikawakobunkan. (In Japanese).
Inada, T. (1986). The formation of Jomon culture. In *Nihon kokogaku 6*. Tokyo: Iwamani shoten. (In Japanese).
Inada, T. (1993). Contact between microblade culture and Jomon culture. *Kokogaku Kenkyu, 40*(2), 21–46. (In Japanese).
Irihirose Village Board of Education, Cave Excavation Team of Uonuma Region. (2004). *Irihirose Village Board of Education Excavation Report (no.1): Kurohime cave site*. Niigata: Irihirose Village Board of Education. (In Japanese).
Iwate Buried Cultural Property Center. (2004). *Iwate buried cultural property Center excavation report (no.437): Hayasakatai site*. Iwate: Iwate Buried Cultural Property Center. (In Japanese).

Izuho, M., & Sato, H. (2008). Landscape evolution and culture changes in the upper Paleolithic of northern Japan. In A. P. Derevianko & M. V. Shunkov (Eds.), *The current issues of Paleolithic studies in Asia; proceedings of the international symposium "the current issues of Paleolithic studies in Asia and contiguous regions"* (pp. 69–77). Novisibirsk: Publishing Department of the Institute of Archaeology and Ethnography SB RAS.

Izuho, M. (2013). Human technological and behavioral adaptation to landscape changes around the last glacial maximum in Japan: A focus on Hokkaido. In K. Graf, C. Kenton, & M. Waters (Eds.), *Paleoamerican odyssey* (pp. 45–64). College Station: Texas A&M University Press.

Kagawa Prefecture Board of Education. (1984). *An excavation report of buried cultural properties accompanied with the construction of Seto big bridge*. Kagawa: Kagawa Prefecture Cultural Property Protection Society. (In Japanese).

Kagoshima City Board of Education. (1990). *Kagoshima City buried cultural property excavation report (10) Yokoi-takenoyama site*. Kagoshima: Kagoshima City Board of Education. (In Japanese).

Kagoshima Prefectural Buried Cultural Property Center. (2005a). *Kagosima prefectural buried cultural property Center excavation report (no.91) Kirikimimitori site I*. Kagoshima: Kagosima buried cultural property center. (In Japanese).

Kagoshima Prefectural Buried Cultural Property Center. (2005b). *Kagosima prefectural buried cultural property center excavation report (No.93) Nagasakobira site*. Kagoshima: Kagosima buried cultural property center. (In Japanese).

Kagoshima Prefectural Buried Cultural Property Center. (2006). *Kagosima prefectural buried cultural property center excavation report (No.96): sankakuyama site cluster (No.3)*. Kagoshima: Kagosima buried cultural property center, Kirishima. (In Japanese).

Kagoshima Prefecture Board of Education. (1981). *Kagoshima prefecture Board of Education Excavation Report (No.16) Kakuriyama site, Kaminokiyama site*. Kagoshima: Prefecture Board of Education. (In Japanese).

Kagoshima Prefecture Board of Education. (1992). *Kagoshima prefecture Board of Education Excavation Report (No.64) Nishimaruo site*. Kagoshima: Kagoshima Prefecture Board of Education. (In Japanese).

Kagoshima Prefecture Aira Town Board of Education. (2005). *Aira town board of education excavation report (No.10): Kenshojo site*. Kagoshima: Kagoshima Prefecture Aira Town Board of Education. (In Japanese).

Kagoshima Prefecture Kaseda City Board of Education. (1999). *Kaseda city board of education excavation report (No.16): Shikazegashira site, Okunano site*. Kagoshima: Kaseda City Board of Education. (In Japanese).

Kagoshima Prefecture Nishino-omote City Board of Education. (1995). *Nishino-omote city board of education excavation report (no.7): Okunonita site, Okuarashi site*. Kagoshima: Kagoshima Prefecture Nishino-omote City Board of Education. (In Japanese).

Kagoshima Prefecture Nishino-omote City Board of Education. (2004). *Nishino-omote city board of education excavation report (no.14): Onigano site*. Kagoshima: Kagoshima Prefecture Nishino-omote City Board of Education. (In Japanese).

Kanagawa Archaeology Foundation. (1998). *Kanagawa archaeology foundation excavation report (no.41): Miyagase sites XV*. Kanagawa: Kanagawa Archaeological Foundation. (In Japanese).

Kanagawa Archaeology Foundation. (1999). *Kanagawa archaeology foundation excavation report (no.49): Yoshioka sites IX*. Kanagawa: Kanagawa Archaeological Foundation. (In Japanese).

Kasakake Town Board of Education. (2003). *Kasakake town board of education excavation report (no. 13): An excavation report of Saishikada-nakajima site Vol.1*. Gunma: Kasakake Town Board of Education. (In Japanese).

Kanomata, Y. (2007). The human adaptation to the environmental change in East Japan from the end of Pleistocene to the beginning of Holocene epoch. *Miyagi Kokogaku, 9*, 1–20. (In Japanese).

Kato Construction Company Research Department of Buried Property. (2004). *Musashino city Gotenyama site Loc. 2N*. Tokyo: Kato Construction Company. (In Japanese).

Kawagoe, T. (1995). Radiocarbon dating of Taishakukyo sites (1). *Bulletin of Taishaku-kyo Sites Research Centre, Graduate School of Letters, Hiroshima University, 10*, 109–117.

Keally, C. T., Taniguchi, Y., & Kuzmin, Y. V. (2003). Understanding the beginnings of pottery technology in Japan and neighbouring East Asia. *The Review of Archaeology, 24*(2), 3–14.

Keio-gijuku University Department of Archaeology. (1993). *Shonan-fujisawa campus site Vol 1*. Tokyo: Keio-gijuku. (In Japanese).

Kishiku Town Board of Education. (1998). *Kishiku town board of education excavation eeport (No.3)*. Nagasaki: Kishiku Town Board of Education. (In Japanese).

Kiyotake Town Board of Education. (2009). *Kiyotake Kamiinoharu Site: Location 5*. Miyazaki: Kiyotake Town Board of Education. (In Japanese).

Kodama, K. (2001). From the Palaeolithic to the Jomon period: Southern Kyushu perspective. *Daiyonki kenkyu, 40*(6), 499–507. (In Japanese with English abstract).

Kobayashi, T. (1981). *Jin site*. Tokyo: Kokugakuin University Archaeological Laboratory. (In Japanese).

Kobayashi, K. (2008). A chronological study on the beginning of sedentism by radiocarbon dating. In Archaeology Group of Chuo University (Ed.), *Hakumon-Kouko-Ronsou III* (pp. 1–28). Tokyo: Archaeology study group of Chuo University. (In Japanese).

Onbe, S., & Kobayashi, K. (2009). Carbon 14 dating of charcoals and shellfish specimens excavated at Kamikuroiwa. *Bulletin of the National Museum of Japanese History, 154*, 504–510.

Kobayashi, K., Sakamoto, M., & Kudo, Y. (Eds.). (2009). *The beginning of Jomon culture*. Chiba: National Museum of Japanese History. (In Japanese).

Kochi Prefecture Buried Cultural Property Center. (2001). *Kochi prefecture buried cultural property center excavation report (no. 63): Okutani-minami site III*. Kochi: Kochi Prefecture Cultural Foundation Buried Cultural Property Center. (In Japanese).

Kudo, Y. (2005). The temporal correspondence between archaeological chronology and environmental changes in the final Pleistocene in eastern Honshu Island. *Daiyonki Kenkyu, 44*(1), 51–64. (In Japanese with English abstract).

Kudo, Y. (2012). *Environment and culture history of the upper Palaeolithic and the Jomon period: High-precision radiocarbon dating and archaeology*. Tokyo: Shinsensha. (In Japanese).

Kudo, Y., Shiraishi, H., & Nakamura, T. (2011). Radiocarbon dating of the pottery adhesions and charcoals of the incipient Jomon period. *Summaries of Research Using AMS at Nagoya University, 22*, 199–210. (In Japanese with English abstract).

Kumamoto Prefecture Board of Education. (2003). *Kumamoto prefecture board of education excavation report (No. 209): Kawayo F site*. Kumamoto: Kumamoto Prefecture Board of Education. (In Japanese).

Kunikita, D., Shevkomud, I., Yoshida, K., Onuki, S., Yamahara, Y., & Matsuzaki, H. (2013). Dating charred remains on pottery and analyzing food habits in the early Neolithic period in northeast Asia. *Radiocarbon, 55*, 1334–1340.

Kurishima, T. (1991). Jomon revolution from the north. *Archaeological Journal, 341*, 8–13. (In Japanese).

Kuzmin, T. V., & Orlova, L. A. (2004). Radiocarbon chronology and environment of woolly mammoth (Mammuthus primigenius Blum.) in northern Asia: Results and perspectives. *Earth-Science Reviews, 68*(1–2), 133–169.

Matsui, H., Tada, R., & Oba, T. (1998). Low-salinity isolation event in the Japan Sea in response to eustatic sea-level drop during LGM: Reconstruction based on salinity-balance model. *Daiyonki Kenkyu, 37*, 221–233.

Matsumuro, T., & Shigeta, T. (2010). A new example of dogu of incipient Jomon in Shiga prefecture: Aidani-kumahara site. *Kokogaku Janaru, 608*, 29–31. (In Japanese).

Mitsuishi, N. (2013). Incipient Jomon period in the western Honshu. In M. Miyoshi (Ed.), *Environmental change and human activity* (pp. 31–46). Mie: Kokogaku-kenkyukai. (In Japanese).

Miyata, E. (2003). Polished arrowheads in earliest Jomon period. *From Jomon no mori, 1*, 29–36. (In Japanese).

Miyazaki Prefecture Archaeological Center. (2001a). *Miyazaki prefectural buried cultural property center excavation report (No.43): Kiwaki site*. Miyazaki: Miyazaki Prefectural Buried Cultural Property Center. (In Japanese).

Miyazaki Prefecture Archaeological Center. (2001b). *Miyazaki prefectural buried cultural property center excavation report (No.44): Matsumoto site, Ideguchi site, Tsukabaru site*. Miyazaki: Miyazaki Prefectural Buried Cultural Property Center. (In Japanese).

Miyoshi, M. (2013). Subsistence and environment during the incipient Jomon period. In M. Miyoshi (Ed.), *Environmental change and human activity* (pp. 1–19). Mie: Kokogaku-kenkyukai. (In Japanese).

Morikawa, M. (2010). Knife shaped tools from Ikariiwa-minamiyama site. In K. Matsufuji (Ed.), *Kokogaku ha nani wo katareruka* (pp. 17–28). Kyoto: Dosisha University Archaeological Series Publishing group. (In Japanese).

Morisaki, K. (2010). *Structural change and regional adaptation of Palaeolithic society*. Tokyo: Rokuichi-shobo. (In Japanese).

Morisaki, K., Izuho, M., & Sato, H. (2010). Upper Paleolithic technological organizations response to landscape changes in northern Japan. In *The initial human habitation of the continental and the insular parts of the North East Asia: The proceedings of the international symposium* (pp. 135–141). Yuzhno-Sakhalinsk: Sakhalin State University.

Morisaki, K., & Sato, H. (2014). Lithic technological and human behavioral diversity before and during the late glacial: A Japanese case study. *Quaternary International, 347*, 200–210.

Morisaki, K., Izuho, M., Terry, K., & Sato, H. (2015). Lithics and climate: Technological responses to landscape change in upper Palaeolithic northern Japan. *Antiquity, 89*(345), 554–572.

Morisaki, K. (2015). Settlement patterns of prehistoric hunter-gatherers in Kyushu region during the terminal Pleistocene. *Daiyonki Kenkyu, 54*(5), 257–270. (In Japanese with English abstract).

Murakami, N., & Onbe, S. (2008). Radiocarbon dating of Tajomon pottery from Torihama Shell mound. In *Bulletin of Fukui prefectural Wakasa history museum* (pp. 19–20). Fukui: Fukui Prefectural Wakasa History Museum.

Nakazato Village Board of Education. (2001). *Nakazato village board of education excavation report (no. 9): Kubodera-minami site*. Niigata: Nakazato Village Board of Education. (In Japanese).

Nakazawa, Y., Izuho, M., Takakura, J., & Yamada, S. (2005). Toward an understanding of technological variability in microblade assemblages in Hokkaido, Japan. *Asian Perspectives, 44*, 276–292.

Nakazawa, Y., Iwase, A., Akai, F., & Izuho, M. (2011). Human responses to the younger Dryas in Japan. *Quaternary International, 242*(2), 416–433.

Nelson, M. C. (1991). The study of technological organization. *Archaeological Method and Theory, 3*, 57–100.

Nishimoto, T. (Ed.) (2009). *Report of grant-in-aid for creative scientific research (16GS0118): The origin of the farming in the Yayoi Period and East Asia: Establishment of high-precision Chronology by Carbon 14 age analysis*. Chiba: National Museum of Japanese History. (In Japanese with English abstract).

Obihiro City Board of Education. (2005). *Obihiro Taisho sites 1*. Obihiro: Obihiro City Board of Education. (In Japanese).

Obihiro City Board of Education. (2006). *Obihiro Taisho sites 2*. Obihiro: Obihiro City Board of Education. (In Japanese).

Odai-yamamoto I site excavation team. (1999). *Archaeological research at the Odai Yamamoto I site*. Tokyo: Odai-yamamoto I Site Excavation Team. (In Japanese).

Okamoto, T. (1979). *On the Mikoshiba-Chojakubo culture. Kenkyu Ronshu V*. Nara: Nara National Research Institute for Cultural Properties. (In Japanese).

Okamura, M. (1997). The formation of Jomon culture in the southern and northern parts of the Japanese Islands. *Daiyonki kenkyu, 36*(5), 319–328. (In Japanese with English abstract).

Onbe, S. (2009). Carbon-14 dating of 'Ento-kei kaigara-mon' pottery. *Newsletter of Jomon Study in South Kyushu, 20*, 141–153. (In Japanese).

Reimer, P. J., Bard, E., Bayliss, A., Warren Beck, J., Blackwell, P. G., Bronk Ramsey, C., Buck, C. E., Cheng, H., Lawrence Edwards, R., Friedrich, M., Grootes, P. M., Guilderson, T. P., Haflidason, H., Hajdas, I., Hatté, C., Heaton, T. J., Hoffmann, T. L., Hogg, A. G., Hughen, K. A., Felix Kaiser, K., Kromer, B., Manning, S. M., Niu, M., Reimer, R. W., Richards, D. A., Marian Scott, E., Southon, J. R., Staff, R. A., Turney, C. S. M., & van der Plicht, J. (2013). Intcal13 and marine 13 radiocarbon age calibration curves 0–50,000 years cal BP. *Radiocarbon, 55*(4), 1869–1887.

Sagawa, M., & Suzuki, M. (2006). *Lithic Assemblage from the Western Terrace of the Hinata Caves Vol. I: Spear Point Workshop in the Incipient Jomon*. Yamagata: Tohoku Gakuin University Sagawa Laboratory, Takahata Town Board of Education, Ukitamu Fudoki no Oka Archaeological Museum. (In Japanese with English abstract).

Saku City Board of Education. (2006). *Tenjin-one Site*. Nagano: Saku City Board of Education. (In Japanese).

Sasebo City Board of Education. (2013). *Preliminary report of Fukui cave National Historic Site*. Nagasaki: Sasebo City Board of Education. (In Japanese).

Sato, H. (2003). The early upper Paleolithic in Hokkaido. *Kodai Bunka, 55*, 181–194. (In Japanese).

Sato, H. (2008). Introduction: The process of Jomonization. In H. Sato (Ed.), *The process of Jomonization* (pp. 1–9). Tokyo: Rokuichi-shobo. (In Japanese).

Sato, H., Yamada, S., & Izuho, M. (2011b). Animal resource and hunting in upper Paleolithic in Japanese Islands. In H. Sato & K. Inuma (Eds.), *No to Hara no Kankyoshi* (pp. 51–71). Tokyo: Bun'ichi Sogo Shuppan. (In Japanese).

Sato, H., Izuho, M., & Morisaki, K. (2011a). Human cultures and environmental changes in the Pleistocene-Holocene transition in the Japanese archipelago. *Quaternary International, 237*, 93–102.

Sato, H., & Tsutsumi, T. (2007). The Japanese microblade industries: Technology, raw material procurement, and adaptations. In Y. V. Kuzmin, S. G. Keates, & C. Shen (Eds.), *Origin and spread of microblade technology in Northern Asia and North America* (pp. 53–78). Vancouver: Archaeology Press, Simon Fraser University.

Shiba, K. (2011). *Study on microblade industries in Kyushu, Japan*. Tokyo: Rokuichishobo. (In Japanese).

Shiba, K., & Obata, H. (Eds.). (2007). *Study of the Palaeolithic culture in the Aso region*. Kumamoto: Kumamoto university archaeological laboratory. (In Japanese).

Shibakawacho Board of Education. (2006). *Oshikakubo, Kubo B Site*. Shizuoka: Shibakawacho Board of Education. (In Japanese).

Shizuoka Prefecture Archaeological Center. (2000). *Shizuoka prefecture archaeological center excavation report (No.122): Ikeda B site*. Shizuoka: Shizuoka Prefecture Archaeological Center. (In Japanese).

Shizuoka Prefecture Archaeological Center. (2009). *Shizuoka prefecture archaeological center excavation report (No.210): Maruokita site*. Shizuoka: Shizuoka Prefecture Archaeological Center. (In Japanese).

Stuiver, M., Grootes, P. M., & Braziunas, T. F. (1995). GISP2 $\delta^{18}O$ climate record of the past 16,500 years and the role of the sun, ocean, and volcanoes. *Quaternary Research, 44*, 341–354.

Sugihara, S., & Ono, S. (1965). Microblade industry at Yasumiba site in Sizuoka Prefecture. *Kokogaku-shukan, 3*(2), 1–33. (In Japanese).

Suzuki, K. (2009). Hamlets and society of Jomon settlements in Tohoku region. In K. Suzuki & Y. Suzuki (Eds.), *Diachronic change of settlement and its regional variation* (pp. 51–94). Tokyo: Yuzankaku.

Takahara, H. (2011). Vegetation history after the last glacial maximum in and around the Japanese archipelago. In H. Takahara & N. Murakami (Eds.), *Kankyoshi wo toraeru giho* (pp. 15–43). Tokyo: Bun'ichi Sogo Shuppan. (In Japanese).

Takahashi, K. (2008). Transition of large mammal faunas in the sea of Japan area during the late Pleistocene. In H. Sato (Ed.), *International symposium on human ecosystem changes in the northern Circum Japan Sea area (NCJSA) in late Pleistocene* (pp. 68–79). Kyoto: Research Institute for Humanity and Nature. (In Japanese and English abstract).

Takahashi, K., & Izuho, M. (2012). Formative history of terrestrial fauna of the Japanese Islands during the Plio-Pleistocene. In A. Ono & M. Izuho (Eds.), *Environmental changes and human occupation in East Asia during OIS3 and OIS2* (pp. 73–85). Oxford: Archaeopress.

Takehiro, F. (2008). Chugoku region. In H. Sato (Ed.), *Jomon ka no kozohendo* (pp. 150–164). Tokyo: Rokuichi-shobo. (In Japanese).

Taniguchi, Y. (2004). Calibrated ^{14}C dates of the oldest pottery in Japanese archipelago and the amount of pottery. *Koukogaku Janaru, 519*, 4–10. (In Japanese).

Taniguchi, Y. (2011). *Reconstruction of the origin of Jomon culture*. Tokyo: Douseisha. (In Japanese).

Taniguchi, Y., & Kawaguchi, J. (2001). ^{14}C dates and calibrated dates of the oldest pottery during the Chnojakubo-Mikoshiba cultural stage. *Daiyonki Kenkyu, 40*(6), 485–498. (In Japanese with English abstract).

Tochigi Archaeological Research Center. (1999). *Tochigi prefecture buried property excavation report (no. 222): Tako-minamihara site*. Tochigi: Tochigi Archaeological Research Center. (In Japanese).

Tochigi Archaeological Research Center. (2003). *Tochigi prefecture buried property excavation report (no. 271): Nozawa site/ Nozawa-ishizuka site*. Tochigi: Tochigi Archaeological Research Center. (In Japanese).

Torihama Shell Midden Research Group. (1987). *Torihama Shell Midden*. Fukui: Fukui Prefecture Board of Education & Wakasa historical museum. (In Japanese).

Tsuji, S. (2004). Environmental history in the earth era. In T. Amuro (Ed.), *Kankyoshi Kenkyu no Kadai* (pp. 40–70). Tokyo: Yoshikawa Kobunkan. (In Japanese).

Yamada, S. (2006). *A study of microblade assemblages in Hokkaido, Japan*. Tokyo: Rokuichi Shobo. (In Japanese).

Yamada, S. (2008). Change and behavioral interpretation of microblade assemblages and technologies in Hokkaido (Japan). In H. Sato (Ed.), *International symposium on human ecosystem changes in the northern Circum Japan sea area (NCJSA) in late Pleistocene* (pp. 115–138). Kyoto: Research Institute for Humanity and Nature. (In Japanese and English).

Yamahara, T. (2008). Terminal Pleistocene and early Holocene in Hokkaido. In H. Sato (Ed.), *The process of Jomonization* (pp. 35–52). Tokyo: Rokuichi-shobo. (In Japanese).

Yamato City Board of Education. (1986). *Yamato city buried cultural property excavation report (no. 21): Tsukimino-kamino sites: Kamino site Loc.1*. Kanagawa: Yamato City Board of Education. (In Japanese).

Yamato Town Board of Education. (2007). *Yamato town board of education excavation report No.1: Takahata-otonohara site, Takahata-maeduru site, Takahata-miyanoshita site*. Kumamoto: Yamato Town Board of Education. (In Japanese).

Yoshida, K., Hara, T., Miyazaki, Y., & Kunikita, D. (2008). The dating of the initial Jomon period: Radiocarbon dating from pottery in the Echigo region. In *The birth of Jomon culture* (pp. 57–63). Tsunan: Tsunan town Board of Education. (In Japanese).

Yuan, D., Cheng, H., Lawrence Edwards, R., Dykoski, C. A., Kelly, M. J., Zhang, M., Qing, J., Lin, Y., Wang, Y., Wu, J., Dorale, J. A., An, Z., & Cai, Y. (2004). Timing, duration, and transitions of the last interglacial Asian monsoon. *Science, 304*, 575–578.

Wang, Y. J., Cheng, H., Edwards, R. L., An, Z. S., Wu, J. Y., Shen, C., & Dorale, J. A. (2001). A high-resolution absolute-dated late Pleistocene monsoon record from Hulu cave, China. *Science, 294*, 2345–2348.

Watanabe, N. (1966). Radiocarbon dates of the Jomon and Yayoi periods. *Daiyonki Kenkyu, 5*(3–4), 157–168. (In Japanese with English abstract).

Setoguchi, N. (1981). Excavation report of Higashi-kurotsuchida site. *Kagoshima-koko, 15*, 1–55. (In Japanese).

Chapter 7
Isolation, Exploration or Seasonal Migration? Investigating Technological Organization in the Late Middle Palaeolithic of Britain During Marine Isotope Stage 3

Rebecca M. Wragg Sykes

Introduction

The Late Middle Palaeolithic archaeological record of Britain has historically been regarded as relatively marginal in discussion on Neanderthals, both literally in terms of its geographic location and also in its capacity to contribute much to our understanding of the lifeways of this species. While analysis and comprehension of the Continental record have become increasingly nuanced over the past two decades, with better recognition of technological diversity, lithic resource management and patterns in mobility (Geneste 1988, 1989, 1991; Féblot-Augustins 1993, 1999, 2009; Dibble 1995; Kuhn 1994, 1995, 2011; Holdaway et al. 1996; Conard and Adler 1997; Roth and Dibble 1998; Gamble 1999; Roebroeks and Gamble 1999; Turq 2001; Delagnes and Meignen 2006; Meignen et al. 2009; Conard and Delagnes 2010; Browne and Wilson 2011, 2013; Conard and Richter 2011; Turq et al. 2013), a similar treatment of the British record has until recently been missing. Despite being numerically very small in comparison to the contemporary Continental archaeological record of, for example, France, the British Late Middle Palaeolithic (LMP) offers a unique situation for learning about this species for three reasons:

- Geographical situation—documents Neanderthal occupation in the very north-western extremes of their range.
- Climatic context—occurs during Marine Isotope Stage (MIS) 3, recognized as a 'failed' interglacial, with highly fluctuating climatic and therefore unpredictable environmental conditions.

R.M. Wragg Sykes (✉)
PACEA UMR-5199, Université de Bordeaux, Bordeaux, France
e-mail: rwraggsykes@gmail.com

E. Robinson, F. Sellet (eds.), *Lithic Technological Organization and Paleoenvironmental Change*, Studies in Human Ecology and Adaptation 9,
DOI 10.1007/978-3-319-64407-3_7

- Cultural adaptation—the British LMP represents a reoccupation following a hiatus in settlement of over 100,000 years and is an unusual archaeological example of colonization known for this species.

Analysis of the entire British LMP archaeological record has now been undertaken for the first time in 20 years (since Coulson 1990), using up-to-date methodologies focusing on the techno-economic characterization of the lithic assemblages (detailed discussion of data in Wragg Sykes 2009, 2010, 2017, in press). A consistent signature of technological organization is apparent, sharing some characteristics with the contemporary northwestern European Continental record but distinctive in others. Attempting to disentangle underlying adaptations that this techno-economic system represents, taking into account the three unique factors described above together with taphonomic complexities, is the concern of this chapter.

The Nature of the British Late Middle Palaeolithic Record

Sites and Chronology

Unlike most Continental contexts, the British LMP is clearly chronologically separated from the Early Middle Palaeolithic by a substantial gap in the archaeological record of over 100 ka, from the end of MIS-7 until MIS-3 (Currant and Jacobi 1997, 2001; Ashton 2002; Ashton and Lewis 2002; Scott 2006, 2010; White et al. 2006; Gilmour et al. 2007; Lewis et al. 2010). Harsh conditions in MIS-6 caused initial depopulation, which was followed by isolation due to raised sea-level fluctuations in MIS-5 interstadials, compounded by an enormous Channel River system and the intense cold of MIS-4 (Gibbard 1988; Antoine et al. 2003; Gibbard and Lautridou 2003). No clear reoccupation in Britain until late MIS-4/early MIS-3 has been identified, despite intensive searching (Lewis et al. 2010)—just two flakes from a late MIS-5 context are known (Wenban-Smith et al. 2010). This stands in contrast to the rich record of Neanderthal occupation across the Channel Plain in northern France and Belgium during MIS-5 and even MIS-4 (Antoine et al. 2006, 2014; Goval 2008; Locht et al. 2010, 2016). Currant and Jacobi (2011) proposed that very rapid sea-level rise in substage 5a prevented the recolonization not only of Neanderthals but also large fauna such as horse, woolly rhinoceros and mammoth, which are similarly missing in late MIS-5 until the MIS-4/3 boundary.

As well as an occupational hiatus, there is also a clear disconnect between the techno-economic character of the British EMP and LMP: the former is heavily focused around varied and flexible Levallois strategies, with virtually no bifaces (Scott 2006, 2010; White et al. 2006). In contrast, the British LMP is based on very different principles of core reduction, tool forms and even techno-economic organization. The archaeological assemblages of interest in this chapter are from 23 assemblages or find spots, selected based on confidence in dating to this period (Table 7.1). Primary lithic analysis and further detail on geological setting can be

Table 7.1 Sites analysed: the number of lithics sampled from extant collections is shown

Site, assemblage size, other archaeological information	Calibrated 14C date range	Dating and methods [*UF* Ultra-filtrated radiocarbon measurement, *AMS* accelerator mass-spectrometry radiocarbon measurement, *OSL* optically stimulated luminescence, *US* Uranium series, *ESR* Electron spin resonance]	Raw materials, artefact types and technological stages of production- see Table caption; bold = non-local		References outside Wragg Sykes (2009, 2010, 2017)
Lynford Quarry, Norfolk Excavated 2002–2003 **N= 310 of c. 2790 total; 738 cores, tools and debitage >20mm** [not all from main organic context]; total including un-sieved microdebitage estimated c. 15,000.	59–72 ka BP [OSL range for channel fill c. 69–51 ka]	OSL on channel cutting organic palaeochannel deposits (containing main archaeological assemblage) 55,000 ± 4000 ka *terminus ante quem.* UF C14 on mammoth within organic palaeochannel >49,700 ka (OxA 11572), 53,700 ± 3100 (OxA 11571). OSL main palaeochannel deposits 64,000 ± 5000 ka; 67,000 ± 5000 ka. OSL near base of deposits 83,000 ± 8000 ka (OxL-1337) *terminus post quem.*	Southern flint	Biface—P, I, M, ?R, D	Boismier et al. (2012)
				Core—Informal W, D	
				Tools—?I, R	
			Northern flint	Biface—D (probably from local glacial source)	
			Quartzite	Hammerstone, probably from local glacial source	

(continued)

Table 7.1 (continued)

<table>
<tr><th>Site, assemblage size, other archaeological information</th><th>Calibrated 14C date range</th><th>Dating and methods [UF Ultra-filtrated radiocarbon measurement, AMS accelerator mass-spectrometry radiocarbon measurement, OSL optically stimulated luminescence, US Uranium series, ESR Electron spin resonance]</th><th colspan="2">Raw materials, artefact types and technological stages of production- see Table caption; bold = non-local</th><th>References outside Wragg Sykes (2009, 2010, 2017)</th></tr>
<tr><td rowspan="7">Pin Hole: Creswell Crags, Derbyshire/Nottinghamshire
Excavated 1870s & 1920s–1930s
N= 50 of 62 (Jacobi 2004) or 118 (White and Pettitt 2011) extant</td><td rowspan="7">39.9–46.4 ka BP
(59.9 kyr max 14C date out of Cal range)</td><td rowspan="7">UF C14 from above level of Upper Palaeolithic artefacts 38,000 ± 2000 (OxA-1470) terminus ante quem.
UF 14C from transition to Middle Palaeolithic artefacts 37,760 ± 340 (OxA-11980).
UF C14 11 measurements from within Middle Palaeolithic, ranging from 43,350 ± 650 (OxA-13592) to 55,900 ± 4000 (OxA-14197; Jacobi et al. 2006 and Higham et al. 2006).
UF C14 from below level of Middle Palaeolithic terminus post quem [sampled twice] 53,400 ± 1700 (OxA-14211) & 50,200 ±1400 (OxA-14212).
TIMS U-series on derived speleothem 63,700 ± 400 terminus post quem.</td><td rowspan="3">Southern flint</td><td>Biface—I, M, R;</td><td rowspan="7">Mello (1875, 1876), Armstrong (1925, 1926, 1928, 1929, 1932, 1935, 1937, 1939, 1956),
Jenkinson (1984)
Jacobi et al. (1998, 2006), Jacobi (2004)
Higham et al. (2006)</td></tr>
<tr><td>Flake—I</td></tr>
<tr><td>Tools—I, W, M</td></tr>
<tr><td rowspan="2">Northern flint</td><td>Flake—I</td></tr>
<tr><td>Tools—?I, M (possibly from regional rather than local glacial), D</td></tr>
<tr><td>Quartzite</td><td>Core—Discoidal, alternate, informal W, M, ?R, D</td></tr>
<tr><td colspan="2">Clay-ironstone: M, R.</td></tr>
<tr><td rowspan="5">Robin Hood’s Cave: Creswell Crags, Derbyshire/ Nottinghamshire
Excavated 1870s, 1880s, 1980s
N= 97 of c. 109 extant; +479 originally excavated</td><td rowspan="5">48.7–53.3 ka BP</td><td rowspan="5">C14 29,300 ± 480 (OxA-3455) Breccia, terminus ante quem.
UF 14C 47,300 ± 1200 (OxA-12772) Sample from spit containing quartzite debitage.
Sample from below lithics and hyaena action in small remnant spit > 52,800 (OxA-12736) ?terminus post quem.
ESR 55,000 ± 4000 Woolly rhinoceros tooth, Red Sand terminus post quem.</td><td colspan="2">Southern flint: I, M, R, D</td><td rowspan="5">Mello (1876, 1877), Dawkins (1877)
Heath (1879)
Laing (1890)
Jenkinson (1984)
Hedges et al. (1994, 1996)
Jacobi and Grun (2003)
Jacobi et al. (1998, 2006)
Jacobi (2004)
Higham et al. (2006)</td></tr>
<tr><td colspan="2">Quartzite: ?P, ?I, M, D</td></tr>
<tr><td colspan="2">Clay-ironstone: I, D</td></tr>
<tr><td colspan="2">Carboniferous chert: ?I</td></tr>
<tr><td colspan="2">Igneous: ?I</td></tr>
</table>

Church Hole, Creswell Crags Derbyshire/Nottinghamshire Excavated 1870s, 1880s, 2000s **N= 15 of c. 47 extant**; +200 originally excavated Cutmarked reindeer tibia from lowest layer (R. Jacobi pers. comm. 2008), previously provenanced only to Creswell Crags.	–	One AMS 14C on bone from lower levels 37,200 ± 1300 (OxA-3417). UF 14C on unstratified hyaena tooth > 40 ka (OxA-14926). Pin Hole MAZ from same deposit as lithics.	Quartzite: W, D	Dawkins (1876) Mello (1877) Heath (1879, 1882) Hedges et al. (1994) Jacobi (2004, 2007) Higham et al. (2006)
Mother Grundy's Parlour, Creswell Crags Derbyshire/ Nottinghamshire Excavated 1880s, 1920s, 1960s **N= 14 of ?14 extant**	–	Pin Hole MAZ from same deposit as lithics.	Clay-ironstone: Biface—I, D Quartzite: W, D	Dawkins and Mello (1879), Armstrong (1925), Campbell (1977)
Ash Tree Cave: Burnhill Grips, Derbyshire Excavated 1950s **N= 28 of c. 30 extant**	Dates out of Cal range; likely >55 ka	UF C14 on fauna from basal clay, below archaeology *terminus post quem* 52,800 ± 3100 (OxA-13802) >54 ka (OxA-13800) >56.5 ka (OxA-13801 bison) >57.7 ka (OxA-15003 bison).	**Southern flint**: I, M, R Northern flint: I, M, R Quartzite: I, M, R Clay-ironstone: I, M, R	Armstrong (1950, 1951, 1952, 1953, 1954, 1955,1956, 1957) Jacobi et al. (2006) Hedges et al. (1994, 1996)
Hyaena Den: Mendips, Somerset Excavated 1860s, 1970s, 1990s **N= 41 of c. 100** including small debitage; original site probably much richer Cutmark on red deer tooth dated to 40,400 ± 1600. Recent excavations found charcoal, large amounts of carbonised bone; ashes noted by early excavators.	Human presence (AMS date, likely too young) 42.9–45.5 ka Cave earth (UF dates) 46.6–54.9 ka	AMS C14 **Cut-marked** red deer tooth 40,400 ± 1600 (OxA-4782) from 1990s excavations. Several UF C14 dates on fauna within cave earth 45,100 ± 1,600 (OxA-13915; red deer) 47,000 ± 1700 BP (OxA-13916; bone) 48,600 ± 1,000 (OxA-13917; hyaena). From silt below cave earth in 1990s excavations 52,700 ± 2000 (OxA-13914; bone) ?*terminus post quem.* OSL on sands below cave earth at cave entrance c. 70 ka. *Terminus post quem.*	**Southern flint**: I, M, R, D Carboniferous chert: I, M, R, D	Dawkins (1862, 1863, 1874) Balch (1914) Tratman et al. (1971), Jacobi and Hawkes (1993) Hedges et al. (1996), Current and Jacobi (2004) Jacobi et al. (2006)

(continued)

Table 7.1 (continued)

Site, assemblage size, other archaeological information	Calibrated 14C date range	Dating and methods [*UF* Ultra-filtrated radiocarbon measurement, *AMS* accelerator mass-spectrometry radiocarbon measurement, *OSL* optically stimulated luminescence, *US* Uranium series, *ESR* Electron spin resonance]	Raw materials, artefact types and technological stages of production- see Table caption; bold = non-local	References outside Wragg Sykes (2009, 2010, 2017)
Rhinoceros Hole: Mendips, Somerset Excavated 1970s, 1990s **N= 4 of 6** [2 Upper Palaeolithic]	No 14C dates Minimum 41 ka BP from U-series	U-series: derived speleothem within layer containing biface and biface debitage c. 51 ka (range 32–71ka, but low thorium). Date for overlying layer gives minimum age of 41 ka.	**Southern flint**: I, M, ?R, D Carboniferous chert: I, M, R	Proctor et al. (1996)
Picken's Hole, Mendips, Somerset Excavated 1960–1970s **N= 37 of 48** [11 natural]	43354–44906 ka	Early conventional 14C measurements on fauna from layer containing artefacts and below, but likely unreliable. UF C14 on woolly rhinoceros 40,550 ± 500 (OxA-10804), >44 ka (OxA-10805) from unit containing Middle Palaeolithic artefacts	**Southern flint**: I, M, R, D Carboniferous chert: I, M, R	Tratman (1964) ApSimon (1986) Jacobi (2007) Jacobi et al. (2006) Wragg Sykes (2016a, b)
Kent's Cavern, Devon Excavated 1890s, 1930s **N= 31 of ?** originally excavated. Few other lithics identified were unavailable, highly likely much debitage especially small not retained during early collections.	43–44.5 ka	Several UF C14 measurements from within cave earth containing Middle Palaeolithic artefacts, the lowest stratigraphically 40,000 ± 700 (OxA-13888) but possibly debris-flow taphonomy. UF 14C measurement on fauna below level of artefacts in one part of cave 49,600 ± 2200 (OxA-14714) ?*terminus post quem.* U-series on small layer of Granular Stalagmite overlying Cave Earth deposit c. >55 ka. ?*terminus ante quem.*	**Southern flint**: I, D Greensand chert: I, D	Vivian (1859), Pengelly (1868, 1869a, b, 1870, 1871, 1872, 1873, 1874, 1875, 1876, 1877, 1878, 1879, 1880) Beynon and Ogilvie (1931, 1932, 1933, 1934, 1935, 1936, 1937, 1938), Beynon et al. (1929), Keith (1926, 1927a, b, 1928, 1929, 1930), Ogilvie and Tebbs (1938), Rogers (1956) Campbell and Sampson (1971) Proctor (1994, 1996), Proctor and Smart (1989), Straw (1996), Higham et al. (2006)

Coygan Cave, Carmarthenshire Excavated 1860s, 1880s, 1910s, 1930s, 1970s **N= 5 of 5** listed in recent publication; possibly more from nineteenth/early twentieth-century excavations Piece of carbonised large mammal bone; excavation noted a 'hearth' at bottom of sequence. *Note: Coygan was the original type site for the MIS-3 MAZ	40.1–45 ka but likely too young (conventional C14)	Conventional C14 measurement from reindeer antler at same level as bifaces 38, 684 + 2713/-2024 (BM-499), almost certainly too young. UF 14C on hyaenas from same general layer ranging from 32,140 ± 250 (OxA-14400) to 43,000 ± 2100 (OxA-14401). U-series: derived speleothem from within same deposit as bifaces 62 ± 2 ka *terminus post quem.*	Re-crystallised rhyolite: Biface—I, M, R, D Rhyolite: Biface—I, D; Flake—D Diorite: Biface—I, D.	Hicks (1867, 1884) Laws (1888) Grant-Dalton (1917) Wardle (1919) Grimes and Cowley (1935), Clegg (1970) Burleigh et al. (1976), Aldhouse Green et al. (1995) Higham et al. (2006)
Little Paxton, Cambridgeshire Collected 1920s, 1940s **N= 236 of 236** extant	–	Pin Hole MAZ from same general gravel layer, plus terrace height.	Southern flint: ?P, I, M, ?R, D Northern flint: Biface—M, D	Paterson and Tebbut (1947) Wymer (1999)
Uphill Quarry Cave 8, Somerset Excavated 1900s **N= 20 of extant.** Some pieces identified as likely Upper Palaeolithic based on techno-typology not included.	–	Pin Hole MAZ, from same deposit as lithics.	**Southern flint**; Biface—I, D Carboniferous chert: Biface—I, D	Wilson and Reynolds (1901) Harrison (1977) Jacobi et al. (2006)
Oldbury, Kent Collected 1880s; excavated 1960s **N= 88 of 464**. A large component of discoidal debitage from the 1960s excavation was unavailable at the time of study.	–	Techno-typology and landscape setting.	Southern flint (poor quality and thermally damaged): Biface—I, M, ?R, D; Core—W, D; Retouched piece—?I, M Eocene flint: Biface—I, D	Collins and Collins (1970) Cook and Jacobi (1998)

(continued)

Table 7.1 (continued)

Site, assemblage size, other archaeological information	Calibrated 14C date range	Dating and methods [*UF* Ultra-filtrated radiocarbon measurement, *AMS* accelerator mass-spectrometry radiocarbon measurement, *OSL* optically stimulated luminescence, *US* Uranium series, *ESR* Electron spin resonance]	Raw materials, artefact types and technological stages of production- see Table caption; bold = non-local	References outside Wragg Sykes (2009, 2010, 2017)
Goat Hole, Paviland, Gower Excavated 1820s, 1910s **N = 10 of ?** originally excavated. Pieces were selected based on previous identification as discoidal cores with some accompanying flakes. Highly likely more material not retained during early twentieth-century excavations.	–	AMS 14C Mid-MIS 3 dates for fauna. Pin Hole MAZ.	**Southern flint:** Core—I, W, D Carboniferous chert: Core—W, D	Buckland (1823) Sollas (1913), Swainston (1999, 2000) Aldhouse-Green (2000), Jacobi and Higham (2008)
Ravencliffe Cave, Derbyshire Excavated 1910s, 1920s **N= 1 of 1** extant; probably some debitage originally excavated judging by publication photo.	–	Pin Hole MAZ	**Southern flint**: Retouched piece—D	Storrs Fox (1908, 1910, 1929), Read 1910)
Fisherton, Wiltshire Collected 1870s. **N= 1 of 1** extant	–	Pin Hole MAZ	Southern flint: Biface—I, D	Stevens, (1870) Delair and Shackley (1978) Green et al. (1983)
Snodland, Kent Collected 1969. **N = 3 of 3**	–	Lithostratigraphy	Southern flint: Biface—I, D	Roe (1981: 261)
Berrymead Priory, Acton Collected 1880s. **N= 1 of 1**	–	Lithostratigraphy	Southern flint: Biface—I, D	Brown (1889) Wymer (1968, 1988, 1999)

Southbourne, Bournemouth Collected 1940s. **N= 1 of 1**	-	Lithostratigraphy	Southern flint: Biface—I, D	Calkin and Green 1949)
Castle Lane, Bournemouth Collected 1940s. **N= 1 of 1**	–	Lithostratigraphy	Southern flint: Biface—I, D	Calkin and Green (1949)
Saham Toney, Norfolk Collected 1970s. **N= 1 of 1**	–	Lithostratigraphy	Southern flint: Biface—I, D	Lawson (1978) Tyldesley (1987)
Sipson, London Excavated 1980s. **N= 1 of 1**	–	Lithostratigraphy	Southern flint: Biface—I, D	Cotton (1984)
Marlow, Buckinghamshire Collected 1930s. **N= 2 of 2**	–	Lithostratigraphy	Southern flint: Biface—I, D	Treacher (1934)

Raw materials present and lithic technological categories present: unless biface or retouched indicated this refers to core and flake production. *P* biface mainly produced in situ, *W* cores worked in situ, *I* artefact imported, *M* artefact maintained in situ, *R* recycling, *D* discard. Bold raw materials indicate non-local stone. See Wragg Sykes (2009, 2017) for full lithic and palaeoenvironmental information on each site. Major chronometric data including calibrated range for archaeological layers shown; ultrafiltrated radiocarbon dates are prioritized over conventional measurements; ESR dates from Pin Hole not included here due to space constraints, see Jacobi et al. (2008). Calibration using OxCal 4.2, IntCal13: Bronk Ramsey (2009)

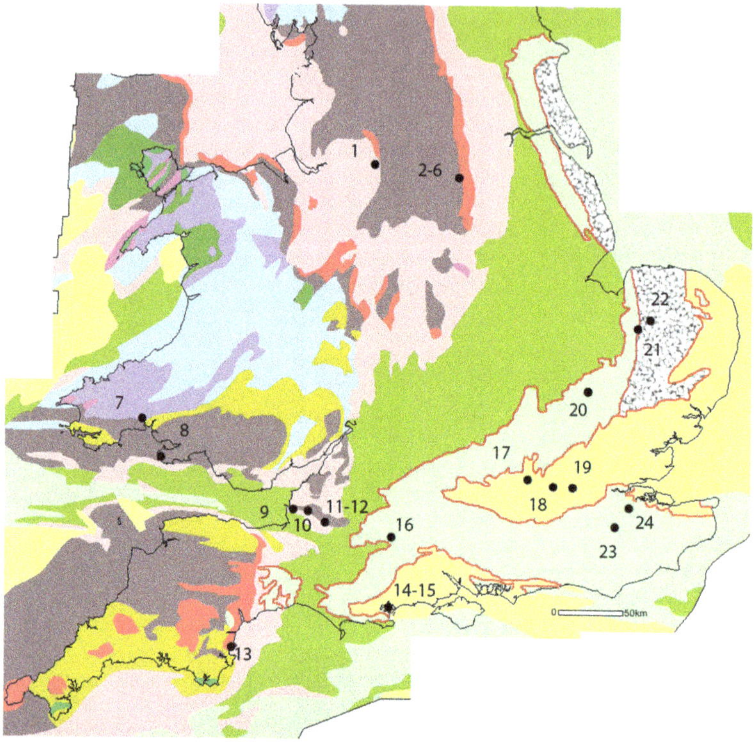

Fig. 7.1 Location of British Late Middle Palaeolithic sites, in relation to geology, including available offshore data (created using data © BGS 2005 NERC). *Pale yellow*, Neogene/Palaeogene; *red outlined pale green*, Cretaceous at surface, including chalk containing southern and northern flint deposits and Greensand containing chert; *red outlined stippled*, Cretaceous currently covered by glacial till; *olive*, Jurassic; *pale pink*, Triassic including quartzite Kidderminster conglomerate; *dark pink*, Permian; *purple*, Carboniferous including cherts; *yellow*, Devonian; *pale blue*, Silurian; *lilac*, Ordovician; *dark green*, Cambrian. *1* Ravencliffe Cave, *2–6* Ash Tree Cave and Creswell Crags caves (Robin Hood Cave, Pin Hole, Church Hole, Mother Grundy's Parlour), *7* Coygan Cave, *8* Paviland (Goat Hole), *9* Uphill Quarry, Cave 8, *10* Picken's Hole, *11–12* Hyaena Den and Rhinoceros Hole, *13* Kent's Cavern, *14–15* Castle Lane and Southbourne, *16* Fisherton, *17* Marlow, *18* Sipson, *19* Berrymead, *20* Little Paxton, *21* Lynford Quarry, *22* Saham Toney, *23* Oldbury, *24* Snodland

found in Wragg Sykes (2009, 2010, 2017, in press). A synthesis of lithic, dating and paleoenvironmental data for each site is presented here in Table 7.1, while Fig. 7.1 shows geographical location. The sites are a mix of cave and open-air (riverine and surface context) locales and were excavated at different times over the past century and a half. Due to the early research history of most sites, they are effectively time-averaged palimpsests within which varying biases operate on artefact collection (larger and more aesthetically distinctive pieces were often prioritized), and therefore caution is applied in the inferences that are drawn. Lynford Quarry is an obvious exception, having been recently excavated to modern standards, but at some other sites, small debitage is also present, for example, where sieves were used (e.g. Ash Tree Cave), indicating less comprehensively biased assemblages. While taking a conservative approach, it is still possible to consider intra- and inter-site patterning in terms of broad features of techno-economics such as raw material preferences, presence/absence of core reduction and tool types and lithic transport.

There is now a much improved chronological understanding of the LMP assemblages, thanks to targeted re-dating, primarily using ultrafiltrated (UF) radiocarbon (Jacobi et al. 1998, 2006; Bronk Ramsey et al. 2004; Higham et al. 2006). Available recent UF radiocarbon measurements and selected other dating information are in Table 7.1, while Fig. 7.2 shows the ultrafiltrated, calibrated radiocarbon measurements from the six best-dated sites. It is clear that the overall chronological range of occupation falls between c. 58–45 ka. At present Lynford Quarry appears to be the

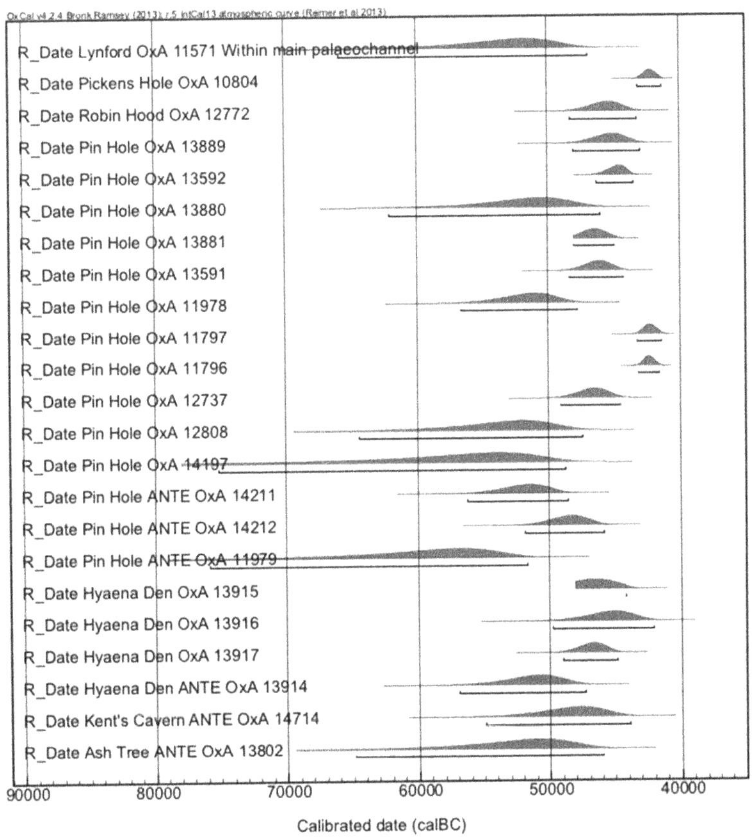

Fig. 7.2 Calibrated ultrafiltrated radiocarbon measurements from British Late Middle Palaeolithic sites, from deposits containing archaeological material. Dates OxA 14197, 12808 and 12737 are from low in the sequence at Pin Hole (based on 2D plots of lithics, Fig. 6, Jacobi et al. (2006)). Dates marked 'ANTE' are stratigraphically below Middle Palaeolithic archaeology at individual sites. See Table 7.1 for further information on uncalibrated measurements, non-radiocarbon dates, materials and references. Calibration run using OxCal, version 4.2 (IntCal13): Bronk Ramsey (2009)

earliest site, with UF radiocarbon results >50 ka (Boismier et al. 2012: 70). OSL dates from the same context vary but suggest an age of c. 59 ka, suggesting that Lynford probably falls within the first interstadial of MIS-3 (taking into account palaeoenvironmental evidence, e.g. Coleoptera; Boismier et al. 2012: 75–94). Other locales have less agreement between dating methods, for example, ESR at Pin Hole suggests c. 67–51 ka (Jacobi et al. 1998), but UF dates sampled from within the vertical range of LMP lithics indicate a younger age range, c. 51–45 ka. Most British LMP sites have less extensively dated assemblages, but their ages still suggest that Neanderthals were present across a relatively wide area of Britain from at least 50–40 ka, including consolidating their presence in the North Midlands (Creswell Crags and Ash Tree Cave), the southwest (Picken's Hole, Rhinoceros Hole and Kent's Cavern) and as far west as Wales (Coygan). Ravencliffe Cave is the most northwestern site, and although consisting of only a single artefact (a typologically Mousterian double convergent scraper), associated biostratigraphic evidence indicates it is certainly MIS-3; therefore the LMP, at least ephemerally, extended into an area of uplands averaging 300 m above sea level (limestone gorges of the Peak District).

Technological Characterization

Turning to the lithic record, in terms of techno-economic character, the British LMP includes a relatively wide range of raw materials (Tables 7.1 and 7.2): ranging from siliceous stones such as flint and various cherts (southern province Cretaceous flint

Table 7.2 Raw materials, split by availability of local vs non-local southern geological province-type flint, the highest quality stone and only material transported outside its geological region

	Local		Non-local	
	Count	Column N %	Count	Column N %
Flint	645	98.5%	61	17.8%
Northern flint	6	.9%	9	2.6%
Green flint	1	.2%		
Carboniferous chert	1	.2%	75	21.9%
Greensand chert	1	.2%	19	5.5%
Quartzite	1	.2%	148	43.1%
Rhyolite			1	.3%
Recrystallized rhyolite			3	.9%
Diorite			1	.3%
Clay-ironstone			22	6.4%
Igneous			1	.3%
Cast			3	.9%
Total	655	100.0%	343	100.0%

Flint identifications between southern and northern types are macroscopic only. Casts were made from originals damaged during the Second World War; original stone not known.

being the highest quality: Mortimore et al. 2001, although it must be emphasized that precise determinations for flint types remain a major under-researched area) to coarser/less homogeneous rocks (quartzite, igneous stones and clay ironstone). Across the different assemblages, there is a striking shared techno-economic signature which is not explained by taphonomic bias. At a basic level, consistent raw material exploitation preferences exist: locally available stone—of whatever quality—tends to be used for most tasks, but only flint appears to have been transported significant distances (>20 km), and this material accounts for 17% of all artefacts from assemblages outside the area where it is geologically available (Table 7.2), although this varies by site. In the larger, more representative assemblages, it ranges from c. 10% at Robin Hood Cave, almost 19% at Picken's Hole and over 30% at Hyaena Den. The maximum transport distance is found at Ravencliffe Cave, although the origin of this flint artefact is not yet identified, and there are potential secondary sources within around 50 km.

Where it has been imported, flint is most frequently found as bifaces (handaxes; Fig. 7.3) and tools (scrapers; Fig. 7.3), and various technological and metric factors indicate they were highly curated and resharpened (Wragg Sykes 2009, 2010, 2017, in press). In several cases, the bifaces were imported and then removed, leaving behind only waste from their maintenance to testify to their presence. Local stone, whether poor or high quality, is used in a more diverse manner, including in situ flake production, as well as for bifaces, scrapers, denticulates and notches. This tool type diversity is not however found in exotic flint: >90% imported retouched tools are scrapers, compared to just over 50% for tools made on locally available stone (including flint in its source region).

The highly selective lithic raw material patterning and unidirectional transport away from the southeast indicate that Neanderthals were considering the different properties of various stones available to them and exploiting this in a complementary system; flint was clearly preferred for particular contexts involving curation and transport. While its northwestward movement could be explained as simply reflecting groups entering Britain from this direction, the fact that no other raw materials are moved away from their own regions indicates that Neanderthals were also applying knowledge about geological resources across the entire landscapes. Alongside raw material selectivity, there is clear evidence that the largest blanks were consistently chosen for retouching; even more than this, scrapers were almost always larger than denticulates or notches both overall and within individual site/raw material contexts (Wragg Sykes 2009, 20117). As scraper retouch offers a greater potential use-life (because, assuming blank form remains equal, less mass is removed during each resharpening using scraper retouch), this points to the selection of large blanks and edge modification specifically with future needs in mind.

Technologically, opportunistic exploitation of cores in an informal manner (following 'migrating' striking platforms: Ashton 1998; White and Ashton 2003) is most common overall. Formal approaches to flake production are used but are limited to non-Levallois, centripetally based methods, discoidtechnology (as defined by Boëda 1993, 1994, 1995; Mourre 2003; Terradas 2003) and alternate reduction, where the previous flake scar is utilized as a striking platform and reduction proceeds

Fig. 7.3 Selection of lithics from the British Late Middle Palaeolithic, demonstrating variation in raw materials and technology. See Wragg Sykes (2009, 2017) for further details. (**a**) flint biface from Lynford Quarry; (**b**) rhyolite biface from Coygan Cave (one of the 'classic' *bout coupé* examples); (**c**) flint biface from Hyaena Den, showing later use as flake source; (**d**) flint biface maintenance flake, with edge retouching, from Pin Hole Cave, Creswell Crags; (**e**) quartzite alternately worked core from Pin Hole, Creswell Crags; (**f**) quartzite discoid core from Robin Hood Cave, Creswell Crags; (**g**) flint convergent scraper from Ravencliffe Cave; (**h**) flint convergent scraper from Pin Hole Cave; (**i**) Carboniferous chert radially worked core from Picken's Hole; (**j**) flint informally worked core from Lynford Quarry

along an edge or perimeter as the core is flipped over (Ashton 1992; Ashton and McNabb 1996; McNabb 2007). A handful of pieces have previously been described as Levallois, but there is no clear, coherent signature of a Levallois technological stratcgy (of any type) within any one assemblage. Jacobi (2004) and White and Pettitt (2011) suggest there is a Levallois blade from Robin Hood Cave; however, it is not regarded here as especially convincing: its longitudinal dorsal scar pattern is unusual in comparison to the rest of the dataset, but this is not conclusive. At Lynford, one core has been described as Levallois (Boismier et al. 2012: 220), but it lacks any platform preparation and the angle between removal surface and platform is not 90°. Furthermore, it is in a much more abraded condition than the vast majority of material from the main paleochannel assemblage and is therefore probably derived. Overall, the wider picture is instead that where formal flake production occurred, it was primarily centripetally organized (Fig. 7.3), either following the discoidal method or alternate reduction. Discoidal technology is found across almost all raw materials/contexts, while alternate reduction is especially common (alongside discoidal technology) on quartzite cobble blanks at the Creswell Crags sites; here it may actually represent early stages of discoidal strategies.

Within the British LMP, much of the flake production appears to have taken place close to where raw materials were available, if not immediately at the source; in common with many Middle Palaeolithic industries, cores do not seem to have been transported long distances. There is however evidence of limited movement at local scales: most sites lack high numbers of decortication flakes, and mismatches between the colours of quartzite cores and flakes at the Creswell Crags caves (e.g. a red core with no red flakes or white flakes with no white cores; Wragg Sykes 2009: 238–9) are not easily explained by any natural process or collector bias. In general, southern flint blanks produced using discoidal reduction were more often moved away from sites than other flakes, but the most mobile objects of all were bifaces and scrapers. As noted above, these are frequently found in contexts indicating long-distance curation; even within local contexts where high-quality stone was easily available, bifaces still appear to have had extended and fragmented trajectories of use, maintenance, repair and recycling, most obviously seen at Lynford (Boismier et al. 2012). Some were used as supports for secondary modifications, such as sections of scraper retouch or notches applied to the margins, and positive correlation between biface size and presence (and amount) of secondary modification suggests selection again related to use-life.

Climatic and Environmental Contexts

A genuine reoccupation of the British landmass seems to begin relatively early in MIS-3. This climatic phase has garnered increasing attention in recent research due to recognition of its unusually unstable nature; while technically an interglacial following MIS-4, various climatic and environmental proxy records consistently show that it never achieved truly temperate conditions found during earlier interstadials such as MIS-5, MIS-7 or MIS-9 (Bond et al. 1993; Dansgaard et al. 1993; Grootes et al. 1993; Shackleton et al. 2000; Svensson et al. 2006; Andersen et al. 2006). The other factor claimed to characterize MIS-3 is a high frequency of rapid and intense climatic oscillations (Dansgaard-Oeschger, D-O events) lasting up to 2000 years.

To some extent, it is possible to track the distinctive climate of MIS-3 in terrestrial palaeoenvironmental records. The most striking feature is the general picture of very little arboreal recovery, likely reflecting cool temperatures (see Fletcher et al. 2010: 2844 for references from northwestern Europe). However, while many pollen sequences from this region point to largely treeless environments dominated by grasses, sedges and diverse herbs, this may be skewed by sample contexts (Willis et al. 2000, Willis and Van Andel 2004; Bos et al. 2009), and there is frequently a low-level of tree pollen present at most British sites—albeit usually interpreted as dwarf varieties (e.g. *Betula nana*) or far-travelled. Very local arboreal refugia may have been present across Europe, with a patchy, shrubby population of at least birch and pine; the typically mosaic nature of the mammoth steppe environments (Guthrie 1982, 1990) of MIS-3 could in themselves have promoted rapid recovery after cold phases or glacials (Stewart and Lister 2001; Willis and Van Andel 2004). Recent results from speleothem samples suggest that some thermophilous taxa were able to survive in MIS-3 Britain within micro-refugia provided by topographically sheltered locations. At Lancaster Hole c. 58–51 ka, while most pollen was pine, there were also large numbers of oak, alder, hazel and willow (Caseldine et al. 2008). Interestingly, analysis of a hyena coprolite from Pin Hole, one of the sites at Creswell Crags, shows very high grassland-related species, with only 1% arboreal pollen, but again this includes oak and hazel—although single grains (Lewis 2010).

In contrast to pollen, Coleopteran evidence overwhelmingly suggests treeless environments, primarily grasslands with sedges, some bogs, heaths and bare ground, frequented by large mammals leaving carrion and dung (Coope 2002). Coope et al. (1961) note the apparent discrepancy between the sometimes mild temperatures and absence of woodland insect faunas; although beetles are sensitive climate indicators, they generally provide very localized pictures of vegetation, presenting the same issue of sample bias as pollen. Some molluscan assemblages seem to point to the existence of more genial microenvironments, for example, stratified material from within the LMP layers at Pin Hole included thermophilous forms (Hunt 1989), again indicating that microenvironments near caves offered much less harsh environments than suggested by wetland/open landscape samples.

There is extensive well-dated evidence for Late Pleistocene mammalian faunal communities in Britain, with specific biostratigraphic 'mammal assemblage zones' (MAZ; Currant and Jacobi 1997, 2001). The most recent reassessment suggests that by late MIS-4 (Brean Down MAZ), an otherwise sparse, cold-adapted fauna representing both remnant species cut-off by high sea levels in MIS-5a and those that could cope with full-glacial conditions was joined by horses, indicating some environmental change and reconnection with the Continent (Currant and Jacobi 2011). Following this, the next MAZ in MIS-3 is dramatically richer in species, and although Pin Hole is its type site, it is also seen at Lynford in probably slightly earlier deposits. The Pin Hole MAZ species appear broadly similar to the rich diversity found across the Eurasian mammoth steppe during this period, including spotted hyenas, horse, mammoth, wolf, woolly rhinoceros, reindeer, giant deer and small carnivores. Red deer are present but seem limited to southern Britain (Currant and Jacobi 2011: 172), which may indicate either regional variation in environment or differences between stadial and interstadial conditions. Overall, the presence of many large herbivores and the carnivore guild that preyed on them is indicative that MIS-3 environments were relatively rich from the beginning of this period and at least as attractive as those further east on the Continent.

The Distinctiveness of the British Late Middle Palaeolithic

In comparison to the archaeological record of Continental Europe, the British LMP is quite clearly of a different scale. While there are not significantly fewer sites compared to Northern France or Belgium, the *richness* of sites in Britain pales in comparison, despite a similarly long history of investigation (also taking into account the differences in geological contexts: less extensive cave resource and deep loess deposits in Britain). This has led some researchers to see the British LMP as something of a disappointment (e.g. Roe 1981; White 2006). The smaller dataset, however, permits comprehensive examination of the entire region, and the reduced numbers of sites and assemblage richness instead should be seen as something requiring explanation, rather than neglect. This is especially relevant given the interesting context of Britain as a geographically peripheral Middle Palaeolithic region and the only unambiguous situation where Neanderthal colonization adaptations following a long hiatus can be examined.

Assemblages from northwestern France and Belgium are not only larger but also technologically more diverse, both in terms of intra- and inter-site variability. Even ignoring the archaeological record of MIS-5/early MIS-4 and restricting discussion to the period equivalent to British reoccupation, the greater technological heterogeneity is noticeable. Only two sites in northern France are dominated by discoidal reduction: Beauvais (c. 59.6–51.6 ka: Locht and Swinnen 1994; Locht et al. 1995, 2016; Michel et al. 1999) and Ormesson (MIS-3, c. 52–41 ka, Bodu et al. 2014). Instead it occurs infrequently as a secondary system alongside dominant and diverse Levallois systems (Locht et al. 2016). Similarly, while present in northern France during this period, bifaces are relatively uncommon (Cliquet and Monnier 1993; Cliquet et al 2001b; Locht and Antoine 2001; Goval 2008; Locht et al. 2010, 2016; Depaepe and Goval 2011; Locht and Depaepe 2011). While a very small number of sites such as the rich workshop at Saint-Amands-les-Eaux (OSL date range ranges c. 52.5–45.8 ka; Deschodt et al. 2006; Feray et al. 2010), and Level 1 at Ploisy (stratigraphically attributed to end MIS-4/start MIS-3: Defaux 2004; Locht et al. 2016) attest to some production of finely worked cordate/triangular bifaces, in the British LMP bifaces appear both much more common and more significant within their respective assemblages (even taking into account taphonomic differences and collector bias). There are some Late Pleistocene assemblages known from Northern

France with high frequencies of bifaces that also include discoidal reduction alongside other methods (Bois-du-Rocher: Launay and Molines 2005; La Vallée de la Vègre: Molines et al. 2001; Saint-Nicolas-d'Attez: Cliquet et al. 2001; Saint-Brice-sous-Rânes: Cliquet et al. 2001, 2009; Saint-Julien-de-la-Liègue: Pinoit 2001; Oosthoven: Ruebens and Van Peer 2011). However, they are often from large palimpsest surface contexts or poorly dated (Cliquet 2001a; Ruebens 2013), and additionally many have been attributed to a different technological biface tradition (Micoquian, rather than MTA, or 'Final Mousterian with bifaces': Locht et al. 2016). Ruebens (2013: table B.1) lists only two examples from northwestern Europe with both bifaces and dominant discoidal reduction, both in Belgium: Grotte du Spy and Trou Magrite.

In summary, while bifaces and discoidal and irregular reduction do occur in Continental northwestern Europe just before (MIS 5d-4), and contemporary with, the British LMP in MIS-3, this combination is very rarely found at individual sites as a complete package. Instead, bifaces are relatively infrequent, and Levallois is dominant. The technological signature of the British LMP is therefore distinctive in comparison, with a consistent focus on bifaces (which appear typologically spatio-temporally including *bout coupés*: Tyldesley 1987; White and Jacobi 2002; Wragg Sykes 2010, 2017; Ruebens and Wragg Sykes 2015), alongside either informal or discoidal core reduction strategies, without other formal flake production systems. It is therefore worth examining the possible reasons for this lack of technological diversity.

Cultural Isolation and Technological Homogeneity

The position of Britain, already at the geographic margins of the Neanderthal world, and further cut-off by the physiographic barrier presented by the Channel River system, could have created a relatively isolated cultural context. Cultural theories involving small, disconnected populations often propose a lack of technological diversity as the maladaptive result of isolation, claiming that frequent social interaction and a large population size are necessary for cultural innovations to be sustained through social learning mechanisms (Cavalli-Sforza and Feldman 1981; Shennan and Steele 1999; Shennan 2001; Henrich 2004; Powell et al. 2009; Premo and Kuhn 2010; Muthukrishna et al. 2013). Neanderthal populations in general are considered to have been within the lowest range known for recent hunter-gatherer groups, if not less, based on various lines of evidence including most recently genetics (Krings et al. 1997, 2000; Briggs et al. 2009; Fabre et al. 2009; Lalueza Fox et al. 2005; Richards and Trinkaus 2009; Bocquet-Appel and Degioanni 2013). If the British LMP colonizing groups became a settled occupation, it could have been even smaller than typical, with low levels of inter-regional interaction—effectively socially islanded due both to the marginal geographical location and the presence of the immense Channel River barrier (Roebroeks et al. 2011). A combination of these factors might have led to loss of cultural richness, visible perhaps in the small size of assemblages, and the low diversity of core reduction methods seen in the British LMP, especially the absence of the Levallois method (more intense preparation and volumetric management during knapping than discoidal or informal reduction), likely requiring robust cultural transmission. However, caution is necessary here, as other authors have recently called into question the link between population size and cultural richness and found that for hunter-gatherer populations as a whole, there is currently no strong evidence that small populations are the main driver behind low diversity (Collard et al. 2013).

Furthermore, when the British LMP archaeological record is examined in detail for evidence of a progressive loss of technology (as far back as current dating resolution allows), there does not seem to be an observable decline in cultural richness over time. If colonizing groups had brought with them a technologically diverse lithic culture, comparable to that on the Continent, we might expect to see at least some sites (especially the oldest) with multiple formal approaches to core reduction. In comparison, northern France does seem to show genuine depopulation during the peak of MIS-4, yet it still includes some technological diversity, with some sites dominated by different Levallois systems (e.g. Fitz-James: Teheux 2000; Savy: Tissoux 2006) or discoidal reduction (Beauvais: Locht and Swinnen 1994; Locht et al. 1995; Ormesson: Bodu et al. 2014; Locht et al. 2016), and others show diversity at the intra-assemblage level, for example, discoidal and Levallois at Corbehem (Tuffreau 1979). In Britain, however, even the earliest sites (Lynford Quarry, Robin Hood Cave and Pin Hole) show no strong evidence for such diversity, and the persistent importance of bifaces is also striking. A final related point is that even if there is a taphonomic effect in play where many assemblages were 'winnowed' by early collection methods, therefore reducing their overall diversity, it is hard to discern a mechanism where only one type of core reduction would be affected. Indeed, in regard to collector bias, one would expect visually distinctive, large Levallois cores and flakes to be preferentially retained.

Colonization, Exploration and Risk

It is possible to take a different perspective on the homogeneous technological signature of the British LMP. Rather than a poor cousin to the Continental Middle Palaeolithic, resulting from cultural degeneration, instead it is possible that the focus on particular core reduction systems and bifaces was adaptive, related to the particular conditions involved in colonization of new landscapes.

Although we see evidence on a large scale for apparent spatio-temporal Neanderthal population fluctuations across their entire range (e.g. van Andel and Davies 2003; Davies and Gollup 2003), it is difficult in the Middle Palaeolithic record to define a clear moment when Neanderthals were moving into a particular landscape that had been entirely unoccupied for a significant time. In the northwestern part of Europe that concerns us, there is a reduction in settlement following MIS-5 (Locht et al. 2010, 2016), but new evidence shows that at least in some areas of France and possibly Germany, there remained a very low frequency of occupation even in MIS-4 (Havrincourt: Antoine et al. 2014; Locht et al. 2016: 13; Beauvais: Locht and Swinnen 1994; Locht et al. 2016; Uthmeier et al. 2011). It is hard therefore to be absolutely certain whether populations became truly locally extinct as has been claimed (Hublin and Roebroeks 2009; Roebroeks et al. 2011). The disappearance after MIS-5 of blade reduction is suggestive of some sort of cultural turnover, but Levallois in various forms remained (Locht 2005; Locht et al. 2010, 2016). Even if there was a genuine hiatus on the Continent, requiring recolonization from either the south or east, the gap in settlement compared to that in Britain would have been considerably less (<10 ka as opposed to 100 ka; Goval 2008; Locht et al. 2016). Conversely, the entire record in Britain between the end of MIS-6 and the end of MIS-4/start of MIS-3 consists of only two flakes in late MIS-5 (Wenban-Smith et al. 2010), therefore offering a much better defined context where visible human occupation is absent for an order of magnitude longer. It provides a unique temporal and geographical situation within the Middle Palaeolithic where we can confidently examine Neanderthal colonization adaptations.

Studies of colonization stemming from biogeographical/human ecology research often focus on the uncertainties of moving into novel landscapes: essentially colonization is risky because the quality, location and reliability of resources are unknown (Keegan and Diamond 1987; Kelly and Todd 1988; Kelly 2003). Additionally, colonizing groups are at the edges of social networks: they not only have fewer chances to benefit from support networks but also risk reproductive isolation. Very few ethnographic examples of hunter-gatherer colonization are known that could provide potential frameworks of reference (Kelly 2003), especially in temperate to subarctic/artic contexts (e.g. Fitzhugh 2004) which might be of most relevance for Neanderthals. Overall, successful colonization appears to depend on factors such as the size and biodiversity of the novel landscapes, the size of the founding population and its ability to maintain links with the source population (Wobst 1978; Keegan and Diamond 1987; Roebroeks 2003). All of these issues are of particular concern for the Middle Palaeolithic where the mammoth steppe environments of MIS-3 were not especially rich (in comparison to tropical contexts: Guthrie 1982, 1990), Neanderthal populations were likely very small (Roebroeks et al. 2011: 114; Bocquet-Appel and Degioanni 2013), and while little is known about the mechanics of intergroup social networks, it is often assumed they were limited (Gamble 1999; Davies and Underdown 2006; although see Wragg Sykes 2012). Recently there has been renewed interest in the discontinuous Neanderthal settlement history of northern France (Goval 2008; Hublin and Roebroeks 2009; Locht et al. 2010, 2016; Roebroeks et al. 2011), but while diachronic lithic variation is discussed, consideration of the role of particular techno-economic adaptations in (re)colonization has been less so.

Although in the British LMP vastly greater timescales are involved than is typical in many geographically focused examinations of colonization, the fact that it was a reoccupation, with groups moving into landscapes not known for thousands of years, is certain. Given the coarseness of Pleistocene chronology and current dating methods, identifying individual sites/assemblages as representing particular colonization phases is impossible, but some broad trends can be suggested. On current evidence Lynford Quarry is the earliest site, and perhaps it is not a coincidence that it is located in north Norfolk (Fig. 7.1), above the confluence of the Thames and Rhine where they formed the Channel River, essentially at the easiest point to enter Britain. Its age, right at the end of MIS-4/ beginning of MIS-3, suggests that colonization occurred before temperatures had peaked during the first warming pulse; although the large herbivore fauna from the site (woolly rhinoceros, horse, mammoth) demonstrates vegetation had recovered from tundra to steppe-grassland. Several other sites also have ages that range up to c. 54 ka (see Table 7.1) including Pin Hole and Robin Hood Cave (both at Creswell) and Hyaena Den. Boismier et al. (2012: 73) also note that the older uncalibrated radiocarbon ages from these sites overlap with the youngest OSL age for the main paleochannel at Lynford.

It is possible that highly volatile climatic and environmental conditions prevalent during early-mid MIS-3 may have forced repeated abandonment and reoccupation of Britain. However the earlier Dansgaard-Oeschger events after c. 60 ka (17–16 and 14) appear to have been reasonably stable, with warmer conditions prevailing for longer (van Andel and Davies 2003), possibly confirmed in some British climate proxies such as spelaeothems from Lancaster Hole in northwest England, where two warm phases occur c. 59.5–57.5 ka and 52–49 ka (Atkinson et al. 2005). It is possible that intervening stadials may have been sufficiently harsh to force out or even terminate groups that entered during the warming events, and at some cave sites there is evidence that hyenas were denning, presumably when humans were absent (Lewis 2010: 269). This could mean that at least some lithic assemblages record successive reoccupations, perhaps after each stadial of MIS-3, for example, at Pin Hole (Jacobi et al. 1998, 2006). However, Lynford at least probably represents only 100 – 1000 years of deposition (Boismier et al. 2012), yet it is striking that the technological signature is so similar to other sites. Could this indicate a colonization adaptation that was successful across diverse landscapes and extended timescales? There is one possible example of a later phase of colonization within the British LMP: at Coygan Cave, Wales, in the far west of the country, three bifaces were made of tough local igneous rocks (Fig. 7.3), but there is no evidence of imported flint bifaces or scrapers, as seen at most other sites where local, lower-quality stone was exploited. Although the assemblage is very small and was excavated early, any flint objects would certainly have been kept. Its geographical position suggests it must have been reached after sites in the east and midlands, and the confident use of hard local stone might hint that greater levels of landscape/resource learning had been accomplished (Kelly 2003; Rockman 2003; Roebroeks 2003; Fitzhugh 2004).

The overall character of the British LMP lithic record, in comparison to the Continental Middle Palaeolithic, can further be considered from a broad colonization perspective. As noted above, the high risk of moving into unfamiliar landscapes is often underlined (Spiess et al. 1998; Rockman 2003; Fitzhugh 2004), and technological adaptations organized around transportability, flexibility and maintainability through long use-lives have been suggested to be especially likely to relate to mitigating risk (Bleed 1986; Torrence 1989; Nelson 1991; Bamforth and Bleed 1997; Andrefsky 2009; although Attenbrow 2004 points out disagreement in details between authors). An example often discussed where these features are seen is the North American early Paleoindian record, which in particular features bifaces (easily transportable, flexible maintainable tools which can also act as raw material sources), and is characterized by raw material selectivity, techno-economic consistency across large regions and generally ephemeral use of sites (Kelly and Todd 1988; Kelly 2003; Barton et al. 2004). Many aspects of the British LMP match these expectations for groups operating in a novel environment where mobility was necessarily high, ranges shifted frequently and resources were not known about in advance. The archaeological record is relatively sparse, even in cave locales, and it is strikingly homogeneous. In terms of lithics, easily transportable and maintainable tools in the form of bifaces and scrapers are a key feature of the British LMP, and the bifaces show evidence of secondary use as flake sources, with at least one retouched biface-working flake (Fig. 7.3D), features seen elsewhere in the MTA (Turq 2001; Soressi 2002). Additionally, the formal core reduction methods are centripetally based—discoidal and alternate reduction—alongside informal migrating core reduction, all of which are much more flexible than non-centripetal Levallois systems (Delagnes 2010) and could be better suited to unfamiliar stone resources. Where used in the British LMP, discoidal reduction seems to reflect a more intensive management of lithic resources, demonstrated by smaller cores with higher scar numbers, compared to informal cores both within and between sites (Wragg Sykes 2009, 2017). On the other hand, this method was also used on lower-quality but immediately available raw materials (e.g. quartzite at Creswell Crags and Carboniferous chert at the Mendips caves), suggesting that the production of particular end products was also a motivation rather than simple conservation of stone.

Although previously regarded as primarily used for expedient exploitation on low-quality stone, discoidal/centripetal technology has increasingly been recognized as a highly adaptable system (Bourguignon and Turq 2003; Peresani 2003a, b; Slimak 2003; Bourguignon et al. 2006; Meignen et al. 2009; Delagnes 2010). It allows economical use of raw materials without requiring high initial investment in core preparation, because the surface is effectively continuously re-prepared if the convexities are correctly maintained. In contrast to non-centripetal Levallois there is also less need to prepare platforms, and as there is no volumetric hierarchization (except at the extreme end of reduction or on very tiny nodules), more of the core surface is utilized (Boëda 1993, 1995; Bourgignon et al. 2006). Even if discoidal reduction produces generally shorter blanks than classic Levallois systems, there are a greater number of immediately useable products per core (Delagnes 2010). Additionally, the products created– such as pseudo-Levallois flakes and core-edge or *débordant* flakes– while nonspecialized, are predictable in general form (Peresani 2003a; Delagnes et al. 2007; Meignen et al. 2009) and have been termed ‘polyvalent’ due to their flexibility (Lemorini et al. 2003; Delagnes 2010). In a sense, discoidal products can be seen as closer to blades in terms of economy and additionally have greater potential use-life as they can be resharpened more easily (Eren et al. 2008; Delagnes 2010). Overall, discoidal reduction provides both the flexibility and reliability required in high-risk situations such as colonization (Kelly and Todd 1988; Torrence 1989; Kelly 2003) and is clearly complementary to the transported toolkit made up of bifaces and scrapers that is common across the British LMP.

Cyclical Landscape Exploitation

The previous discussion demonstrates that the overall technological strategy seen across the British LMP (a combination of curated bifaces/scrapers and discoidal/informal flake production) offers flexibility, reliability and maintainability in one complementary system, which makes sense in a high-risk colonization situation. However, the stimuli for Neanderthals entering Britain should be further explored as a possible influence on technology: without considering the broader techno-economic system of mobility, we risk missing the wider picture. Perhaps the features of the British LMP that were especially well suited to the high risks of colonization (non-hierarchical core reduction systems, curated biface and scraper toolkits) were in fact a result of novel landscape-scale adaptations for subsistence strategies targeting unpredictable mobile fauna within the fluctuating climate and environments of MIS-3.

Recently, the major Middle Palaeolithic industries have been reconsidered using a holistic functional and techno-economic perspective in relation to mobility and hunting systems (Delagnes 2010; Delagnes and Rendu 2011). These studies underline the fact that the MTA—which the British LMP is generally regarded as a variant of (Soressi 2002; Wragg Sykes 2009, 2017; White and Pettitt 2011; Ruebens 2013; Ruebens and Wragg Sykes 2015)—has no clear correlation with particular faunal taxa, unlike, for example, Quina industries which appear targeted at reindeer. MTA sites from southwest France include a mix of species (bison, horse and red deer: Delagnes and Rendu 2011), but in the few northern French contexts, there is less faunal data available in association with this industry. Overall the MTA is regarded as a system organized around nonspecialized meat procurement by highly mobile groups, with bifaces providing a durable, easily transportable butchery toolkit plus raw material source in the form of maintenance flakes. While the MTA's long-lived and mobile bifaces are its techno-economic fulcrum, core reduction is highly variable (including Levallois, semi-rotating for elongated blanks or discoidal). This is in contrast to other Mousterian industries such as Discoid-Denticulate and Quina, which are interpreted as adaptations to a similarly highly mobile hunting systems, but species-targeted (Delagnes 2010; Delagnes and Rendu 2011).

The British LMP certainly features highly mobile toolkits focused on bifaces, but as described above, unlike the Continental MTA, it has a restricted range of accompanying core reduction strategies. Given the repetition of this patterning across most sites and raw materials, this approach seems to be as important as the bifaces to the technological system. Delagnes and Rendu (2011) see the core reduction processes within the Continental MTA as a complementary component to the transported bifaces with their successive phases of resharpening and use. As noted above, discoidal reduction requires low investment, yet produces easily transportable blanks, more versatile than Levallois and laminar systems (which require greater initial investment and create less flexible products). The British LMP seems, therefore, to combine features of the Continental MTA (biface-focused curated toolkits, in the sense of Binford 1979) with a distinctive mix of informal and non-Levallois formal core reduction, plus resharpening promoting long use-life (i.e. scraper retouch). If the MTA is adapted to high levels of mobility, but is not aimed at targeting migratory herds or particular species (Delagnes and Rendu 2011), then it might instead be understood as an adaptation to a particular kind of open landscape exploitation. While there are many MTA assemblages from cave sites, Delagnes (2010) has noted the high frequency of open-air locales and suggests this may reflect increasingly seasonal hunting strategies requiring higher levels of mobility and extended techno-economic organization across larger areas of the landscape (Delagnes and Meignen 2006; Delagnes et al. 2007).

As described earlier, some of the most dramatic and rapid fluctuations in temperature known from the entire Quaternary occurred during MIS-3. Disrupted responses of weather and ecosystems would have been noticeable during the lifetimes of individuals and certainly within the memory of several generations in a group. While sometimes colder than typical interglacials, greater summer insolation levels were present in early MIS-3 (peaking around 60 ka), which could have caused increased seasonality in resource availability (Van Meerbeeck et al. 2009). While some have portrayed Britain in MIS-3 as an unpromising wasteland (White 2006), the faunal resources present were just as rich as elsewhere on the European mammoth steppe and were more than adequate for other large carnivores such as spotted hyenas. Interestingly, palaeoclimatic indicators for Lynford do not suggest especially warm conditions (Boismier et al. 2012), which may suggest that the quality of environment, rather than temperature, was most important. Furthermore, Britain did offer something different to the northern European plains, being a large region of topographically diverse terrain (likely with a less extreme continental climate). Herbivores appear to have quickly taken advantage of this new ecological dynamic, and reindeer at least may have been doing so seasonally (Wragg Sykes 2009, 2017), perhaps part of a broader trend of increased migratory behaviour in MIS-4/3 (Delagnes and Rendu 2011). It is notable therefore that some of the rare evidence for Neanderthal faunal processing from the British LMP involves reindeer (see Table 7.1 for all faunal associations).

In such conditions of shifting environments and resources, the most valuable skill enabling survival would have been adaptability, and Neanderthal groups equipped with techno-economic systems that supported high levels of residential and range mobility would have been able to quickly adopt a strategy of expanded activity ranges across the open landscapes of MIS-3, rather than attempting to diversify their subsistence base, potentially requiring new skills (Barton et al. 2004). We currently lack direct data for Neanderthal territory size, the main inferences being drawn from the proxy of lithic raw material transfer distances. Based on the assumption that maximum known distances are likely to be most informative about the scales of traversed landscapes (if exchange was not taking place), there is a clear increase in territorial size between the EMP and LMP (Féblot-Augustins 1993, 1999, 2009). The distances involved (100–300 km) make it feasible that northern French groups operating in high-mobility hunting systems could have incorporated Britain into their territories, even as far west as Coygan Cave in Wales, meaning that the colonization may have been an expansion, rather than a migration. Some sort of cyclical, non-residential use of the landscapes of Britain within a hunting context might explain its techno-economic patterning. The importance of bifaces in almost all assemblages makes sense if primary activities were related to carcass processing, given that recent use-wear results have pointed to this as a major task associated with these tools (Claud 2008). The high levels of mobility indicated by curated and transported toolkits can also be understood in a hunting-focused context, as for the MTA more widely. Even the unidirectional raw material movement might be related to a large-scale landscape exploitation: a similar pattern can be seen in North American Folsom contexts, with strong regionality in raw material transport, including outward movement of high-quality stone, without a corresponding returning of other stone types, which has been interpreted as resulting from seasonal long-distance hunting (Hofman 2003; Jones et al. 2003). While the scales of lithic transport in the Folsom are apparently greater than seen in the British LMP (although see Jones et al. 2012), such a system organized around exploiting fauna does offer another possibility for explaining the ephemeral nature of the MIS-3 Neanderthal presence and the distinctive techno-economic signature found there. Faunal evidence from France suggests that at least some Neanderthals were hunting in a strategic manner, using communal organization and possibly also storing surplus (Rendu et al. 2012). Perhaps groups occupying northern France, Belgium or even the north Channel Plain were only temporary visitors to Britain, following the expansion of herds and using a suite of lithic adaptations that were suited at once to exploring relatively unfamiliar landscapes and making the most of transient resources found there. It may be that one way of dealing with the MIS-3 world meant shifting away entirely from being geographically based to an itinerant lifestyle permitting exploitation of short-term opportunities across large areas, thanks to a highly mobile, flexible technology and ecological knowledge.

Conclusions

Investigation into the British Late Middle Palaeolithic has to a large extent previously been focused on determining chronology and describing the character of individual assemblages, often within limited and typologically-focused frameworks. Consequently, the wider relevance of the region to discussion on Neanderthal behaviour has been limited, and it has been viewed as of marginal importance. Based on comprehensive analysis of the best-dated sites from this period, and taking into account taphonomic biases, a new perspective is now possible. It is argued here that the relatively small size of the British LMP record, alongside its distinctive techno-economic character (dominated by bifaces and a focus on discoidal and informal core reduction), are explicable when viewed as an adaptation providing flexibility and reliability through high levels of mobility and easily transportable, maintainable toolkits (Torrence 1989). Such a techno-economic strategy would have enabled the colonization of Britain's MIS-3 landscapes, following an absence of over 100,000 years. Until (and if) further refinements in chronology become available, it is not possible to determine whether the LMP colonization involved migration and settlement by groups from northwest Europe or if instead these landscapes were exploited in a more transitory manner, as a MIS-4/3 western extension of spatially-expanded hunting systems. What is clear is that the British LMP supports wider evidence that Neanderthals were actively and systematically managing lithic resources within varied systems of mobility (Delagnes 2010; Delagnes and Rendu 2011; Kuhn 2011).

The British record stands out in its consistent techno-economic character, and the persistence over some 10,000 years of this particular approach to technological organization, which would have been especially well adapted to high-risk situations, whether colonization per se; high mobility for targeted, seasonal hunting; or a combination of both. The British LMP suggests that the environments of MIS-3 should not be principally viewed as problematic for Neanderthals (e.g. Finlayson and Carrión 2007), when the archaeological record shows that they were able to not only recover from the MIS-4 glaciation across their previous range of northern France but also expand into entirely new territory. While climate and environments were undoubtedly fundamental drivers in Neanderthal behavioural variation, some level of adaptive innovation is also clear in terms of the developing diversity in techno-economic organization during MIS-5, through to the MIS-3 colonization of Britain.

Acknowledgements My thanks to the editors for including my chapter, for their patience in its final stages, and especially to Erick Robinson for over a decade of friendship and good debates. Having originally conceived the idea for an edited volume on lithic technology and environmental change with Erick, circumstances meant I was unable to attend the SAA symposium or continue as Co-editor. It is wonderful to see how the idea has progressed to a tangible book, full of exciting contributions from around the globe. I am grateful to the many colleagues who have discussed this research during and since my PhD thesis, in particular, Roger Jacobi, Karen Ruebens and especially, since 2013, Brad Gravina. The original research was primarily undertaken at the University of Sheffield, funded by a University Research Studentship. Subsequent development of this paper was supported by a Marie Curie Intra-European Fellowship at PACEA, UMR-5199, Université de Bordeaux (European 7th Framework, 2013-15, Grant 301693). I am grateful to all staff and colleagues there for welcoming me into such a stimulating research environment.

I thank all the institutions and individuals who provided access to the collections studied (full list in Wragg Sykes 2009). Any errors or omissions are my responsibility.

References

Aldhouse-Green, S. (Ed.). (2000). *Paviland Cave and the 'Red Lady': A definitive report*. Bristol: Western Academic and Specialist Press.

Aldhouse-Green, S., Scott, K., Schwarcz, H., Grün, R., Housley, R., Rae, A., Bevins, R., & Rednap, M. (1995). Coygan Cave Laugharne, south Wales. A Mousterian site and hyaena den: A report on the University of Cambridge excavations. *Proceedings of the Prehistoric Society, 61*, 37–79.

Andersen, K. K., Svensson, A., Johnsen, S. J., Rasmussen, S. O., Bigler, M., Rothlisberger, R., Ruth, U., Siggaard-Andersen, M.-L., Steffensen, J. P., Dahl Jensen, D., Vinther, B. M., & Clausen, H. B. (2006). The Greenland ice core chronology 2005, 15–42 ka. Part 1: Constructing the time scale. *Quaternary Science Reviews, 25*(23–24), 3246–3257.

Andrefsky, W. (2009). The analysis of stone tool procurement, production and maintenance. *Journal of Archaeological Research, 17*, 65–103.

Antoine, P., Coutard, J.-P., Gibbard, P., Hallegouet, B., Lautridou, J.-P., & Ozouf, J.-C. (2003). The Pleistocene rivers of the English Channel region. *Journal of Quaternary Science, 18*(3–4), 227–243.

Antoine, P., Limondin-Lozouet, N., Auguste, P., Locht, J.-L., Galheb, B., Reyss, J.-L., Escude, E., Carbonel, P., Mercier, N., Bahain, J.-J., Falguères, C., & Voinchet, P. (2006). Le tuf de Caours (Somme, France) : mise en évidence d'une séquence eemienne et d'un site paléolithique associé. *Quaternaire, 17*(4), 281–320.

Antoine, P., Goval, E., Jamet, G., Coutard, S., Moine, O., Hérisson, D., Auguste, P., Guérin, G., Lagroix, F., Schmidt, E., Robert, V., Debenham, N., Meszner, S., & Bahain, J.-J. (2014). Les séquences loessiques pléistocène supérieur d'Havrincourt (Pas-de-Calais, France) : stratigraphie, paléolenvironnements, géochronologie et occupations paléolithiques. *Quaternaire, 25*(4), 321–368.

ApSimon, A. P. (1986). Picken's Hole, Compton Bishop, Somerset; early Devensian bear and wolf den, and Middle Devensian hyena den and Palaeolithic site. In S. N. Collcutt (Ed.), *The Palaeolithic of Britain and its nearest neighbours: Recent trends* (pp. 55–56). Sheffield: University of Sheffield.

Armstrong, A. L. (1925). Excavations at Mother Grundy's Parlour, Creswell Crags, Derbyshire 1924. *Journal of the Royal Anthropological Institute, 55*, 146–178.

Armstrong, A. L. (1926). Excavations at Creswell Crags, Derbyshire, 1924-26. *Transactions of the Hunter Archaeological Society, 3*(2), 116–122.

Armstrong, A. L. (1928). Excavations in the Pin Hole Cave, Creswell Crags Derbyshire. *Proceedings of the Prehistoric Society, 6*, 330–334.

Armstrong, A. L. (1929). Excavations at Creswell Crags, the Pin Hole cave 1925-28. *Transactions of the Hunter Archaeological Society, 3*, 332–334.

Armstrong, A. L. (1932). Excavations at Creswell crags, Derbyshire. 1928-32. The Pin Hole Cave. *Transactions of the Hunter Archaeological Society, 4*, 178–184.

Armstrong, A. L. (1935). Report of Committee for the Archaeological Exploration of Derbyshire Caves, Interim Report 12. *Derbyshire Archaeological Journal, 55*, 85–92.

Armstrong, A. L. (1937). Excavations at Creswell Crags, the Pin Hole Cave. *Transactions of the Hunter Archaeological Society, 4*, 178–118.

Armstrong, A. L. (1939). Excavations in Derbyshire during 1938. Palaeolithic. Creswell crags, Whaley Rockshelter No.2 and Trent Valley Gravels. *Derbyshire Archaeological Journal, 59*, 81–82.

Armstrong, A. L. (1950). Exploration of prehistoric sites in East Derbyshire. *Derbyshire Archaeological Journal, 70*, 88–91.

Armstrong, A. L. (1951). Exploration of prehistoric sites in East Derbyshire. *Derbyshire Archaeological Journal, 71*, 66–68.

Armstrong, A. L. (1952). Exploration of prehistoric sites in East Derbyshire. *Derbyshire Archaeological Journal, 72*, 133–135.

Armstrong, A. L. (1953). Exploration of prehistoric sites in East Derbyshire. *Derbyshire Archaeological Journal, 73*, 102–104.

Armstrong, A. L. (1954). Exploration of prehistoric sites in East Derbyshire. *Derbyshire Archaeological Journal, 74*, 156–157.

Armstrong, A. L. (1955). Exploration of prehistoric sites in East Derbyshire. *Derbyshire Archaeological Journal, 75*, 145–150.

Armstrong, A. L. (1956). Exploration of prehistoric sites in East Derbyshire. *Derbyshire Archaeological Journal, 76*, 133–137.

Armstrong, A. L. (1957). Report on the excavation of ash tree cave, near Whitwell. *Derbyshire, Derbyshire Archaeological Journal, 76*, 57–74.

Ashton, N. M. (1992). The high lodge flint industries. In N. M. Ashton, J. Cook, S. G. Lewis, & J. Rose (Eds.), *Excavations at high lodge, G. De G. Sieveking 1962–1968, J. Cook 1988* (pp. 124–163). London: British Museum Press.

Ashton, N. M. (1998). The technology of the flint assemblages. In N. M. Ashton, S. G. Lewis, & S. Parfitt (Eds.), *Excavations at the Lower Palaeolithic site at East Farm, Barnham, Suffolk 1989-94, Occasional paper* (Vol. 125, pp. 183–204). London: British Museum Press.

Ashton, N. M. (2002). Absence of humans in Britain during the last interglacial period (Oxygen Isotope Stage 5e). In W. Roebroeks & A. Tuffreau (Eds.), *Le Dernier Interglaciare et les Occupations Humaines du Paléolithique Moyen* (pp. 93–103). Lille: Université des Sciences et Technologies de Lille, CREP Publication No. 8.

Ashton, N. M., & McNabb, J. (1996). Appendix 1: Methodology of flint analysis. In B. Conway, J. McNabb, & N. M. Ashton (Eds.), *Excavations at Barnfield Pit, Swanscombe, 1968–72, Occasional Paper* (Vol. 94, pp. 241–246). London: British Museum Press.

Ashton, N. M., & Lewis, S. (2002). Deserted Britain: Declining populations in the British Late Middle Pleistocene. *Antiquity, 76*, 388–396.

Atkinson, T. C., Rowe, P. J., & Dennis, P. F. (2005). Isotopic composition of Palaeo-precipitation from a British speleothem: The temporal slope of δ18O *vs.* mean annual temperature over a 100 ka timescale. *Geophysical Research Abstracts, 7*, 06961.

Attenbrow, V. J. (2004). *What's changing? Population size or land–use patterns? The archaeology of Upper Mangrove Creek, Sydney Basin, Terra australis No 21*. Canberra: Pandanus Press, ANU.

Balch, H. E. (1914). *Wookey hole: Its caves and cave dwellers*. London: Oxford University Press.

Bamforth, D. B., & Bleed, P. (1997). Technology, flaked stone technology, and risk. In C. M. Barton & G. A. Clark (Eds.), *Rediscovering Darwin: Evolutionary theory in archaeological explanation, Archaeological Papers of the American Anthropological Association* (Vol. 7, pp. 267–290). Arlington: American Anthropological Association.

Barton, C. M., Schmich, S., & James, S. R. (2004). The ecology of human colonization in pristine landscapes. In C. M. Barton, G. A. Clark, D. R. Yesner, & G. A. Pearson (Eds.), *The settlement of the American continents: A multidisciplinary approach to human biogeography* (pp. 138–161). Tucson: University of Arizona Press.

Beynon, F., & Ogilvie, A. H. (1931). Kent's Cavern Report. *Report of the British Association for the Advancement of Science*, 273–274.

Beynon, F., & Ogilvie, A. H. (1932). Kent's Cavern Report. *Report of the British Association for the Advancement of Science*, 289–290.

Beynon, F., & Ogilvie, A. H. (1933). Kent's Cavern Report. *Report of the British Association for the Advancement of Science*, 301–302.

Beynon, F., & Ogilvie, A. H. (1934). Kent's Cavern Report. *Report of the British Association for the Advancement of Science*, 258.

Beynon, F., & Ogilvie, A. H. (1935). Kent's Cavern Report. *Report of the British Association for the Advancement of Science*, 334–335.

Beynon, F., & Ogilvie, A. H. (1936). Kent's Cavern Report. *Report of the British Association for the Advancement of Science*, 303–304.

Beynon, F., & Ogilvie, A. H. (1937). Kent's Cavern Report. *Report of the British Association for the Advancement of Science, 299*.

Beynon, F., & Ogilvie, A. H. (1938). Kent's Cavern Report. *Report of the British Association for the Advancement of Science, 347.*

Beynon, F., Dowie, H. G., & Ogilvie, A. H. (1929). Report on the excavations in Kent's Cavern, 1926-9. *Transactions and Proceedings, Torquay Natural History Society, 5*(3), 237–242.

Binford, L. R. (1979). Organization and formation processes: Looking at curated technologies. *Journal of Anthropological Research, 35*(3), 255–273.

Bleed, P. (1986). The optimal design of hunting weapons: Maintainability or reliability. *American Antiquity, 51*, 737–747.

Bocquet-Appel, J.-P., & Degioanni, A. (2013). Neanderthal demographic estimates. *Current Anthropology, 54*(8), S202–S213.

Bodu, P., Salomon, H., Leroyer, M., Naton, H. G., Lacarriere, J., & Dessoles, M. (2014). An open-air site from the recent Middle Palaeolithic in the Paris Basin (France): Les Bossats at Ormesson (Seine-et-Marne). *Quaternary International, 331*, 39–59.

Boëda, E. (1993). Le débitage discoïde et le débitage Levallois récurrent centripète. *Bulletin de la Société Préhistorique Française, 90*(6), 392–404.

Boëda, E. (1994). *Le concept Levallois: Variabilité des Méthodes, CNRS CRA monograph* (Vol. 9). Paris: Ed. du CNRS.

Boëda, E. (1995). Levallois: A volumetric reconstruction, methods, a technique. In H. L. Dibble & O. Bar-Yosef (Eds.), *The definition and interpretation of Levallois technology* (pp. 41–68). Madison: Prehistory Press.

Boismier, W., Gamble, C., & Coward, F. (2012). *Neanderthals among mammoths: Excavations at Lynford Quarry, Norfolk*. Swindon: English Heritage.

Bond, G. C., Broecker, W. D., Johnsen, S., McManus, J., Labeyrie, L., Jouzel, J., & Botani, G. (1993). Correlations between climate records from North Atlantic sediments and Greenland ice. *Nature, 365*, 143–147.

Bos, J. A. A., Helmens, K. F., Bohnkce, S. J. P., Seppä, H., & Birks, H. J. B. (2009). Fauna, vegetation and climate at Sokli, north-eastern Fennoscandia , during the Weichselian Middle Pleniglacial. *Boreas, 38*(2), 1–14.

Bourguignon, L., & Turq, A. (2003). Une chaîne opératoire de débitage discoïde sur éclats du Moustérien à denticulés aquitain. In M. Peresani (Ed.), *Discoid Lithic technology. Advances and implications, BAR International Series* (Vol. 1120, pp. 131–152). Oxford: Archaeopress.

Bourguignon, L., Delagnes, A., & Meignen, L. (2006). Systèmes de production lithique, gestion des outillages et territoires au Paléolithique moyen: où se trouve la complexité? In *Normes Techniques et Pratiques Sociales: de la Simplicité des Outillages Pré- et Protohistoriques, XXVIe Rencontres Internationales d'Archéologie et d'Histoire d'Antibes* (pp. 75–86). Antibes: Editions APDCA.

Briggs, A. W., Good, J. M., Green, R. E., Krause, J., Maricic, T., Stenzel, U., Lalueza-Fox, C., et al. (2009). Targeted retrieval and analysis of five Neandertal mtDNA genomics. *Science, 325*(5938), 318–321.

Bronk Ramsey, C. (2009). Bayesian analysis of radiocarbon dates. *Radiocarbon, 51*(1), 337–360.

Bronk Ramsey, C., Higham, T., Bowles, A., & Hedges, R. (2004). Improvements to the pretreatment of bone at Oxford. *Radiocarbon, 46*, 155–163.

Brown, J. A. (1889). Working sites and inhabited land surfaces of the Palaeolithic period in the Thames Valley, etc. *Transactions of the Middlesex Natural History Society*, 3–36.

Browne, C. L., & Wilson, L. (2011). Resource selection of lithic raw materials in the Middle Palaeolithic in southern France. *Journal of Human Evolution, 61*, 597–608.

Browne, C. L., & Wilson, L. (2013). Evaluating inputs to models of hominin raw material selection: Map resolution and path choices. *Journal of Archaeological Science, 40*(11), 3955–3962.

Buckland, W. (1823). *Reliquiae Diluvianae*. London: John Murray.

Burleigh, R., Hewson, A., & Meeks, N. (1976). British Museum natural radiocarbon measurements VIII. *Radiocarbon, 18*, 16–42.

Calkin, J. B., & Green, J. F. N. (1949). Palaeoliths and terraces near Bournemouth. *Proceedings of the Prehistoric Society, 15*, 21–37.

Campbell, J. (1969). Excavations at Creswell Crags. *Derbyshire Archaeological Journal, 89*, 47–58.
Campbell, J. (1977). *The Upper Palaeolithic of Britain: A study of man and nature in the late ice age*. Oxford: Clarendon Press.
Campbell, J., & Sampson, C. G. (1971). *A new Analysis of Kent's Cavern, Devonshire, England, University of Oregon Anthropological Papers* (Vol. 3). Eugene, OR: University of Oregon.
Caseldine, C. J., McGarry, S. F., Baker, A., Hawkesworth, C., & Smart, P. L. (2008). Late quaternary speleothem pollen in the British Isles. *Journal of Quaternary Science, 23*(2), 193–200.
Cavalli-Sforza, L., & Feldman, M. (1981). *Cultural transmission and evolution: A quantitative approach*. Princeton: Princeton University Press.
Claud E. (2008). *Le Statut fonctionnel des bifaces au Paléolithique moyen récent dans le sud-ouest de la France. Etude tracéologique intégrée des outillages des sites de La Graulet, La Conne de Bergerac, Combe Brune 2, Fonseigner et Chez-Pinaud/Jonzac*. Unpublished PhD thesis, Université Bordeaux I.
Clegg, J. (1970). Excavations at Coygan Cave, near Laugharne. First draft report. *The Carmarthen Antiquary, 5*, 13–20.
Cliquet, D. (Ed.). (2001a). *Les industries à outils bifaciaux du Paléolithique moyen d'Europe occidentale. Actes de la table-ronde internationale organisée à Caen (Basse-Normandie – France) (14 et 15 October 1999), ERAUL* (Vol. 98). Liège: Université de Liège.
Cliquet, D. (2001b). Le Paléolithique moyen à outils bifaciaux en Normandie: état des connaissances. In D. Cliquet (Ed.), *Les industries à outils bifaciaux du Paléolithique moyen d'Europe occidentale (pp. 115–127). Actes de la table-ronde internationale organisée à Caen (Basse-Normandie – France) (14 et 15 October 1999), ERAUL* (Vol. 98). Liège: Université de Liège.
Cliquet, D., & Monnier, J. L. (1993). Signification et evolution du Paléolithique moyen recent armoricain. *Bulletin de la Société Préhistorique Française, 90*(4), 275–282.
Cliquet, D., Lautridou, J.-P., Rivard, J.-J., Alix, P., Gosselin, R., & Lorren, P. (2001). Les industries à outils bifaciaux du Paléolithique moyen en Normandie armoricaine: l'exemple du site de Saint-Brice-sous-Rânes (Orne – France). In D. Cliquet (Ed.), *Les industries à outils bifaciaux du Paléolithique moyen d'Europe occidentale: Actes de la table-ronde internationale organisée à Caen (Basse-Normandie – France) (14 et 15 October 1999), ERAUL* (Vol. 98, pp. 93–106). Liège: Université de Liège.
Cliquet, D., Mercier, N., Lautridou, J.-P., Alix, P., Beugnier, V., Bianchini, R., Caspar, J.-P., Coutard, S., Lasseur, E., Lorren, P., Gosselin, R., Rivard, J.-J., & Valladas, H. (2009). Un atelier de production et de consommation d'outils bifaciaux de la fin du Paléolithique moyen à Saint-Brice-sous-Rânes (Orne, France) dans son contexte environnemental. *Quaternaire, 20*(3), 361–379.
Collard, M., Buchanan, B., & O'Brien, M. J. (2013). Population size as an explanation for patterns in the Paleolithic archaeological record: More caution is needed. *Current Anthropology, 54*(S8), S388–S396.
Collins, A., & Collins, D. (1970). Cultural evidence from Oldbury. *Bulletin of the Institute of Archaeology, 8-9*, 151–176.
Conard, N. J., & Adler, D. (1997). Lithic reduction and hominid behaviour in the Middle Palaeolithic of the Rhineland. *Journal of Anthropological Research, 53*, 147–176.
Conard, N. J., & Delagnes, A. (Eds.). (2010). *Settlement dynamics of the Middle Palaeolithic and Middle Stone Age* (Vol. 3). Tubingen: Kerns Verlag.
Conard, N. J., & Richter, J. (Eds.). (2011). *Neanderthal lifeways, subsistence and technology: On hundred fifty years of Neanderthal study*. New York: Springer.
Cook, J., & Jacobi, R. (1998). Discoidal core technology in the Palaeolithic at Oldbury, Kent. In N. Ashton, F. Healy, & P. Pettitt (Eds.), *Stone Age Archaeology: Essays in Honour of John Wymer* (pp. 124–136). Oxford: Oxbow.
Coope, G. R. (2002). Changes in the thermal climate in Northwestern Europe during Marine Isotope Stage 3, estimated from fossil insect assemblages. *Quaternary Research, 57*, 401–408.
Coope, G. R., Shotton, F. W., & Strachan, I. (1961). A Late Pleistocene fauna and flora from Upton Warren, Worcestershire. *Philosophical Transactions of the Royal Society B, 244*, 379–417.
Cotton, J. (1984). A Mousterian handaxe from West London. *Lithics, 5*, 40–43.

Coulson, S. (1990). *Middle Palaeolithic Industries of Great Britain*. Bonn: Hallos.
Currant, A., & Jacobi, R. (1997). Vertebrate faunas from the British Late Pleistocene and the chronology of human settlement. *Quaternary Newsletter, 82*, 1–8.
Currant, A., & Jacobi, R. (2001). A formal mammalian biostratigraphy for the Late Pleistocene of Britain. *Quaternary Science Reviews, 20*, 1707–1716.
Currant, A., & Jacobi, R. M. (2004). A Middle Devensian mammalian assemblage from the Hyaena Den, Wookey Hole, Somerset. In *The Quaternary mammals of Southern and Eastern England –Field guide*. London: Quaternary Research Association.
Currant, A., & Jacobi, R. M. (2011). The mammal faunas of the British Late Pleistocene. In N. Ashton, S. Lewis, & C. Stringer (Eds.), *Ancient human occupation of Britain, Developments in quaternary science* (Vol. 14, pp. 165–180). London: Elsevier.
Dansgaard, W., Johnsen, S. J., Clausen, H. B., Dahl-Jensen, D., Gundestrup, N. S., Hammer, C. U., Hvidberg, C. S., Steffensen, J. P., Sveinbjörnsdottir, H., Jouzel, J., & Bond, G. (1993). Evidence for general instability of past climate from a 250-kyr ice-core record. *Nature, 364*, 218–220.
Davies, W., & Gollup, P. (2003). The human presence in Europe during the Last Glacial Period II: Climate tolerance and climate preferences of mid- and late Glacial hominids. In T. H. Van Andel & W. Davies (Eds.), *Neanderthals and modern humans in the European landscape during the Last Glaciation: Archaeological results of the Stage 3 Project* (pp. 131–146). Cambridge: McDonald Institute Monographs.
Davies, R., & Underdown, S. J. (2006). The Neanderthals: A social synthesis. *Cambridge Archaeological Journal, 16*(2), 145–164.
Dawkins, W. B. (1862). On a Hyaena Den at Wookey Hole, near Wells. *Quarterly Journal of the Geological Society of London, 18*, 115–126.
Dawkins, W. B. (1863). On a Hyaena Den at Wookey Hole, near Wells. *Quarterly Journal of the Geological Society of London, 19*, 260–274.
Dawkins, W. B. (1874). *Cave hunting*. London: MacMillan.
Dawkins, W. B. (1876). On the mammalia and traces of man in Robin Hood's Cave. *Quarterly Journal of the Geological Society of London, 32*, 240–259.
Dawkins, W. B. (1877). On the mammal fauna of the caves of Creswell Crags. *Quarterly Journal of the Geological Society of London, 33*, 589–612.
Dawkins, W. B., & Mello, J. M. (1879). Further discoveries at Cresswell Crags. *Quarterly Journal of the Geological Society of London, 35*, 724–734.
Defaux, F. (2004). "Ploisy, le Bras de Fer": Amiens: Rapport de fouille, INRAP, Service Régionale d'Archéologie de Picardie.
Delagnes, A. (2010). Du rôle structurant de la mobilité dans les systèmes techniques du Paléolithique moyen. In N. Conard & A. Delagnes (Eds.), *Settlement dynamics of the Middle Paleolithic and Middle Stone Age* (Vol. 3, pp. 373–396). Tübingen: Kerns Verlag.
Delagnes, A., & Meignen, L. (2006). Diversity of lithic production systems during the Middle Paleolithic in France. In E. Hovers & S. Kuhn (Eds.), *Transitions before the transition. Evolution and stability in the Middle Paleolithic and Middle Stone Age* (pp. 85–107). New York: Springer.
Delagnes, A., & Rendu, W. (2011). Shifts in Neandertal mobility, technology and subsistence strategies in western France. *Journal of Archaeological Science, 38*, 1771–1783.
Delagnes, A., Jaubert, J., & Meignen, L. (2007). *Les technocomplexes du Paléolithique moyen en Europe occidentale dans leur cadre diachronique et geographique. Les Néandertaliens. Biologie et cultures, Documents préhistoriques* (Vol. 23, pp. 213–229). Paris: Éditions du CTHS.
Delair, J. B., & Shackley, M. L. (1978). The Fisherton Brickpits: Their stratigraphy and fossil contents. *Wiltshire Natural History Society Magazine, 74*, 3–19.
Depaepe, P., & Goval, E. (2011). Regards portes sur les travaux de François Bordes en France septentrionale. In *Francois Bordes et la préhistoire, 134e congres CTHS – Célèbres ou obscurs*. Hommes et femmes dans leurs territoires et leur histoire, 21/24 Avril 2009, pp. 255–265.
Deschodt, L., Feray, P., & Vallin, L. (2006). Saint-Amand-les-Eaux (59) "Bas du Mont des Bruyeres". *Rapport de Diagnostique Archéologique, 20*.
Dibble, H. (1995). Middle Paleolithic scraper reduction: Background, clarification, and review. *Journal of Archaeological Method and Theory, 2*(4), 299–368.

Eren, M. I., Greenspan, A. G., & Sampson, C. G. (2008). Are Upper Paleolithic blade cores more productive than Middle Paleolithic discoidal cores? A replication experiment. *Journal of Human Evolution, 55*(6), 952–961.

Fabre, V., Condemi, S., & Degioanni, A. (2009). Genetic evidence of regional groups among Neanderthals. *PloS One, 4*(4), e5151.

Féblot-Augustins, J. J. (1993). Mobility strategies in the Late Middle Palaeolithic of central Europe and western Europe: Elements of stability and variability. *Journal of Anthropological Archaeology, 12*, 211–265.

Féblot-Augustins, J. J. (1999). Raw material transport patterns and settlement systems in the European Lower and Middle Palaeolithic: Continuity, change and variability. In W. Roebroeks & C. Gamble (Eds.), *The Middle Palaeolithic occupation of Europe* (pp. 193–214). Leiden: University of Leiden.

Féblot-Augustins, J. J. (2009). Revisiting European Upper Paleolithic raw material transfers: The demise of the cultural ecological paradigm? In B. Adams & B. S. Blades (Eds.), *Lithic materials and Paleolithic Societies* (pp. 25–46). Chichester: Wiley-Blackwell.

Feray, P., Deschodt, L., Lantoine, J., & Schwenninger, J.-L. (2010). Le site paléolithique moustérien de tradition acheuléenne (MTA) de Saint-Amand-les-Eaux (nord de la France). Résultats préliminaires. Available at : https://www.academia.edu/3097458/Le_site_paleolithique_MTA_de_Saint-Amand-les-eaux_nord_de_la_France_.Resultats_preliminaires._Saint-Amand-les-Eaux_MTA_palaeolithic_site_North_of_France_._First_results_

Finlayson, C., & Carrión, J. S. (2007). Rapid ecological turnover and its impact on Neanderthal and other human populations. *Trends in Ecology and Evolution, 22*(4), 213–222.

Fitzhugh, B. (2004). Colonizing the Kodiak Archipelago: Trends in raw material use and lithic technologies at the Tanginak Spring Site. *Arctic Anthropology, 41*(1), 14–40.

Fletcher, W. J., Sánchez Goñi, M. F., Allen, J. R. M., Cheddadi, R., Combourieu-Nebout, N., Huntley, B., Lawson, I., Londeix, L., Magri, D., Margari, V., Muller, U., Naughton, F., Novenko, E., Roucoux, K., & Tzedakis, P. (2010). Millennial-scale variability during the last glacial in vegetation record from Europe. *Quaternary Science Review, 29*, 2839–2864.

Gamble, C. S. (1999). *The Palaeolithic Societies of Europe*. Cambridge: Cambridge University Press.

Geneste, J. M. (1988). Systèmes d'approvisionement en matières premières lithique. In M. Otte (Ed.), *L'Homme de Néanderthal, La Mutation* (Vol. 3, pp. 61–70). Liège: Université de Liège. ERAUL 35.

Geneste, J.-M. (1989). Economie des resources lithiques dans le Moustérien du sud-ouest de la France. In M. Otte (Ed.), *L'homme de Néandertal, La subsistence* (Vol. 6, pp. 75–97). Liège: Université de Liège. ERAUL 33.

Geneste, J. M. (1991). Systèmes technique de production lithique: variations techno-économique dans les processus de realization des outillages Paléolithique. *Techniques et culture, 17-18*, 1–35.

Gibbard, P. L. (1988). The history of the great northwest European rivers during the past three million years. *Philosophical Transactions of the Royal Society of London Series B, 318*, 559–602.

Gibbard, P. L., & Lautridou, J. P. (2003). The Quaternary history of the English Channel: An introduction. *Journal of Quaternary Science, 18*, 195–199.

Gilmour, M., Currant, A., Jacobi, R., & Stringer, C. (2007). Recent TIMS dating results from British Late Pleistocene vertebrate faunal localities: Context and interpretation. *Journal of Quaternary Science, 22*(8), 793–800.

Goval, E. (2008). *Definitions, Analyses et Caracterisations des Territoires des Neandertaliens au Weichselien Ancien en France Septentrionale: Approches Technologiques et Spatiales des Industries Lithiques, Élargissement au Nord-Ouest de l'Europe*. Unpublished PhD thesis, Université de Lille I.

Grant-Dalton, S. (1917). *Untitled note in an Inventory of the Ancient Monuments of Wales and Monmouthshire: V County of Carmarthenshire*. London: Royal Commission on the Historical Monuments of England.

Green, C. P., Keen, D. H., McGregor, D. F., Robinson, J. E., & Williams, R. B. W. (1983). Stratigraphy and environmental significance of Pleistocene deposits at Fisherton, near Salisbury, Wiltshire. *Proceedings of the Geologists Association, 94*, 17–22.

Grimes, W. F., & Cowley, L. F. (1935). Coygan Cave, Llansadyrnin, Carmarthenshire. *Archaeologia Cambrensis, 90*, 95–111.

Grootes, P. M., Stuiver, M., White, J. W., Johnsen, S., & Jouzel, J. (1993). Comparison of oxygen isotope records from the GISP2 and GRIP Greenland ice cores. *Nature, 366*, 552–554.

Guthrie, R. D. (1982). Mammals of the mammoth steppe as palaeoenvironmental indicators. In D. M. Hopkins, J. V. Matthews, C. E. Schweger, & S. B. Young (Eds.), *Palaeoecology of Beringia* (pp. 307–326). New York: Academic.

Guthrie, R. D. (1990). *Frozen fauna of the mammoth steppe*. Chicago: Chicago University Press.

Harrison, R. A. (1977). The Uphill Quarry Caves, Weston-Super-Mare. A reappraisal. *Proceedings of University of Bristol Spelaeology Society, 14*(3), 233–354.

Heath, T. (1879). *An abstract description and history of the Bone Caves of Creswell Crags*. Derby: Wilson and Ellis.

Hedges, R. M., Pettitt, P. B., Bronk Ramsey, C., & van Klinken, G. J. (1994). Radiocarbon dates from the Oxford AMS system: Datelist 18. *Archaeometry, 36*(2), 337–374.

Hedges, R. M., Pettitt, P. B., Bronk Ramsey, C., & van Klinken, G. J. (1996). Radiocarbon dates from the Oxford AMS system: Datelist 22. *Archaeometry, 38*(2), 391–415.

Henrich, J. (2004). Demography and cultural evolution: Why adaptive cultural processes produced maladaptive losses in Tasmania. *American Antiquity, 69*, 197–218.

Hicks, H. (1867). Discovery of a Hyaena Den, near Laugharne, Carmarthenshire. *Geological Magazine, 4*, 307–309.

Hicks, H. (1884). On some recent researches in born caves in Wales. *Proceedings of the Geological Association, 9*, 1–42.

Higham, T., Jacobi, R. M., & Bronk Ramsey, C. (2006). AMS radiocarbon dating of ancient bone using ultrafiltration. *Radiocarbon, 48*(2), 179–195.

Hofman, J. (2003). Tethered to stone or free to move: Folsom biface technology in regional perspective. In M. Soressi & H. L. Dibble (Eds.), *Multiple approaches to the study of biface technology* (pp. 229–250). Philadelphia: University of Pennsylvania Museum of Archaeology and Anthropology.

Holdaway, S., McPherron, S., & Roth, B. (1996). Notched tool reuse and raw material availability in French Middle Palaeolithic sites. *American Antiquity, 61*(2), 377–387.

Hublin, J.-J., & Roebroeks, W. (2009). Ebb and flow or regional extinctions? On the character of Neanderthal occupation of northern environments. *Comptes Rendues Palevol, 8*, 503–509.

Hunt, C. O. (1989). Molluscs from A. L. Armstrong's excavations in Pin Hole Cave, Creswell crags. *Cave Science, 16*(3), 97–100.

Jacobi, R. M. (2004). Some observations on the non-flint lithics from Creswell Crags. *Lithics, 25*, 39–64.

Jacobi, R. M. (2007). The Stone Age Archaeology of Church Hole, Creswell Crags, Nottinghamshire. In P. Pettitt, P. Bahn, & S. Ripoll (Eds.), *Palaeolithic Cave Art at Creswell Crags in European Context* (pp. 75–111). Oxford: Oxford University Press.

Jacobi, R. M., & Hawkes, C. J. (1993). Archaeological notes: Work at the Hyaena Den Wookey hole. *Proceedings of University of Bristol Spelaeology Society, 19*(3), 369–371.

Jacobi, R. M., & Grun, R. (2003). ESR dates from Robin Hood Cave, Creswell Crags, Derbyshire, UK, and the age of its Neanderthal occupation. *Quaternary Newsletter of the Quaternary Research Association, 100*, 1–12.

Jacobi, R. M., & Higham, T. (2008). The "Red Lady" ages gracefully: New ultrafiltration AMS determinations from Paviland. *Journal of Human Evolution, 55*, 898–907.

Jacobi, R. M., Rowe, P. J., Gilmour, M. A., Rainer, G., & Atkinson, T. C. (1998). Radiometric dating of the Middle Palaeolithic tool industry and associated fauna of Pin Hole Cave, Creswell Crags, England. *Journal of Quaternary Science, 13*(1), 29–42.

Jacobi, R. M., Higham, T., & Bronk Ramsey, C. (2006). AMS radiocarbon dating of Middle and Upper Palaeolithic bone in the British Isles: Improved reliability using ultrafiltration. *Journal of Quaternary Science, 21*(5), 557–573.

Jenkinson, R. D. S. (1984). *Creswell Crags. Late Pleistocene Sites in the East Midlands, British archaeological reports* (Vol. 122). Oxford: B.A.R.

Jones, G. T., Beck, C., Jones, E. E., & Hughes, R. E. (2003). Lithic source use and Paleoarchaic foraging territories in the Great Basin. *American Antiquity, 68*(1), 5–38.

Jones, G. T., Fontes, L. M., Horowitz, R. A., Beck, C., & Bailey, D. G. (2012). Reconsidering Paleoarchaic mobility in the central Great Basin. *American Antiquity, 77*, 351–367.

Keegan, W. F., & Diamond, J. (1987). Colonization of islands by humans: A biogeographical perspective. *Advances in Archaeological Method and Theory, 10*, 49–92.

Keith, A. (1926). Kent's Cavern Report. *Report of the British Association for the Advancement of Science*, 326–328.

Keith, A. (1927a). Kent's Cavern Report. *Report of the British Association for the Advancement of Science*, 303–306.

Keith, A. (1927b). Report on a fragment of a human jaw. *Transactions and Proceedings, Torquay Natural History Society, 5*(1), 1–2.

Keith, A. (1928). Kent's Cavern Report. *Report of the British Association for the Advancement of Science*, 434–436.

Keith, A. (1929). Kent's Cavern Report. *Report of the British Association for the Advancement of Science*, 265–266.

Keith, A. (1930). Kent's Cavern Report. *Report of the British Association for the Advancement of Science*, 266–267.

Kelly, R. L. (2003). Colonization of new land by hunter-gatherers: Expectations and implications based on ethnographic data. In M. Rockman & J. Steele (Eds.), *Colonization of unfamiliar landscapes: The archaeology of adaptation* (pp. 44–58). London: Routledge.

Kelly, R. L., & Todd, L. C. (1988). Coming into the country: Early Paleoindian hunting and mobility. *American Antiquity, 53*, 231–244.

Krings, M., Stone, A., Schmitz, R. W., Krainitzki, H., Stoneking, M., von Haeseler, A., Grossschmidt, K., Possnert, G., Paunovic, M., & Pääbo, S. (1997). Neandertal DNA sequences and the origin of modern humans. *Cell, 90*, 19–30.

Krings, M., Capelli, C., Tschentscher, F., Geisert, H., Meyer, S., von Haeseler, A., Grossschmidt, K., Possnert, G., Paunovic, M., & Paabo, S. (2000). A view of Neandertal genetic diversity. *Nature Genetics, 26*, 144–146.

Kuhn, S. L. (1994). A formal approach to the design and assembly of mobile toolkits. *American Antiquity, 59*(3), 426–442.

Kuhn, S. L. (1995). *Mousterian lithic technology*. Princeton: Princeton University Press.

Kuhn, S. L. (2011). Neanderthal technoeconomics: An assessment and suggestions for future developments. In N. J. Conard & J. Richter (Eds.), *Neanderthal lifeways, subsistence and technology: On hundred fifty years of Neanderthal study* (pp. 99–110). New York: Springer.

Laing, R. (1890). On the bone caves of Creswell and discovery of an extinct Pleistocene feline (*Felis brevirostris*) new to Great Britain. *Report of the British Association for the Advancement of Science, 1889*, 582–584.

Lalueza-Fox, C., Sampietro, M., Caramelli, M., Puder, Y., Lari, M., Calafell, F., Martinez-Maza, C., et al. (2005). Neanderthal evolutionary genetics: Mitochondrial DNA data from the Iberian peninsula. *Molecular Biology and Evolution, 22*, 1077–1081.

Launay, S., & Molines, N. (2005). Le Moustérien à outils bifaciaux dans le Massif Armoricain: première lecture technologique. *Bulletin de la Société Préhistorique Française, 102*(4), 837–853.

Laws, E. (1888). *The history of Little England beyond Wales*. London: George Bell and Sons.

Lawson, A. J. (1978). A hand-axe from Little Cressingham. *East Anglian Archaeology, 8*, 1–8.

Lemorini, C., Peresani, M., Rossetti, P., Malerba, G., & Giacobini, G. (2003). Techno- morphological and use-wear functional analysis: An integrated approach to the study of a discoid industry. In M. Peresani (Ed.), *Discoid lithic technology: Advancements and implications, BAR International Series* (Vol. 1120, pp. 257–275). Oxford: Archaeopress.

Lewis, M. (2010). Pleistocene hyaena coprolite palynology in Britain: Implications for the environments of early humans. In N. Ashton, S. Lewis, & C. Stringer (Eds.), *Ancient human occupation of Britain* (pp. 263–278). London: Elsevier.

Lewis, S., Ashton, N., & Jacobi, R. (2010). Testing human presence during the last interglacial (MIS 5e): A review of the British evidence. In N. Ashton, S. Lewis, & C. Stringer (Eds.), *Ancient human occupation of Britain* (pp. 125–164). London: Elsevier.

Locht, J.-L. (2005). Le Paléolithique Moyen en Picardie : état de la recherche. *Revue Archéologique de Picardie, 3*(4), 29–38.

Locht, J.-L., & Swinnen, C. (1994). Le débitage discoïde du gisement de Beauvais (Oise): aspects de la chaîne opératoire au travers de quelques remontages. *Paléo, 6*, 89–104.

Locht, J.-L., & Antoine, P. (2001). Caractérisation techno-typologique et position chronostratigraphique de plusiers industries à rares bifaces ou amincissements bifaciaux du nord de la France. In D. Cliquet (Ed.), *Les industries à outils bifaciaux du Paléolithique moyen d'Europe occidentale. Actes de la table-ronde internationale organisée à Caen (Basse-Normandie – France) (14 et 15 October 1999), ERAUL* (Vol. 98, pp. 129–134). Liège: Université de Liège.

Locht, J.-L., & Depaepe, P. (2011). Regards sur le Paléolithique moyen de France septentrionale et de Belgique. In *Le Paleolithique moyen en Belgique. Mélanges Marguerite Ulrix-Closset, Bulletin des Chercheurs de La Wallonie, hors-serie, 4*, 229–237.

Locht, J.-L., Swinnen, C., Antoine, P., Auguste, P., Pathou-Mathis, M., Depaepe, P., Falguères, C., Laurent, M., Bahain, J.-J., & Mathys, P. (1995). Le gisement paléolithique moyen de Beauvais (Oise). *Bulletin de la Société Préhistorique Francaise, 92*(2), 213–226.

Locht, J.-L., Goval, E., & Antoine, P. (2010). Reconstructing Middle Palaeolithic homind behaviour during OIS 5 in northern France. In N. J. Conard & A. Delagnes (Eds.), *Settlement dynamics of the Middle Palaeolithic and Middle Stone Age* (Vol. 3, pp. 329–355). Tubingen: Kerns Verlag.

Locht, J.-L., Hérisson, D., Goval, E., Cliquet, D., Huet, B., Coutard, S., Antoine, P., & Feray, P. (2016). Timescales, space and culture during the Middle Palaeolithic in northwestern France. *Quaternary International, 411A*, 129–148.

McNabb, J. (2007). *The British Lower Palaeolithic: Stones in Contention*. London: Routledge.

Meignen, L., Delagnes, A., & Bourgignon, L. (2009). Patterns of lithic material procurement and transformation during the Middle Palaeolithic in Western Europe. In B. Adams & B. S. Blades (Eds.), *Lithic materials and Palaeolithic societies* (pp. 15–24). London: Blackwell.

Mello, J. M. (1875). On some bone caves in Creswell Crags. *Quarterly Journal of the Geological Society of London, 31*, 679–691.

Mello, J. M. (1876). On the bone caves of Creswell. *Quarterly Journal of the Geological Society of London, 32*, 240–259.

Mello, J. M. (1877). The bone caves of Creswell Crags. *Quarterly Journal of the Geological Society of London, 33*, 579–588.

Michel, V., Locht, J. L., Antoine, P., Masaoudi, H., Falgueres, C., & Yokoyama, Y. (1999). U/ Th/ ESR combined dating of faunal remains from the Mousterian Beauvais open-air site, France. *Comptes Rendus de l'Academie des Sciences Serie II Sciences de la Terre et des Planetes, 329*(5), 369–375.

Molines, N., Hinguant, S., & Monnier, J.-L. (2001). Le Paléolithique Moyen à outils bifaciaux dans l'ouest de la France: synthèse des données anciennes et récentes. In D. Cliquet (Ed.), *Les industries à outils bifaciaux du Paléolithique moyen d'Europe occidentale. Actes de la table-ronde internationale organisée à Caen (Basse-Normandie – France) (14 et 15 October 1999), ERAUL* (Vol. 98, pp. 107–112). Liège: Université de Liège.

Mortimore, R. N., Wood, C. J., & Gallois, R. W. (2001). British Upper Cretaceous stratigraphy. *Geological Conservation Review Series, 23*.

Mourre, V. (2003). Discoïd ou pas discoid? Réflexions sur la pertinence de critères techniques définissant le débitage discoïd. In M. Peresani (Ed.), *Discoid lithic technology: Advances and implications* (pp. 1–18). Oxford: British Archaeological Reports.

Muthukrishna, M., Shulman, B. W., Vasilescu, V., & Henrich, J. (2013). Sociality influences cultural complexity. *Proceeding of the Royal Society B, 281*(1774).

Nelson, M. C. (1991). The study of technological organisation. In M. B. Schiffer (Ed.), *Archaeological method and theory, Vol. 3* (pp. 57–100). Tucson: University of Arizona Press.
Ogilvie, A. H., & Tebbs, B. N. (1938). Kent's Cavern. *Report of the British Association for the Advancement of Science, 347.*
Paterson, T. T., & Tebbutt, C. F. (1947). Studies in the Palaeolithic succession in England no III palaeoliths from St Neots, Huntingdonshire. *Proceedings of the Prehistoric Society, 13*, 37–46.
Pengelly, W. (1868). The literature of Kent's Cavern, Torquay, prior to 1859. *Transactions of Devonshire Association, 2*(2), 469–522.
Pengelly, W. (1869a). The literature of Kent's Cavern, part II. *Transactions of Devonshire Association, 3*, 191–482.
Pengelly, W. (1869b). Fifth report of the committee for exploring Kent's Cavern, Devonshire. *Report of the British Association for the Advancement of Science*, 189–208.
Pengelly, W. (1870). Sixth report of the committee for exploring Kent's Cavern, Devonshire. *Report of the British Association for the Advancement of Science*, 189–208.
Pengelly, W. (1871). Seventh report of the committee for exploring Kent's Cavern, Devonshire. *Report of the British Association for the Advancement of Science*, 16–29.
Pengelly, W. (1872). Eighth report of the committee for exploring Kent's Cavern, Devonshire. *Report of the British Association for the Advancement of Science*, 1–14.
Pengelly, W. (1873). Ninth report of the committee for exploring Kent's Cavern, Devonshire. *Report of the British Association for the Advancement of Science*, 28–47.
Pengelly, W. (1874). Tenth report of the committee for exploring Kent's Cavern, Devonshire. *Report of the British Association for the Advancement of Science*, 198–209.
Pengelly, W. (1875). Eleventh report of the committee for exploring Kent's Cavern, Devonshire. *Report of the British Association for the Advancement of Science*, 1–17.
Pengelly, W. (1876). Twelfth report of the committee for exploring Kent's Cavern, Devonshire. *Report of the British Association for the Advancement of Science*, 1–13.
Pengelly, W. (1877). Thirteenth report of the committee for exploring Kent's Cavern, Devonshire. *Report of the British Association for the Advancement of Science*, 1–8.
Pengelly, W. (1878). Fourteenth report of the committee for exploring Kent's Cavern, Devonshire. *Report of the British Association for the Advancement of Science*, 1–8.
Pengelly, W. (1879). Fifteenth report of the committee for exploring Kent's Cavern, Devonshire. *Report of the British Association for the Advancement of Science*, 1–7.
Pengelly, W. (1880). Sixteenth and final report of the committee for exploring Kent's Cavern, Devonshire. *Report of the British Association for the Advancement of Science*, 1–7.
Peresani, M. (Ed.). (2003a). *Discoid lithic technology. Advances and implications, BAR International Series* (Vol. 1120). Oxford: Archaeopress.
Peresani, M. (2003b). Preface. In M. Peresani (Ed.), *Discoid lithic technology. Advances and implications, BAR International Series* (Vol. 1120, pp. i–iv). Oxford: Archaeopress.
Piniot, L. (2001). Analyse typo-technologique du gisement de Bois l'Abbé (Saint Julien de la Liègue, Eure). In D. Cliquet (Ed.), *Les industries à outils bifaciaux du Paléolithique moyen d'Europe occidentale. Actes de la table-ronde internationale organisée à Caen (Basse-Normandie – France) (14 et 15 octobre 1999), 2001, ERAUL* (Vol. 98, pp. 85–91). Liège: Université de Liège.
Powell, A., Shennan, S. J., & Thomas, M. G. (2009). Late Pleistocene demography and the appearance of modern human behaviour. *Science, 324*, 1298–1301.
Premo, L. S., & Kuhn, S. L. (2010). Modeling effects of local extinctions on culture change and diversity in the Paleolithic. *PloS One, 5*(12), 1–10.
Proctor C. J. (1994). *A British Pleistocene chronology based on U-series and ESR dating of speleothems.* Unpublished Ph.D. Thesis, University of Bristol.
Proctor, C. J., & Smart, P. L. (1989). A new survey of Kent's Cavern, Devon. *Proceedings of University of Bristol Spelaeology Society, 18*(3), 422–429.
Proctor, C. J., Collcutt, S. N., Currant, A. P., Hawkes, C. J., Roe, D. A., & Smart, P. L. (1996). A report on the excavations at Rhinoceros Hole, Wookey. *Proceedings of University of Bristol Spelaeology Society, 20*(3), 237–262.
Read, C. (1910). Ravencliffe cave. II. *Derbyshire Archaeological Journal, 32*, 147–151.

Rendu, W., Costamagno, S., Meignen, L., & Soulier, M.-C. (2012). Monospecific faunal spectra in Mousterian contexts: Implications for social behavior. *Quaternary International, 247*, 50–58.

Richards, M. P., & Trinkaus, E. (2009). Isotopic evidence for the diets of European Neanderthals and early modern humans. *Proceedings of the National Academy of Sciences, 106*(38), 16034–16039.

Rockman, M. H. (2003) *Landscape learning in the Late Glacial Recolonization of Britain*. Unpublished Ph.D. Thesis, University of Arizona.

Roe, D. A. (1981). *The Lower and Middle Palaeolithic periods in Britain*. London: Routledge and Kegan Paul.

Roebroeks, W. (2003). Landscape learning and the earliest peopling of Europe. In M. Rockman & S. Steele (Eds.), *Colonization of unfamiliar landscapes: The archaeology of adaptation* (pp. 99–115). London: Routledge.

Roebroeks, W., & Gamble, C. (1999). *The Middle Palaeolithic occupation of Europe*. Leiden: Faculty of Archaeology, University of Leiden.

Roebroeks, W., Hublin, J.-J., & MacDonald, K. (2011). Continuities and discontinuities in Neandertal presence – A closer look at northwestern Europe. In N. Ashton, S. Lewis, & C. Stringer (Eds.), *Ancient human occupation of Britain* (pp. 111–121). London: Elsevier.

Rogers, E. H. (1956). Stratification of the cave earth in Kent's Cavern. *Proceedings of the Devon Archaeological Exploration Society, 5*, 68–92.

Roth, B., & Dibble, H. L. (1998). The production and transport of blanks and tools at the French Middle Paleolithic site of Combe-Capelle Bas. *American Antiquity, 63*(1), 47–62.

Ruebens, K. (2013). Regional behaviour among late Neanderthal groups in Western Europe: A comparative assessment of late Middle Palaeolithic bifacial tool variability. *Journal of Human Evolution, 65*(4), 341–362.

Ruebens, K., & Van Peer, P. (2011). A Middle Palaeolithic site with small bifaces at Oosthoven-Heieinde (Northern Belgium). In M. Toussaint, K. Di Modica, & S. Pirson (Eds.), *Le Paléolithique moyen en Belgique. Mélanges Marguerite Ulrix-Closset, ERAUL 128, Bulletin des Chercheurs de la Wallonie Hors-série No 4* (pp. 353–359). Liège: Université de Liège.

Ruebens, K., & Wragg Sykes, R. M. (2015). Spatio-temporal variation in late Middle Palaeolithic Neanderthal behaviour: British bout coupé handaxes as a case study. *Quaternary International, 411*, 305–326.

Scott, B. (2006) *The Early Middle Palaeolithic of Britain; origins, technology and landscape*. Unpublished Ph.D. Thesis, University of Durham.

Scott, B. (2010). *Becoming Neanderthals. The Earlier British Middle Palaeolithic*. Oxford: Oxbow Books.

Shennan, S. J. (2001). Demography and cultural innovation: A model and some implications for the emergence of modern human culture. *Cambridge Archaeological Journal, 11*, 5–16.

Shennan, S. J., & Steele, J. (1999). Cultural learning in hominids: A behavioural ecological approach. In H. Box & K. Gibson (Eds.), *Mammalian social learning* (pp. 367–388). Cambridge: Cambridge University Press.

Slimak, L. (2003). Les débitages discoïdes moustériens: évaluation d'un concept technologique. In L. Slimak & Y. Giraud (Eds.), *Circulations sur Pleusieurs de Kilometres Durant le Paléolithique Moyen, Contribution à la Connaissance des Sociétés Néandertaliennes. Comptes Rendues Palevol, 6*, 359–368.

Sollas, W. J. (1913). Paviland Cave: An Aurignacian station in Wales. *Journal of the Royal Anthropological Institute, 43*, 325–374.

Soressi, M. (2002). *Le Moustérien deTtradition Acheuléenne du Sud-Ouest de la France: Discussion sur la Signification du Faciès à Partir de l'Étude Compare de Quatre Sites: Pech-de-L'Azé I, Le Moustier, La Rochette et la Grotte XVI*. Unpublished Ph.D. Thesis, l'Université de Bordeaux I.

Spiess, A., Wilson, D., & Bradley, J. (1998). Paleoindian occupation in the New England Maritimes region: Beyond cultural ecology. *Archaeology of Eastern North America, 26*, 201–264.

Stevens, E. T. (1870). *Flint chips: A guide to pre-historic archaeology, as illustrated by the collection in the Blackmore Museum, Salisbury*. London: Bell and Daldy.

Stewart, J. R., & Lister, A. M. (2001). Cryptic northern refugia and the origins of modern biota. *Trends in Ecology and Evolution, 16*, 608–613.
Storrs Fox, W. (1908). [No title]. *Proceedings of the Society of Antiquaries, 22*, 129–139.
Storrs Fox, W. (1910). Ravencliffe Cave. I. *Derbyshire Archaeological Journal, 32*, 71–146.
Storrs Fox, W. (1929). Ravencliffe Cave. *Derbyshire Archaeological Journal, 3*, 71–78.
Straw, A. (1996). The Quaternary record of Kent's Cavern – A brief reminder and update. *Quaternary Newsletter, 80*, 17–25.
Svensson, A., Andersen, K. K., Bigler, M., Clausen, H. B., Dahl-Jensen, D., Davies, S. M., Johnsen, S. J., Muscheler, R., Rasmussen, S. O., Rothlisberger, R., Steffensen, J. P., & Vinther, B. M. (2006). The Greenland ice core chronology 2005, 15–42 ka. Part 2: Comparison to other records. *Quaternary Science Reviews, 25*(23–24), 3258–3267.
Swainston, S. (1999). Unlocking the inhospitable. In W. Davies & R. Charles (Eds.), *Dorothy Garrod and the progress of the Palaeolithic* (pp. 41–56). Oxford: Oxbow Books.
Swainston, S. (2000). The lithic artefacts from Paviland. In S. Aldhouse-Green (Ed.), *Paviland Cave and the 'Red Lady': A definitive report* (pp. 95–113). Bristol: Western Academic and Specialist Press.
Teheux, E. (2000). Observations préliminaires sur le site de La Minette à Fitz-James (Oise). *Archéopages, 1*, 30–37.
Terradas, X. (2003). Discoid flaking method: Conception and technological variability. In M. Peresani (Ed.), *Discoid lithic technology. Advances and implications, BAR international series* (Vol. 1120, pp. 19–32). Oxford: Archaeopress.
Tissoux, H., 2006. La séquence loessique pléistocène supérieur de Savy (Aisne France): stratigraphie, datations et occupations paléolithiques. *Quaternaire, 1, 3* , 269–275.
Torrence, R. (1989). Retooling: Towards a behavioral theory of stone tools. In R. Torrence (Ed.), *Time, energy and stone tools* (pp. 57–66). Cambridge: Cambridge University Press.
Tratman, E. K. (1964). Picken's Hole, Crook Peak, Somerset. A Pleistocene site. Preliminary note. *Proceedings of the University of Bristol Speleological Society, 10*, 112–115.
Tratman, E. K., Donovan, D. T., & Campbell, J. B. (1971). The Hyaena Den (Wookey Hole), Mendip Hills, Somerset. *Proceedings of the University of Bristol Speleological Society, 12*(3), 245–279.
Treacher, L. (1934). Field meeting in the Marlow district. *Proceedings of the Geologists' Association, 45*, 107–108.
Tuffreau, A. (1979). Le gisement moustérien du château d'eau de Corbehem (Pas-de-Calais). *Gallia Préhistoire, 2*, 371–389.
Turq, A. (2001). Réflexions sur le biface dans quelques sites Paléolithique Ancien-Moyen en grottes ou abri du Nord-Est du Bassin Aquitain. In D. Cliquet (Ed.), *Les industries à outils bifaciaux du Paléolithique moyen d'Europe occidentale. Actes de la table-ronde internationale organisée à Caen (Basse-Normandie France) (14 et 15 October 1999), 2001, ERAUL* (Vol. 98, pp. 141–148). Liège: Université de Liège.
Turq, A., Wilbroeks, W., Bourgignon, L., & Faivre, J.-P. (2013). The fragmented character of Middle Palaeolithic stone tool technology. *Journal of Human Evolution, 65*(5), 645–655.
Tyldesley, J. A. (1987). *The Bout Coupé Handaxe. A typological problem* (Vol. 170). Oxford: British Archaeological Reports.
Uthmeier, T., Kels, H., Schirmer, W., & Böhner, U. (2011). Neanderthals in the cold: Middle Palaeolithic sites from the open cast mine of Garzweiler, Nordrhein-Westfalen (Germany). In N. Conard & J. Richter (Eds.), *Neanderthal lifeways, subsistence and technology: On hundred fifty years of Neanderthal study* (pp. 25–41). New York: Springer.
van Andel, T. H., & Davies, W. (Eds.). (2003). *Neanderthals and modern humans in the European landscape during the last glaciation: Archaeological results of the stage 3 project*. Cambridge: McDonald Institute Monographs.
Van Meerbeeck, C. J., Renssen, H., & Roche, D. N. (2009). How did Marine Isotope Stage 3 and Last Glacial Maximum climates differ? Perspectives from equilibrium simulations. *Climate of the Past, 5*, 33–51.

Vivian, E. (1859). *Cavern researches, or, discoveries of organic remains, and of British and Roman Reliques, in the Caves of Kent's Hole, Anstis Cove, Chudleigh, and Berry Head; by the Late Rev. J. MacEnery, F.G.S.* London: Marshall.

Wardle, F. C. (1919). The Coygan Bone Cave. *Transactions of the Carmarthenshire Antiquarian Society, 13*, 50–52.

Wenban-Smith, F., Bates, M., & Schwenninger, J.-L. (2010). Early Devensian occupation (MIS 5d-5b) at Dartford, Southeast England. *Journal of Quaternary Science, 25*(8), 1193–1199.

White, M. J. (2006). Things to do in Doggerland when you're dead: Surviving OIS3 at the northwestern-most fringe of Middle Palaeolithic Europe. *World Archaeology, 38*(4), 547–575.

White, M. J., & Jacobi, R. M. (2002). Two sides to every story: *bout coupé* handaxes revisited. *Oxford Journal of Archaeology, 21*(2), 109–133.

White, M. J., & Ashton, N. (2003). Lower Palaeolithic core technology and the origins of the Levallois method in North-Western Europe. *Current Anthropology, 44*(4), 598–609.

White, M. J., Scott, B., & Ashton, N. (2006). The Early Middle Palaeolithic in Britain: Archaeology, human settlement history and technological organisation. *Journal of Quaternary Science, 21*(5), 525–542.

White, M. J., & Pettitt, P. B. (2011). The British Late Middle Palaeolithic: An interpretative synthesis of Neanderthal occupation at the northwestern edge of the Pleistocene world. *Journal of World Prehistory, 24*(1), 25–97.

Willis, K. J., Rudener, E., & Sümegi, P. (2000). The full-glacial forests of central and southeastern Europe. *Quaternary Research, 53*, 203–213.

Willis, K. J., & van Andel, T. H. (2004). Trees or no trees? The environments of central and eastern Europe during the Last Glaciation. *Quaternary Science Reviews, 23*, 2369–2387.

Wilson, E., & Reynolds, S. H. (1901). The Uphill Bone Caves. *Proceedings of the Bristol Natural History Society, 9*(3), 152–160.

Wobst, M. H. (1978). Boundary conditions for Paleolithic social systems: A simulation approach. *American Antiquity, 39*(2), 147–178.

Wragg Sykes, R. M. (2009). *The British Mousterian: Late Neanderthal Archaeology in Landscape Context*. Unpublished Ph.D. Thesis, University of Sheffield. Available at https://univ-bordeaux.academia.edu/RebeccaWraggSykes

Wragg Sykes, R. M. (2010). Beyond bout coupés: The dynamic role of bifaces in the British Mousterian. *Lithics, 31*, 20–32.

Wragg Sykes, R. M. (in press). The Mousterian lithic assemblage from Picken's Hole, Compton Bishop, Somerset. *Proceedings of the University of Bristol Spelaeological Society*.

Wragg Sykes, R. M. 2012. Neanderthals 2.0? Evidence for expanded social networks, ethnic diversity and encultured landscapes in the Late Middle Palaeolithic. *Unravelling the Palaeolithic–10 Years of Research at the Centre for the Archaeology of Human Origins (CAHO, University of Southampton)* (pp. 73–84).

Wragg Sykes, R. M. (2017). Neanderthals in the Outermost West: Technological adaptation in the Late Middle Palaeolithic (re)-colonization of Britain, Marine Isotope Stage 4/3. *Quaternary International, 433*(Part B), 4–32.

Wymer, J. J. (1968). *Lower Palaeolithic Archaeology in Britain*. London: John Baker Publishers.

Wymer, J. J. (1988). Palaeolithic archaeology and the British Quaternary Sequence. *Quaternary Science Reviews, 7*, 79–98.

Wymer, J. J. (1999). *The lower Palaeolithic occupation of Britain*. London: Wessex Archaeology and English Heritage.

Chapter 8
Proto-Aurignacian Lithic Technology, Mobility, and Human Niche Construction: A Case Study from Riparo Bombrini, Italy

Julien Riel-Salvatore and Fabio Negrino

Introduction

There is in anthropological archaeology a long history of research stretching to Steward and beyond that has sought to establish links between hunter-gatherer adaptations and their environments. In fact, recent synthetic analyses of the lifeways of ethnographic forager groups conducted by archaeologists (e.g., Kelly 1995; Binford 2001) have made some of these links quite clear, especially as they relate to issues of mobility and concomitant technological organization strategies. Given the ubiquity of arguments about how forager adaptations are deeply rooted in their local ecologies in the archaeological literature, it has even been said that the ability to quickly change technological patterns is a defining feature of what it means to be a "modern" human (Mellars 2005: 13). Considering this, it is surprising that most discussions of variability in the Aurignacian have largely limited themselves to defining culture historical facies of the techno-complex rather than trying to understand, first, whether there is internal variability within individual facies and, if so, what the root causes of this variability might be.

The Proto-Aurignacian (also known as archaic Aurignacian or Aurignacian 0) is now widely agreed to be one of the earliest manifestations of the Aurignacian and, by extension, of putatively modern human adaptations (Hublin 2015; Benazzi et al. 2015). Documented in different parts of Europe, the earliest Proto-Aurignacian is now generally thought to be that of the Balzi Rossi site complex in the region of Liguria (NW Italy), where it has been identified in multiple levels at the sites of Riparo Mochi and Riparo Bombrini (cf. Szmidt et al. 2010; Anderson et al. 2015).

J. Riel-Salvatore (✉)
Département d'anthropologie, Université de Montréal, Montreal, Canada
e-mail: julien.riel-salvatore@umontreal.ca

F. Negrino
Dipartimento di antichità, filosofia, storia e geografia, Università di Genova, Genova, Italy

E. Robinson, F. Sellet (eds.), *Lithic Technological Organization and Paleoenvironmental Change*, Studies in Human Ecology and Adaptation 9,
DOI 10.1007/978-3-319-64407-3_8

However, due to the history of how the published Proto-Aurignacian assemblages from Mochi were recovered and subsequently analyzed, they have only ever been studied as a whole (Laplace 1977; Palma di Cesnola 1993; Kuhn and Stiner 1998). This has impeded explorations of the internal variability of the Proto-Aurignacian at that site. The identification of two distinct Proto-Aurignacian layers at Riparo Bombrini during recent excavations thus offers a unique opportunity to tackle this issue head-on and to provide one of the first empirical analyses of the internal variability of the earliest manifestations of the Aurignacian in Europe.

The fundamental question this paper addresses is whether internal variability in the Proto-Aurignacian at the site of Riparo Bombrini can be correlated in any meaningful way to paleoenvironmental change. Clarifying this question will help shed light on what other factors might be worth investigating to explain this variability, if ecological factors only had a negligible impact. After a description of the site and its research history, the archaeological record of Proto-Aurignacian sequence is described in detail. This is followed by an in-depth discussion of the variability documented in the lithic assemblages in the site's two Proto-Aurignacian assemblages, and how this correlates to other sequences where multiple Proto-Aurignacian levels are known. The lithic and paleoenvironmental datasets are then combined, to discuss whether the two can be shown to covary in a coherent manner. The paper closes with a discussion of what the patterns evidenced at Bombrini mean for our understanding of the Proto-Aurignacian specifically, and more generally the dispersal of Homo sapiens in that part of Europe. This discussion includes considerations of previous models of diffusion and explores whether the internal variability evident in the Proto-Aurignacian can be reconciled with recent scenarios that posit that some facets of the archaeological record can provide evidence of human niche construction in the Pleistocene.

Riparo Bombrini: An Overview

Riparo Bombrini has recently come to the fore in discussions of the Middle-Upper Paleolithic transition. This is due to it having yielded some of the most recent Mousterian deposits known in Eurasia (Higham et al. 2014) as well as very early Proto-Aurignacian deposits (Benazzi et al. 2015) just slightly later than those from Riparo Mochi where the earliest Proto-Aurignacian is documented (Douka et al. 2012). Additionally, in contrast to Mochi where the Proto-Aurignacian levels have always been studied as a whole (Kuhn and Stiner 1998; Alhaique et al. 2000; Bertola et al. 2013), Riparo Bombrini comprises two distinct and well-dated Proto-Aurignacian levels. This permits an empirical evaluation of the internal variability of this techno-complex, which has only rarely been possible with most other Italian Proto-Aurignacian assemblages for a variety of reasons.

Riparo Bombrini is a collapsed rock-shelter and part of the Balzi Rossi site complex in coastal Liguria, immediately next to the French border (Fig. 8.1). Several decades after its initial discovery in the late nineteenth century, Cardini (1938)

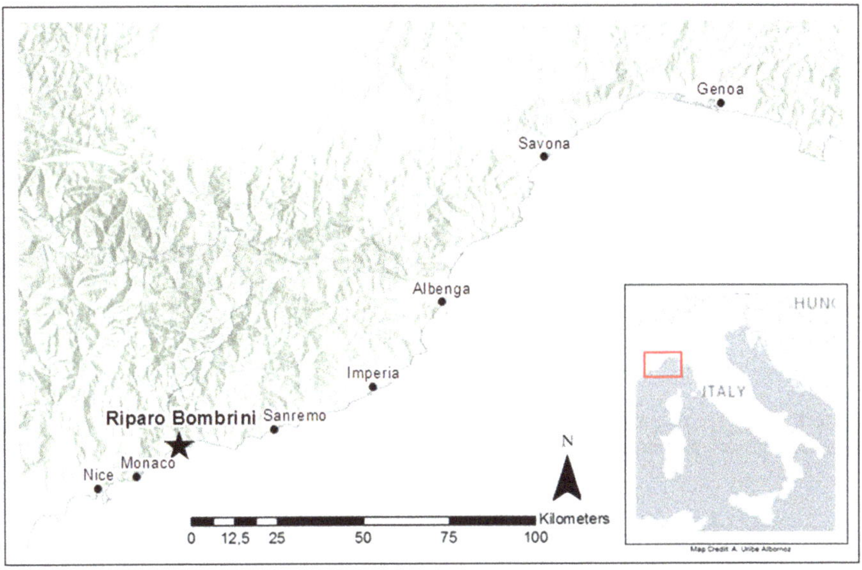

Fig. 8.1 Location of Riparo Bombrini, in Liguria near the French-Italian border

excavated a test pit near the base of the cliff that revealed the presence of hearths and of a lithic assemblage rich in bladelets, subsequently attributed to the Aurignacian *sensu lato*. In 1976, an excavation undertaken to build a walkway linking the Balzi Rossi Museum to the other caves of the site complex identified a longer stratigraphy that included Late Mousterian and Proto-Aurignacian levels, both of which yielded abundant lithic and faunal artifacts, along with several ornaments, ochre chunks, and decorated objects in the Proto-Aurignacian (Vicino 1984). Vicino's excavations also brought to light a deciduous incisor attribute to a *Homo sapiens* individual (Formicola 1984, 1989; Benazzi et al. 2015), making it one of the few early Upper Paleolithic modern human remains found in a reliable context.

From 2002 to 2005, the site was excavated anew by Negrino and colleagues to clarify the context of the Middle-Upper Paleolithic transition in this region (Del Lucchese et al. 2004; Negrino 2005; Holt et al. 2006; Bietti and Negrino 2008; Del Lucchese and Negrino 2008). Their work identified three sedimentary macro-units (Fig. 8.2), one of which comprises two distinct Proto-Aurignacian layers (levels A1 and A2), while the other two (MS 1–2 and M1–7) document the Late Mousterian. While some of the recently excavated material is still under study, some completed analyses have demonstrated the presence of distinct activity areas in the Mousterian levels (Riel-Salvatore et al. 2013) and that both the Mousterian and Proto-Aurignacian industries appear to show an ability to adjust mobility strategies along something akin to a forager-collector continuum (Riel-Salvatore 2007, 2010). In contrast, raw material provisioning strategies differ markedly, with the Mousterian showing a predominantly local and circum-local procurement pattern with very rare exotic elements, while the Proto-Aurignacian shows up to 20% of exotic lithotypes and little use of even high-quality circum-local stone (Riel-Salvatore 2007; Riel-

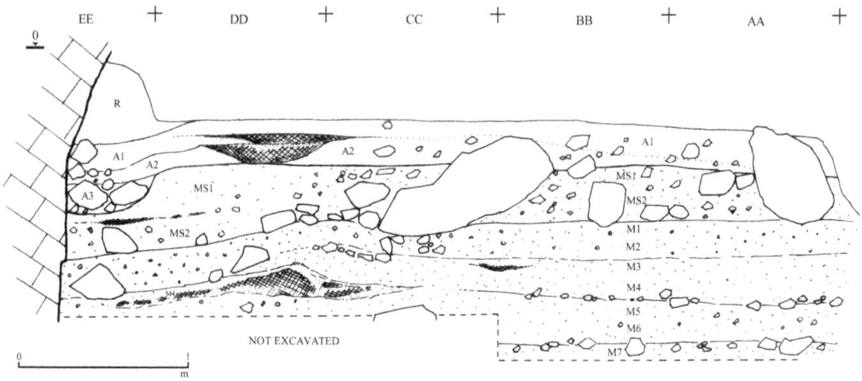

Fig. 8.2 Stratigraphic profile of Riparo Bombrini. *A* Proto-Aurignacian levels, *MS* semi-sterile Mousterian levels, *M* Mousterian levels, *R* disturbed deposits (Drawn by Fabio Negrino)

Salvatore and Negrino 2009; Bertola et al. 2013). In addition the Late Mousterian levels show low-frequency presence of ochre (the exact significance of which remains to be explored) and of shellfish exploitation (Riel-Salvatore et al. 2013), while the Proto-Aurignacian levels show conspicuous evidence of ochre, as well as of incised steatite, notched bird bones, abundant shell ornaments, and an osseous industry comprising awls, needles, and points (Bertola et al. 2013, cf. Vicino 1984).

It should be noted that, as mentioned above, Riparo Bombrini is part of the Balzi Rossi site complex. As such, in the Paleolithic, it was at the eastern end of what was likely a large talus slope that opened in front of Grotta del Caviglione, a large and conspicuous wedge-shaped cave that gouges the Balzi Rossi cliff, and also spread westward to include what is today Riparo Mochi. Thus, the recognition of these different points as different sites is, in part, an artifact of both geomorphology and the history of human landscape development in the region over the past 150 years. It is thus very likely that this paleo-talus area was not considered as distinct occupation sites by its Paleolithic occupants and the site nomenclature is to be understood as an accident of history. This also means that people occupying Riparo Bombrini more than 40 ka ago may have also concurrently occupied Riparo Mochi (there was likely no neat and definite spatial separation between the two areas when they were occupied in the Proto-Aurignacian). That said, because the two rock-shelters are separated by some 50 m and would have sat at opposite ends of a single talus slope that peaked at Grotta del Caviglione (an imposing visual landmark), a case can be made that they were also not part of a single occupation site. It is more likely that they were occupied as alternative spots when people visited the Balzi Rossi and that they captured slightly different palimpsests of the proto-Aurignacian. Additionally, because they have been (and continue to be) excavated as distinct sites, using different methods, and at different times, their records are not directly comparable. Studying them separately is, from a strictly logistical standpoint, the best strategy under current circumstances. While this may make the settlement patterns presented below difficult to evaluate, the distinct position of Riparo Bombrini and Riparo Mochi relative to Caviglione's entrance and their location at two opposite ends of its paleo-talus mean that it is probably safe, for the moment, to develop interpretations about the record of each site separately, even if only to provide empirically grounded and explicitly testable working hypotheses on which to base future work on the Proto-Aurignacian at the Balzi Rossi. Ongoing fieldwork at both sites will in fact soon permit this, as well as an integrated synthesis of the earliest modern human behavioral and technological strategies at the site.

In sum, Riparo Bombrini provides a unique, high-resolution view into the internal dynamics of the Proto-Aurignacian as a behavioral package, certainly one of the best records currently available in Italy. Having two distinct and well-dated layers attributed to that techno-complex and associated with faunal remains also allows an evaluation of whether (and how) modern humans reacted to climatic variability as they first settled Western Europe.

The Paleoenvironmental Record

The Proto-Aurignacian levels at Riparo Bombrini comprise a 10–20 cm accumulation of yellowish clayey loam containing abundant angular clasts and large vault blocks (see Fig. 8.2), indicating that the Proto-Aurignacian accumulated under conditions that were colder than those that characterized the underlying Mousterian (Bertola et al. 2013; Holt et al. n.d.). The paleoenvironmental record from Riparo Bombrini is based on its radiocarbon chronology (which permits correlations to Greenland Ice Core records), the faunal assemblages recovered during the 2002–2005 excavations, as well as palynological data on samples collected during the same period. Forthcoming microfaunal data will soon expand the lines of evidence on which to base these reconstructions although preliminary analyses broadly conform to the reconstructions proposed below (Holt et al. n.d.).

The available radiocarbon dates for the Proto-Aurignacian at Riparo Bombrini indicate that Level A2 (the earliest one, which yielded the human tooth) dates to roughly 41.3–39.1 cal BP; this interval corresponds to the cold period immediately preceding Heinrich Event 4, based on both the NGRIP Ice Core and the Monticchio lake core (Meese et al. 1997; Allen et al. 2000; cf. van Andel et al. 2003). In contrast, Level A1 dates to 38.3–35.9 cal BP associated with comparatively warmer, though overall still cold and arid conditions (Table 8.1). Thus, the Proto-Aurignacian at Bombrini appears to continue without interruption through the Phlegrean Fields eruption of ca. 40 ka (cf. Giaccio et al. 2006). This agrees with the detailed chronostratigraphy of Level G at Riparo Mochi (Douka et al. 2012), which is comparable to that of Bombrini, where Level A3 probably corresponds to the basal proto-Aurignacian at Mochi (Bertola et al. 2013). The fact that the Proto-Aurignacian both pre- and postdates the age of the Campanian Ignimbrite (i.e., 40,012 GISP2 cal BP; Giaccio et al. 2006) is in itself a significant observation that we will return to later in this analysis (see also Lowe et al. 2012). With reference to the CI tephra, it is important to stress that it has not been identified at Riparo Bombrini, either micro- or macroscopically, meaning that there may be a degree of ambiguity in discussions related to that chrono-stratigraphic marker and the volcanic eruption

Table 8.1 Select AMS radiocarbon dates for the Proto-Aurignacian levels from Riparo Bombrini (from Benazzi et al. 2015)

Sample #	Level	Age	Error	Cal age	Cal error
S-EVA 29022	A1	34,030	260	39,592	937
S-EVA 29026	A1	33,220	240	37,710	679
S-EVA 29023	A1	32,750	230	37,267	669
S-EVA 29021	A1	32,210	150	36,393	447
S-EVA 29017	A2	35,600	310	40,496	845
S-EVA 29015	A2	34,810	280	39,945	825

Calibrated using CalPal (http://www.calpal-online.de/).

Table 8.2 Faunal spectra of Level A1 (*NISP* = 70) and A2 (*NISP* = 27)

Species	A1	A2
Sus scrofa	10%	2.7%
Bison priscus	1.4%	–
Bos/bison	7.1%	21.6%
Capra ibex	12.8%	–
R. rupicapra	7.1%	–
Caprines	31.4%	32.4%
D. dama	2.9%	2.7%
Cervus elaphus	14.2%	16.2%
C. capreolus	4.2%	–
Cervids	7.1%	8.1%
Equids	–	13.5%
Rhinoceros sp.	–	2.7%
V. vulpes	1.4%	–

Data from personal communication by Almudena Arellano

that created it. That said, this marker is absent from other well-studied proto-Aurignacian stratigraphies in northern Italy, including Mochi and Fumane, indicating that it likely did not accumulate in that part of the peninsula, in spite of the important general climatic impact of the eruption.

The faunal assemblages from Bombrini provide another datum to evaluate the chrono-climatic data. While the faunal spectrum documented in the two levels is largely comparable, some of its details bear out the climatic distinction outlined above (Table 8.2). These data obviously need to be taken with circumspection in terms of their ecological implications, given the low NISPs and the potential bias introduced by hunter preferences; however, they can nonetheless provide one of a number of independent checks to the trend drawn from the chrono-climatic data alone. With those caveats in mind, it is interesting to note, in Level A2, the presence of equids and rhinoceros, along with greater frequencies of bovids, which could indicate more open, potentially even steppe-like conditions near the site at that time. In contrast, Level A1 yields much higher frequencies of boar and roe deer, indicating a greater forest cover and overall milder conditions, although the significant presence of ibex nonetheless attests to relatively rigorous conditions. Overall, however, the radiocarbon chronology and faunal spectrum agree to suggest colder conditions in Level A2 than in Level A1.

These trends are further supported by the palynological analysis of Arobba and Caramiello (2009).[1] While that analysis considered the ecology of the Aurignacian as a whole (per Arobba 1984) and attests to an overall trend of aridification near the site during the Proto-Aurignacian, the data presented by the authors allow for a finer-grained reading of the pollen spectra documented in the two Proto-Aurignacian sublevels, with Level A1 being associated with higher proportions of oak and lower frequencies of pine than Level A2, as well as with the only evidence for buckthorn (*Rhamnus*), a temperate/subtropical species.

It should be noted that these data contrast with the overall frequencies of arboreal pollen, which is markedly lower in A1 than A2 (13.4% vs. 31.3%). This apparent discrepancy is likely due to the low overall absolute pollen frequency (11–100 p/g) at the site noted by the authors themselves (Arobba and Caramiello 2009: 43); it will also be the focus of renewed investigation during renewed excavations at Riparo Bombrini conducted by the authors that began in 2015.

In sum, the trends from the pollen data dovetail neatly those from the site's chrono-climatic and faunal records, giving a good deal of confidence in the reconstruction proposed here of Level A2 being associated with colder conditions than Level A1, which is associated with somewhat less rigorous and more forested conditions. Given the sedimentary continuity between the two levels, we can thus infer that there was a warming trend over the course of the Proto-Aurignacian at Riparo Bombrini. This sets the stage for an evaluation of whether changes in a major component of the Proto-Aurignacian behavioral package—lithic technology—can be correlated to this pattern.

Proto-Aurignacian Lithic Technology at Riparo Bombrini

First identified by Laplace (1966) over 50 years ago based on a seminal study of the Mochi lithic assemblage, the Proto-Aurignacian has over the last two decades been recognized as a distinctive and very early expression of the Aurignacian phenomenon (Bon 2006). Often mistakenly presented as an essentially Mediterranean phenomenon (cf. Mellars 2006), the Proto-Aurignacian has also more recently been documented as Far West at the site of Isturitz, in the northern Pyrenees (Szmidt et al. 2010), and potentially as Far East as the Bulgarian site Kozarnika (Tsanova et al. 2012). While some researchers have argued for a clear chronological attribution of the Proto-Aurignacian (e.g., Banks et al. 2013), it is important to underscore that, in

[1] Here, we use our finalized reconstruction of the site stratigraphy to assign Arobba and Caramiello's sample no. four to Level A2 rather than Level A3, the latter being found only in a limited area against the back wall of the shelter (cf. Bertola et al. 2013, Riel-Salvatore et al. 2013). As such, sample nos. four and eight should be considered as reflecting the pollen record of Level A2.

the Italian peninsula, the Proto-Aurignacian essentially is "the Aurignacian," being found at multiple sites in layers dating well after the 39.9 cal BP, cutoff used by Banks and colleagues (e.g., Riparo Bombrini [Benazzi et al. 2015], Riparo Mochi [Douka et al. 2012], Grotta di Fumane [Higham et al. 2009]). Thus, at least in Italy, it is best to consider the Proto-Aurignacian as an archaeological and behavioral adaptation rather than as strictly a chronological phase.

As such, the archaeological signature of the Proto-Aurignacian is generally agreed to be characterized by the overwhelming dominance of bladelet technology in the lithic industry, a relatively scant and unstandardized osseous technology, the conspicuous presence of personal ornaments made on a wide range of materials, and very long-distance lithic raw material transfers (Kuhn 2002; Anderson et al. 2015). While in France, the production sequences documented at a single site are often truncated, they are present in their entirety in some Italian cases (e.g., Fumane; see Broglio et al. 2005), while in the Balzi Rossi, some of the initial preparation clearly occurred at nearby sources (Bertola et al. 2013; Grimaldi et al. 2014). Likewise, the Italian sites reveal uneven frequencies of exotic raw materials, with Riparo Bombrini and Riparo Mochi showing transfers over several hundred kilometers, while Fumane and sites in southern Italy (i.e., Serino, Castelcivita, Paglicci) were characterized by essentially local and circum-local raw material acquisition (Riel-Salvatore and Negrino 2009). Enlarging the geographical focus reveals that the Proto-Aurignacian of Grotte Mandrin, in the Rhône Valley (Slimak et al. 2006), shows transfers over less than 100 km, while at Grotte de l'Observatoire (Monaco), the pattern is extremely similar to that of the Balzi Rossi (Porraz et al. 2010).

Just as the dependence on exotic raw material varies rather significantly, the importance of bladelet technology also varies considerably in the Italian peninsula, being rather underwhelming in the south where they account for an average of ca. 25% of the retouched pieces, in contrast to the north, where they account for ca. 66% (Riel-Salvatore 2010). The south also is associated with markedly lower number of personal ornaments and ochre than the north. It is thus against this backdrop of relative techno-complex heterogeneity that the variability in the Bombrini Proto-Aurignacian must be understood.

The Proto-Aurignacian assemblages from Riparo Bombrini comprise several thousand pieces. They are dominated by shapeless debris less than 1.5 cm in maximal dimension that is excluded from consideration here. This section first presents a description of the technology and typology of these stone tool assemblages, followed by a discussion of artifact curation and raw material procurement patterns. Finally, given the importance of bladelets in Proto-Aurignacian lithic technology, a sample of bladelets from both levels is analyzed to provide more detailed data about the differences between the two levels.

By way of introduction, previous analyses have suggested that the two assemblages display distinct behavioral strategies (Riel-Salvatore 2007, 2010). Specifically, it has been argued on the basis of an assemblage-scale analysis of retouch frequencies and artifact densities that, as a whole, the Proto-Aurignacian is a more curated technology than the Mousterian, likely a reflection of more frequent moves across the landscape. In spite of this general tendency, it was also proposed that Level A2 displayed an expedient lithic technological organization reflecting a logistical land-use strategy, while Level A1 displayed a more curated lithic technological system congruent with a shorter-term residential mobility pattern.

Additionally, it was observed that the Proto-Aurignacian at Riparo Bombrini drew on a range of far-flung sources to provision itself with needed lithic raw material; these sources range several hundred kilometers east and west of the site and form a significant proportion of the overall assemblage (Negrino 2002; Riel-Salvatore and Negrino 2009). While these analyses provide preliminary insights into the distinctiveness of the Proto-Aurignacian as an adaptation and suggest that there is some internal variability within it, these trends have yet to be verified by analyses based on large samples from both levels. Thus, in addition to assessing correlations between paleoenvironmental and archaeological variability, this study also promises to test some of these preliminary observations based on analyses of more representative artifact samples from both levels.

Technology and Typology

Typologically, both assemblages are heavily dominated by various kinds of retouched bladelets, especially Dufour bladelets (Dufour subtype) with the typical curved (though not twisted) profile and semi/steep alternating marginal retouch (Fig. 8.3; Table 8.3). Upper Paleolithic forms, such as endscrapers, burins, and splintered pieces, are present but rare, in contrast to notched and denticulated flakes that make up over 10% of the Level A1 retouched tools. While notched/denticulated elements may be the result of postdepositional processes rather than purposeful manufacture, the fact that they are largely restricted to flake blanks and that they appear in noticeably different frequencies across the two levels suggests that there

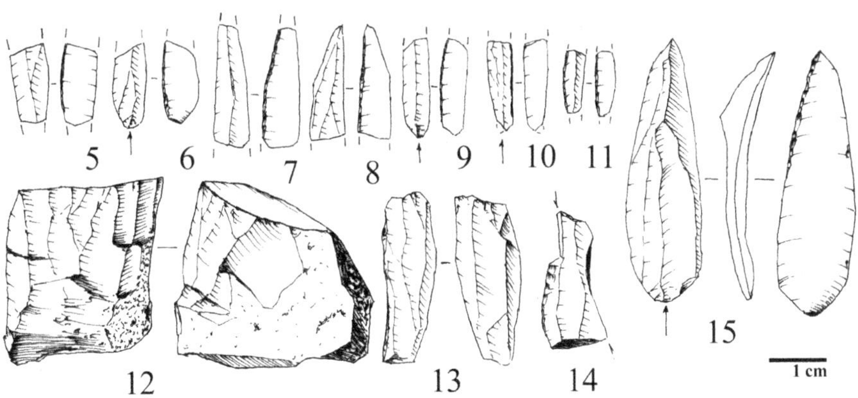

Fig. 8.3 Proto-Aurignacian lithics from Riparo Bombrini. 1–7, Dufour bladelets; 8–9, cores; 10, burin; 11, point

Table 8.3 Typological inventories of the Proto-Aurignacian assemblages from Riparo Bombrini

Retouched tool type	A1	%	A2	%
Dufour bladelets	34	59.7	142	71
Denticulated bladelet	2	3.5	6	3
Retouched bladelet	1	1.8	2	1
Steeply retouched bladelet	0	0	3	1.5
Marginally retouched bladelet	1	1.8	2	1
Bliaterally retouched bladelet	0	0	1	0.5
Endscraper, simple	2	3.5	2	1
Endscraper, marginal	0	0	6	3
Burin	1	1.8	4	2
Dihedral burin	0	0	4	2
Perforator	1	1.8	0	0
Truncation	0	0	1	0.5
Sidescraper	0	0	1	0.5
Sidescraper, marginal	2	3.5	2	1
Sidescraper, atypical	0	0	1	0.5
Notch	3	5.3	5	2.5
Denticulate	4	7	6	3
Splintered piece	1	1.8	6	3
Marginally retouched blade	1	1.8	1	0.5
Denticulated blade	1	1.8	1	0.5
Retouched flake	0	0	4	2
Marginally retouched flake	3	5.3	0	0
Total	*57*	*100*	*200*	*100*

is nonetheless some significance to their different frequency in A1 and A2. Cores are quite rare at Bombrini, with a single amorphous specimen for Level A1, while Level A2 yielded nine cores showcasing a wider range of core forms, including prismatic, bladelet, and bidirectional examples. These patterns echo those identified at Mochi (Kuhn and Stiner 1998; Negrino 2002), the main difference being that blades are very rare at Bombrini, be it in the retouched tools or in the debitage. Lithic production at Bombrini was clearly focused on bladelets (Table 8.3).

From a technological standpoint, then, two distinct *chaînes opératoires* are documented in the Bombrini Proto-Aurignacian. The first, and principal, one is geared toward the production of bladelets from predominantly unidirectional prismatic cores; carenated cores are altogether absent in this context. The resulting bladelets were then generally steeply retouched, often in alternating bilateral fashion resulting in classic Dufour bladelets. Given the absence of blades and larger blade cores at Bombrini, bladelet production appears to have been a definite goal rather than simply the tail end of laminar production on cores of ever-decreasing dimensions. The second *chaîne opératoire* is less coherent and geared toward flake production (including some elongated blade-like blanks), raising the possibility that flake production was a secondary product of blade production as core reduction advanced. These unstandardized supports were occasionally retouched to make notches, denticulates, sidescrapers, and splintered pieces.

Table 8.4 Basic breakdown of Riparo Bombrini Proto-Aurignacian lithic assemblages

Level	Total lithics	Total > 1.5 cm	Retouched tools	Cores	Debitage	Unretouched bladelets	Technological elements
A1	947	448	57	1	390	225	35
A2	7027	2395	202	9	2184	1209	126

Overall, while the typological inventory of Level A1 is poorer than that of Level A2 (Table 8.3), the differences are fairly trivial and are likely the result of the larger sample of retouched tools recovered in A2 rather than any significant behavioral distinction between the two assemblages. Likewise, even though there are almost no cores in Level A1, the fact that both assemblages are dominated by bladelets and flakes indicates that the *chaînes opératoires* deployed in both assemblages were essentially the same.

Assemblage Curation

Techno-typological uniformity can nonetheless hide other significant axes of behavioral variability. For instance, these largely qualitative descriptive data provide little, if any, insights into the technological organization of given assemblages. To this end, we use here a method employed elsewhere to identify how some parameters of lithic assemblages (i.e., their position along a theoretical curation continuum) covaried in predictable manners with the land-use strategies of prehistoric foragers that is based on comparing artifact volumetric densities and retouch frequencies (Barton 1998; Riel-Salvatore and Barton 2004, 2007; Riel-Salvatore 2007, 2010; Riel-Salvatore et al. 2008; Barton et al. 2011; Barton and Riel-Salvatore 2014; see also Clark 2008; Kuhn 2004; Kuhn and Clark 2015). Since plotting the results of this method makes for easier interpretation and since two points by definition make a line, the resulting analysis also plots Bombrini's Late Mousterian assemblages to help contextualize the Proto-Aurignacian pattern (cf. Riel-Salvatore 2010).

Table 8.4 presents the gross lithic volumetric density figures (i.e., the total number of pieces extrapolated per cubic meter of excavated sediment; see Riel-Salvatore and Barton 2004), as well as the usable blank density of each level, which excludes lithics under 1.5 cm in maximum dimension (except bladelets, all of which were counted). The usable total lithic ratio indicates that Level A1 yielded 14% more usable products than Level A2. Excluding bladelets yields much lower and more comparable ratios. This pattern highlights two things. First, bladelet production seems to result in much higher frequencies of usable products suggesting that, outside of bladelet production, the Proto-Aurignacian produces much more shatter than usable flakes. Second, that being the case, the discrepancy in overall usable blank density indicates important differences in bladelet production in the two levels, with Level A1 indicating a less wasteful strategy.

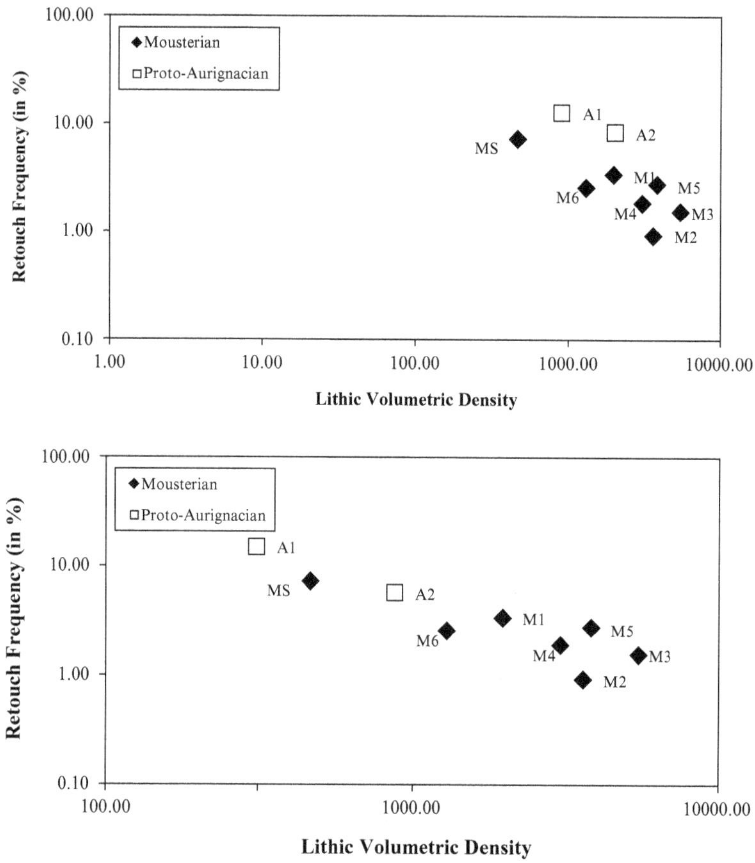

Fig. 8.4 Whole assemblage graphs showing the negative relationship between lithic volumetric density and the frequency of retouch. *Top*: All pieces confounded; *Bottom*: Excluding bladelets

Calculating lithic volumetric densities and frequencies of retouched pieces from the data presented in Table 8.4 for all assemblages at Bombrini shows that the Proto-Aurignacian (mean % retouch = 10.58) is more intensely retouched than the Mousterian (mean = 2.53%), although this is again largely due to the prevalence of retouched bladelets in the Proto-Aurignacian. Following Riel-Salvatore and Barton (2004), the relationship between volumetric density (corrected for length of accumulation, per Riel-Salvatore 2007) and retouch frequency of each assemblage was plotted on a logged-axis scatterplot, with the expectation that they should be correlated negatively and thus yield insights into the mobility strategies of the Riparo Bombrini toolmakers (Fig. 8.4a). From this analysis, the predicted relationship clearly holds, but two separate clusters cleaving along industrial lines can be seen on the graph, with Mousterian assemblages being denser and less retouched than in the Proto-Aurignacian. In both industries, however, assemblages are organized along an expedient-curated ratio. That being the case, the assemblage from Level A2 is more expedient than that from Level A1, indicating a significant behavioral distinction between the two assemblages not picked up on in terms of their techno-typological makeup.

Interestingly, removing the bladelet component of the Proto-Aurignacian to make it more directly comparable to the Late Mousterian simply—and unsurprisingly, since it strips the Proto-Aurignacian of its distinctiveness—strengthens the adherence of the Bombrini sequence to the expected relationship (Fig. 8.4b). The net effect in this case is to deflate Proto-Aurignacian artifact densities and to slightly increase and decrease the retouch frequency of A1 and A2, respectively. This reinforces the observation above that the lithic production of the Proto-Aurignacian was aimed at the large-scale production of bladelets, resulting in systematically higher densities of knapped stone, more of which was retouched than in solely flake-based industries. Figure 8.4b, however, also shows that even in its non-bladelet dimension, Proto-Aurignacian lithic resources were managed rather differently than in the Mousterian.

To sum up, then, it appears that a more residential land-use strategy was in place in Level A1, while Level A2 shows a more logistical strategy (sensu Riel-Salvatore and Barton 2004). These were reflected by stone tool assemblages that were more curated and more expedient, respectively. Since different land-use strategies are likely to have been characterized by different resource management strategies, we now turn to a consideration of raw material procurement patterns to explore this possibility.

Raw Material Procurement

Raw material procurement is an interesting dimension of forager adaptation in coastal Liguria, given the general dearth of sources of fine-grained siliceous stone in the region. At the Balzi Rossi, local raw materials (i.e., found within 10 km of the site) include heterogeneous flint from the nearby Ciotti conglomerate (del Lucchese et al. 2001–2002; Negrino et al. 2006), limestone, and coarse quartzite. Circum-local lithotypes include distinctive, fine-grained quartzite and silicified limestone found near modern-day Sanremo, some 20 km away (Negrino 2002); these have occasionally been mislabeled as Perinaldo flint. Exotic raw materials include a series of fine-grained, homogeneous flints from southern France, rhyolite from the Esterel region of France, fine-grained jaspers from eastern Liguria and/or Emilia-Romagna, a gray flint from the Apennines, and "Scaglia" flint found in the modern-day Marche region in outcrops found more than 350 km away as the crow flies from Riparo Bombrini (Negrino 2002; Negrino and Starnini 2003).

There are interesting differences in raw material procurement strategies across the two Proto-Aurignacian levels (Table 8.5). While there is no fundamental difference in the proportion of debris to debitage (i.e., in "usable" pieces) and the frequency of exotic raw material is broadly similar in the two levels, in Level A1, exotic debris is almost twice as frequent as exotic debitage, while they are equally

Table 8.5 Proto-Aurignacian raw material use patterns, per lithic category

	A1			A2		
Lithic category	L	CL	E	L	CL	E
Debitage >1.5cm	93.85	0	5.38	91.87	0.47	7.54
Débitage <1.5cm	89.36	1.81	8.84	90.29	1.12	8.43
Technological elements	85.71	0	14.29	78.57	0.79	20.63
Unretouched bladelets	71.56	4.44	24	80.23	2.07	17.45
Dufour bladelets	64.71	5.88	29.41	69.01	5.63	23.94
Retouched tools and cores	84	0	16	68.12	2.9	28.99

L local, *CL* circum-local, *E* exotic. Counts for each category can be found in Table 8.4

represented in A2. This difference is statistically significant ($z = -2.059, p = 0.0395$) in Level A1 but not Level A2, which suggests that the management of exotic lithotypes differed in the two assemblages. For one thing, the A2 assemblage indicates that lithic production took place on site, something that is further suggested by the higher frequency in A2 of exotic lithotypes among technological elements, that is to say, those pieces involved in core preparation (e.g., core rejuvenation flakes, crest blades, etc.) or edge rejuvenation (e.g., burin spalls). In contrast, the A1 assemblage contains significantly more exotic debris than debitage, suggesting that exotic material might have been worked while passing through Riparo Bombrini, mainly as retouching curated chipped stone implements or in the production of blanks on exotic material that were subsequently removed from the site. Considering the small size of the debris under consideration here which can easily be affected by site formation processes like runoff, it is worth noting that there is no evidence of such processes in the sediment of Level A1, which in fact is associated with overall more arid conditions that Level A2 (see above). This means that the pattern just discussed is most likely anthropic in nature and thus behaviorally meaningful.

Such a pattern is in keeping with the observation that the A2 assemblage shows a more expedient lithic organization that likely represents the occupation of Riparo Bombrini as a logistical base camp. Under this mobility regimen, distant resources were the subject of targeted forays and brought back to the site where they were stockpiled, creating conditions of effective availability leading to a more profligate use of them. The pattern for A1 is also in keeping with its interpretation as a more curated assemblage, likely representing occupation of Bombrini as part of a more residential mobility strategy.

This pattern is further corroborated by an inspection of raw material representation in bladelets, cores, and retouched tools (Table 8.5). While the frequency of retouched bladelets is broadly comparable in both Proto-Aurignacian assemblages (though slightly higher in A2), the bladelets from Level A2 (especially unretouched ones) are more frequently made on local lithotypes than those from Level A1. This, again, accords well with the idea of A2 having been occupied predominantly as a long-term base camp. In such a context, locally available raw material could have been accumulated at the site and worked to retool multicomponent tools comprising bladelets to a greater extent than in contexts where the site was occupied as part of a more residentially organized land-use strategy. Under that context, allochthonous

Table 8.6 Counts of retouched and unretouched pieces in the Proto-Aurignacian assemblages of Riparo Bombrini

	A1			A2		
	Retouched	Total	% Retouch	Retouched	Total	% Retouch
Bladelets	34	259	13.13	142	1351	10.51
Other debitage	23	189	12.17	60	1044	5.75

raw material of higher quality would have preferentially been used because it displayed "effective availability" (*sensu* Riel-Salvatore and Barton 2004) as a result of being carried around the landscape and therefore being dependably reliable for unexpected bladelet production events.

That allochthonous material was treated in this way is reinforced by the fact that the only core on exotic material (a prismatic bladelet core on French flint) is found in Level A2. This suggests that cores on exogenous lithotypes were only discarded when the site was used as a logistical base camp in the lowermost Proto-Aurignacian. Combined with the higher tool diversity (see Table 8.3) and with the much higher frequency of discarded retouched tools on exotic material in A2 relative to A1, this suggests that the A2 assemblage reflects a site occupation modality defined by the occurrence of a wide range of tasks along with discard and retooling of packages of raw material, local or not, having fallen below a certain threshold of utility. That this took place under a context of effective lithic abundance is further attested to by the fact that, despite being absolutely more numerous in assemblage A2 than in A1, retouched stone tools only account for about 5% of A2, while they account for just under 15% of A1, if considered separately from bladelets (Table 8.6).

It is also interesting to note that, while the majority of the Proto-Aurignacian assemblages at Bombrini were made on local materials (i.e., Ciotti flint), exotic stone is nonetheless documented in all of the technological categories of the assemblages. The rare circum-local lithotypes are found mostly in the form of bladelets (retouched and unretouched). Thus, while the makers of the Proto-Aurignacian depended, for external infusions of goods, on a far-flung network, their use of circum-local (i.e., close but not immediately available near the site) resources appears to have been rather more limited. Since circum-local material is almost only found in the form of bladelets, it may be that this material was procured only opportunistically. In short, most lithic raw material was procured locally and supplemented by an important fraction of exotic lithotypes; circum-local stone was not the focus of any sustained activity, maybe in part due to the difficulty of producing bladelets from raw materials such as quartzite.

Bladelet Technology

Be it in terms of the technology, typology, or assemblage curation patterns, it appears that bladelets were an especially central part of Proto-Aurignacian technology. We therefore analyzed a subset of bladelets from both levels (152 for Level A1, and 318 for Level A2, of which 3 and 35 were complete, respectively) to see if there

Table 8.7 Average dimensions (in mm) of the retouched and unretouched components of Proto-Aurignacian levels at Riparo Bombrini

	Length	Width	Thickness
A1 Unretouched	12.3	7.3	2.0
A1 Retouched	14.4	8.1	2.6
A2 Unretouched	13.5	7.6	2.1
A2 Retouched	15.4	7.0	2.0

are additional differences between them that can be teased out. Looking first at their dimensions (Table 8.7), bladelets in Level A1 appear to be slightly shorter than those in Level A2, an observation that holds for both retouched and unretouched subsets. The retouched bladelets of A2 are also slightly narrower, though this probably is due to bilaterally retouched pieces having only being found in that subset, though this leaves unanswered the question of why they are noticeably thinner than the A1 subset. While none of these differences is statistically significant, they are nonetheless suggestive of differences between the two assemblages that should be further analyzed using the complete bladelet assemblages from Bombrini to see if this bears out these differences.

Another relevant attribute to understand how bladelets were produced and curated is cortex, as its presence may indicate an early stage of the production sequence. Retouched bladelets in both assemblages show an almost complete absence of cortex (Fig. 8.5). This is unsurprising if one considers them to be the desired end product of the prismatic core bladelet *chaîne opératoire*. In contrast, while the same general pattern holds for the unretouched bladelets in Level A2, the A1 unretouched bladelets are much more frequently cortical (though not significantly so). This suggests, somewhat counter-intuitively, that the earlier stages of bladelet production are more prevalent in this assemblage. Alternatively, this may be due to the high reliance on poor-quality local raw material in this sample, which may have required greater *façonnage* in order to exploit properly.

Turning to the evidence for retooling (another potential indicator of different degrees of assemblage curation), Table 8.8 shows that in both assemblages, complete bladelets are much more frequently unretouched and mesial segments account for about one-third of the broken pieces of all subsets (for the purposes of this part of the analysis, we ignore the retouched bladelets because of their extremely low number), while basal fragments are the most abundant and tip fragments are the least frequent. That said, tip fragments (i.e., missing the base) are most frequent in the A2 retouched bladelet subset, likely reflecting a strategy where only the distal portions of bladelets were brought back to camp, possibly as broken parts of armatures stuck in prey animals. Thus, their discard at the site is perhaps less reflective of active retooling than that of the basal fragments (i.e., missing the tip), that dominate the broken unretouched pieces in both levels. Discarding pieces with the tip missing indicates that the pieces were most likely

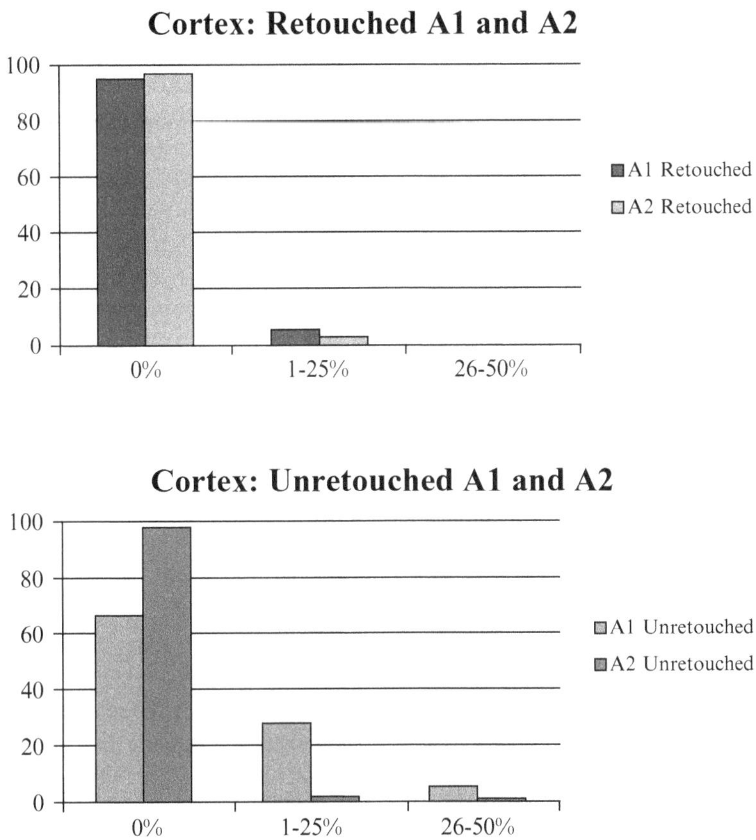

Fig. 8.5 Frequencies of cortex categories on retouched (*top*) and unretouched (*bottom*) bladelets

Table 8.8 Count of complete and broken retouched and unretouched bladelets in Levels A1 and A2

	A1 ($N = 152$)		A2 ($N = 318$)	
	Ret.	Unret.	Ret.	Unret.
Complete N	1	35	3	77
Broken N	2	114	32	206
Basal fragments	0	0.47	0.41	0.57
Tip fragments	0	0.18	0.25	0.13
Mesial fragments	1.00	0.35	0.34	0.3

The frequency of different broken pieces is presented in the bottom three rows of the table as a proportion of the total number of broken pieces broken off site but retained in whatever haft they were inserted into and brought back to the site where they were finally discarded. This would agree with a scenario whereby retooling was done mostly on site using local raw materials, while exotic raw materials were more often incorporated into tools during logistical task forays away from Riparo Bombrini. While we do not yet know with certainty how bladelets were hafted in the Bombrini Proto-Aurignacian, in contrast to the situation at Fumane, they are unlikely to have served as point or dart tips (see discussion in Bertola et al. 2013). However, if other studies of Dufour bladelet breakage (e.g., O'Farrell 2005; Pelegrin and O'Farrell 2005; cf. Hays and Lucas 2001) can be taken as indicators, it is likely that they were used as inset part of composite tools, which may also help explain the large proportion of (intentionally made?) mesial fragments in all bladelet subsets. In this context, interpreting the differential breakage pattern of distinct bladelet parts would be less straightforward than if they had been used as weapon tips, but nonetheless certainly significant as well as commensurate with the scenario just outlined.

Summary

The above analyses present the following picture of the differences between the two Proto-Aurignacian assemblages at Bombrini. While both assemblages are dominated by bladelets and the *chaînes opératoires* employed to produce bladelets are the same, overall Level A1 attests to a lithic organizational strategy that emphasized curation as a result of the more residential land-use strategy in use in that level. This is further supported by raw material data that show a greater exploitation of local resources and by the bladelet data that indicate that stouter but shorter bladelets were being produced in that level and that more retooling took place in that level than in A2. This suggests that retooling activities waited until Proto-Aurignacian foragers reached the site, replenished their raw material supplies with local lithotypes (usually Ciotti flint), discarded broken armatures, and replaced them with new ones manufactured there, prior to their next move. In contrast, Level A2 displays patterns suggestive of the site's use as a logistical base camp, provisioned in needed resources from some distance away, as suggested by raw material procurement patterns. Likewise, bladelet production at the site produced longer, narrower bladelets, usually meant to replace armatures that had broken and had mostly been discarded at task sites away from the camp. The presence of greater numbers of cores in this level also indicates that production was a greater focus of activity than in Level A1, where it is likely that bladelets were produced from cores that were subsequently removed from the site, as foragers left with them for the next leg of their yearly rounds.

Variability in Proto-Aurignacian Lithic Technology and Paleoenvironmental Conditions

Two main conclusions emerge from the analysis of the lithic and paleoenvironmental data presented above. First, in spite of its techno-typological uniformity, there was variability in Proto-Aurignacian adaptations in Levels A1 and A2 at Riparo Bombrini. Second, the two levels are also associated with distinct paleoenvironmental regimes. Based on these two sets of information, it is certainly possible to propose a correlation between environmental change and human behavior: Under colder conditions, a more logistical strategy was adopted, with Riparo Bombrini serving as a long-term base camp provisioned by material procured some distance away. This is reflected in the lithic record by a more expedient assemblage. In contrast, the more curated assemblage found under warmer conditions reflects a more residential land-use strategy. On the basis of this correlation, then, it would be tempting to argue for a straightforward relationship, wherein increases in temperatures results in higher mobility, which in turn caused Proto-Aurignacian foragers to curate their lithic industry more (and depend slightly less on exotic raw material except for the production of bladelets, arguably the most portable and flexible component of the lithic assemblages discussed in this study). We would argue, however, that it is premature to make such definitive statements, mainly due to the fact that we are dealing here with only two assemblages which will by default create a kind of conceptual "line," linked by the two data points presented here. This is complicated in the case of Riparo Bombrini by the presence of contemporary Proto-Aurignacian deposits at Riparo Mochi very close by and by the fact that that site's record may either complement or have strongly conditioned the record analyzed here. Thus, if anything, we should take the trend highlighted in the Proto-Aurignacian at Riparo Bombrini as a working hypothesis to be tested, using assemblages from other Proto-Aurignacian sites containing multiple levels. Ideally, the approach employed in this study will provide a template for future comparable or complementary studies at other sites having yielded multiple Proto-Aurignacian levels. This would allow a test of the expectation of diachronic, paleoenvironmentally mediated internal variability in the Proto-Aurignacian, to address whether the behavioral flexibility proposed in this paper is manifest across the geographical range of that techno-complex.

That being said, it is also important to underscore that this study demonstrates that there is some degree of behavioral variability within the Proto-Aurignacian. Therefore, finding explanations for what pushed this variability in one direction or another would seem to be key to future research. This is in contrast to other analyses that characterize variants of the Aurignacian as internally very stable and indeed draw on this stability as an explanation for the Aurignacian's success (a view rightly criticized as opportunistic and simplistic by Roebroeks and Corbey (2000)). Because the timespan during which the Proto-Aurignacian is documented also comprises instances of very dramatic paleoenvironmental change, we must at least consider the potential for causal links between the two.

A few observations stemming from the present analysis should therefore be stressed. First, at Bombrini, the Proto-Aurignacian survived the climatic turbulence precipitated by the eruption of the Phlegrean Fields. Chronological data indicate that, in contrast to southern Italy where the CI tephra caps the Proto-Aurignacian in all known sequences (except perhaps Paglicci), this perduration is also documented at Mochi (Douka et al. 2012) and at Fumane (Higham et al. 2009). Thus, in areas perhaps less directly affected by this catastrophic event, the Proto-Aurignacian proved to be a resilient behavioral adaptation to large-scale paleoenvironmental change (Lowe et al. 2012). If the data from Riparo Bombrini are any indication, it would seem that in the centuries immediately following it, foragers shifted to a more residential mobility strategy to better cope with the ecological changes triggered by HE4. What should be highlighted, however, is that the Proto-Aurignacian's internal variability may well have been a key defining feature of its lasting success in the EUP of northern Italy. Documenting whether this was the case elsewhere would therefore seem to be an important question to investigate in the future.

Even if the links between behavioral variability and environmental variability remain to be ascertained, the data presented here nonetheless allow a few conclusive observations to be made and permit a critical evaluation of other models about the nature of the internal variability of the Aurignacian generally and the implications of this for the Proto-Aurignacian more specifically. For instance, one of the few researchers to explicitly acknowledge a significant degree of internal variability in the Aurignacian has been Davies (2001). He attributes this variability principally to the nature of different phases of the settlement of Europe by modern human populations (but cf. Anderson et al. 2015). Under this scenario (see also Davies 2007), the earliest assemblages would correspond to the "pioneer" phase of settlement and would correspondingly be small, low density, less typologically diverse, made predominantly on local raw materials. This is the opposite of the pattern seen at Bombrini, where the earliest Aurignacian of Level A2 is the denser and more diverse of the two Proto-Aurignacian occupations, and is also associated with a wider typological diversity and greater dependence on exotic raw materials. While this observation must be tempered with the fact that this level may not represent the very earliest Proto-Aurignacian at the Balzi Rossi (cf. Douka et al. 2012), it is likely sufficiently early to broadly correspond to it. This would therefore suggest that the pioneer/developed dichotomy is insufficient to account for all of the variability documented in the techno-complex. In a like manner, a thought-provoking recent study by Anderson et al. (2015) proposes that the Proto-Aurignacian was essentially a comparatively short-lived "pioneering" adaptation by *Homo sapiens* across Europe. This interpretation is, however, based on a view of the techno-complex as quite homogeneous across its range. It also ignores its longer chronology in the Italian peninsula than in France. This perspective unfortunately glosses over some of the regional dynamics described earlier, notably the sharp contrasts between Proto-Aurignacian assemblages from northern and southern Italy, and does not explicitly address the kind of internal variability described in this paper (and its possible links to paleoenvironmental variability). As such, while issues related to the dynamics of a colonization of new territories by new populations certainly need to be considered, several other dimensions need to be given conceptual room to understand the Proto-Aurignacian as a fully developed adaptive system.

The links between the Proto-Aurignacian and climatic factors have remained largely unexplored. Although Banks et al. (2013) have recently proposed that the Proto-Aurignacian was associated with a smaller ecological niche than the subsequent Early Aurignacian, that has rightly been criticized on methodological grounds and because of issues with sample selection (Higham et al. 2013; Ronchitelli et al. 2013), notably for the fact that it largely ignored data from the Balzi Rossi, and recently published new data have shown that the chronological boundaries used by the authors are likely incorrect (e.g., Nigst et al. 2014). Banks et al. (2013) nonetheless raise the thought-provoking possibility that there was a well-defined niche for the Proto-Aurignacian that contrasts with those of other facies of the Aurignacian.

Interestingly, the data from Riparo Bombrini would seem to support some dimensions of this idea while invalidating others. In terms of according with it, the two Proto-Aurignacian assemblages at Bombrini are associated with the same overall territory, as defined by the geography of raw material procurement. In both levels, the breadth of this geography is the same, with the same sources being exploited. Thus, it would appear that, in spite of climatic changes and the impact they had on faunal and plant resources (see above), the overall social geography of the Proto-Aurignacian remained stable. What appears to have changed is how foragers navigated and exploited that landscape under different conditions. As such, then, one could make the case that the Proto-Aurignacian was a behavioral response inherently suited to dealing with variability in conditions within given areas, in this case the biogeographical corridor of Liguria which is defined by a narrow band of coastal plains at the foot of the Maritime Alps and a very steep coastal shelf that insured that the topography that foragers encountered even under very different paleoclimatic regimes would have been quite similar. What would have changed in the face of climatic fluctuations, then, is not the overall behavioral package, but rather what facets of it were emphasized in different contexts.

Thus, the landscape was not simply the physical background over which the Proto-Aurignacian was overlaid. Rather, that landscape was actively negotiated and reconceptualized in the face of changing conditions, with a single site, in this case Riparo Bombrini, having the potential to serve as a different kind of node in the networks Proto-Aurignacian foragers depended on for survival. Therefore, it could be argued that the Proto-Aurignacian niche in Liguria was not simply the result of a compromise between human needs and available resources, but that by repositioning themselves on the landscape strategically, humans continually constructed and shaped their niche. A corollary of this reasoning is that the variability documented in the Proto-Aurignacian at Bombrini provides evidence of deliberate prehistoric human niche construction (cf. Riel-Salvatore 2010), since there appears to have been willful modification of how the geography and resources of its setting was manipulated in patterned ways.

Where these data depart from Banks et al. (2013)'s view is that in this context, the Proto-Aurignacian can be shown to be a flexible set of conditions, some of which could be purposefully manipulated by humans, as opposed to simply the backdrop against which it existed and as a factor of which its inherent complexity can be defined.

Acknowledgments This research was funded by the National Science Foundation (BCS 0526073), Université de Montréal, the University of Colorado Denver, and the Social Sciences and Humanities Research Council of Canada. Data on the bladelets analyzed here was collected with the help of Crystal Simms, Claudine Gravel-Miguel, and Enza Spinapolice; Crystal Simms also conducted a first compilation of some of the bladelet data. We kindly thank an anonymous reviewer who gave detailed, pointed, and extremely useful feedback on an earlier draft of this paper. Finally, we wish to thank Erick Robinson and Fred Sellet for inviting us to contribute this chapter and, especially, for their patience and encouragement during the process.

References

Alhaique, F., Bietti, A., Del Lucchese, A., Grimaldi, S., Manzi, G., Martini, S., Negrino, F., & Recchi, A. (2000). Biological and cultural variability at the Neandertal/modern human transition in Italian archeological sites. *Rivista di Antropologia, 78*, 105–116.

Allen, J. R. M., Watts, W. A., & Huntley, B. (2000). Weichselian Palynostratigraphy, Palaeovegetation and Palaeoenvironment: The record from Lago Grande di Monticchio, southern Italy. *Quaternary International, 74*, 91–110.

Anderson, L., Bon, F., Bordes, J. G., Pasquini, A., Slimak, L., & Teyssandier, N. (2015). Relier des espaces, construire de nouveaux réseaux : aux origines du Protoaurignacien et des débuts du Paléolithique supérieur en Europe occidentale.

Arobba, D. (1984). Analisi pollinica dei sedimenti del Riparo Bombrini (Balzi Rossi, Ventimiglia). *Rivista Ingauna e Intemelia, XXXIX*(3–4), 13–19.

Arobba, D., & Caramiello, R. (2009). Analisi paleobotanica sui sedimneti del Riparo Bombrini (Balzi Rossi, Ventimiglia). *Bulletin du Musée d'Anthropologie préhistorique de Monaco, 49*, 41–48.

Banks, W. E., d'Errico, F., & Zilhão, J. (2013). Human-climate interaction during the early upper Paleolithic: Testing the hypothesis of an adaptive shift between the proto-Aurignacian and the early Aurignacian. *Journal of Human Evolution, 64*, 39–55.

Barton, C. M. (1998). Looking back from the world's end: Paleolithic settlement and mobility at Gibraltar. In J. L. Sanchidrián Torti & M. D. Simón Vallejo (Eds.), *Las Culturas del Pleistoceno Superior en Andalucía* (pp. 13–23). Málaga: Patronato de la Cueva de Nerja.

Barton, C. M., & Riel-Salvatore, J. (2014). The formation of lithic assemblages. *Journal of Archaeological Science, 46*, 334–352.

Barton, C. M., Riel-Salvatore, J., Anderies, J. M., & Popescu, G. (2011). Modeling Human Ecodynamics and Biocultural Interactions in the Late Pleistocene of Western Eurasia. *Human Ecology, 39*, 705–725.

Benazzi, S., Slon, V., Talamo, S., Negrino, F., Peresani, M., Bailey, S. E., Sawyer, S., Panetta, D., Vicino, G., Starnini, E., Mannino, M. A., Salvadori, P. A., Meyer, M., Pääbo, S., & Hublin, J.-J. (2015). The makers of the Protoaurignacian and implications for Neandertal extinction. *Science, 348*, 793–796.

Bertola, S., Broglio, A., Cristiani, E., De Stefani, M., Gurioli, F., Negrino, F., Romandini, M., & Vanhaeren, M. (2013). La diffusione del primo Aurignaziano a sud dell'arco alpino. *Preistoria Alpina, 47*, 123–152.

Bietti, A., & Negrino, F. (2008). L'Aurignacien et le Gravettien du Riparo Mochi, l'Aurignacien du Riparo Bombrini: comparaisons et nouvelles perspectives. *Arch. Inst. Paléontologie Humaine, mém., 39*, 133–140.

Binford, L. R. (2001). *Constructing frames of reference: An analytical method for archaeological theory building using ethnographic and environmental data sets*. Berkely: University of California Press.

Bon, F. (2006). A brief overview of Aurignacian cultures in the context of the industries of the transition from the middle to the upper Paleolithic. In O. Bar-Yosef & J. Zilhão (Eds.), *Towards a definition of the Aurignacian* (pp. 133–144). Lisbon: Instituto Português de Arqueologia.

Broglio, A., Bertola, S., De Stefani, M., Marini, D., Lemorini, C., & Rossetti, P. (2005). La production lamellaire et les armatures lamellaires de l'Aurignacien ancien de la grotte de Fumane (Monts Lessini, Vénétie). In F. Le Brun-Ricalens (Ed.), *Productions lamellaires à l'Aurignacien* (pp. 415–436). Luxembourg: MNHA, Archéologiques.

Cardini, L. (1938). Recenti scavi dell'Istituto Italiano di Paleontologia Umana alla Barma Grande di Grimaldi. *Archivio per l'Antropologia e la Etnologia, 68*, 385–389.

Clark, A. E. (2008). *Changes in Occupation Intensity during the Lower and Middle Paleolithic at Tabun Cave, Israel*. Unpublished M.A. Thesis, University of Arizona, Tucson, Arizona.

Davies, W. (2001). A very model of a modern human industry: New perspectives on the origins and spread of the Aurignacian in Europe. *Proceedings of the Prehistoric Society, 67*, 195–217.

Davies, W. (2007). Re-evaluating the Aurignacian as an expression of modern human mobility dispersal. In P. Mellars, K. Boyle, O. Bar-Yosef, & C. B. Stringer (Eds.), *Rethinking the human revolution: New behavioural and biological perspectives on the origin and dispersal of modern humans* (pp. 263–274). Cambridge: McDonald Institute for Archaeological Research.

Del Lucchese, A., Martini, S., Negrino, F., & Ottomano, C. (2001-2002). 'I Ciotti' (Mortola Superiore, Ventimiglia, Imperia). Una località di approvvigionamento della materia prima per la Scheggiatura durante il Paleolitico. *Bullettino di Paletnologia Italiana, 91-92*, 1–26.

Del Lucchese, A., Formicola, V., Holt, B., Negrino, F., & Vicino, G. (2004). Riparo Bombrini, Balzi Rossi (Ventimiglia, Imperia). Notizie preliminari degli scavi 2002-2004. *Ligures, 2*, 287–289.

Del Lucchese, A., & Negrino, F. (2008). Riparo Bombrini Balzi Rossi (Ventimiglia, Imperia). *Archeologia in Liguria, N.S. 1*, 238–240.

Douka, K., Grimaldi, S., Boschian, G., Del Lucchese, A., & Higham, T. F. G. (2012). A new chronostratigraphic framework for the upper Palaeolithic of Riparo Mochi (Italy). *Journal of Human Evolution, 62*, 286–299.

Formicola, V. (1984). Un Incisivo Umano dal Deposito Aurignaziano del Riparo Bombrini ai Balzi Rossi. *Rivista Ingauna e Intemelia, 39*, 11–12.

Formicola, V. (1989). Early Aurignacian deciduous incisor from Riparo Bombrini at Blazi Rossi (Grimaldi, Italy). *Rivista di Antropologia, 67*, 287–292.

Giaccio, B., Hajdas, I., Peresani, M., Fedele, F. G., & Isaia, R. (2006). The Campanian ignimbrite tephra and its relevance for the timing of the middle to upper Paleolithic shift. In N. J. Conard (Ed.), *When Neanderthals and modern humans met* (pp. 343–375). Tübingen: Kerns Verlag.

Grimaldi, S., Porraz, G., & Santaniello, F. (2014). Raw material procurement and land use in the northern Mediterranean arc: Insight from the first proto-Aurignacian of Riparo Mochi (Balzi Rossi, Italy). *Quartär, 61*, 113–127.

Hays, M., & Lucas, G. (2001). Experimental investigations of Aurignacian Dufour bladelets. In M. Hays & P. Thaker (Eds.), *Questioning the answers: Resolving fundamental problems of the early upper Paleolithic* (pp. 109–116). Oxford: British Archaeological Reports 105.

Higham, T., Brock, F., Peresani, M., Broglio, A., Wood, R., & Douka, K. (2009). Problems with radiocarbon dating the middle to upper Palaeolithic transition in Italy. *Quaternary Science Reviews, 28*, 1257–1267.

Higham, T., Wood, R., Moreau, L., Conard, N., & Bronk Ramsey, C. (2013). Comments on "human-climate interaction during the early upper Paleolithic: Testing the hypothesis of an adaptive shift between the proto-Aurignacian and the early Aurignacian" by banks et al. *Journal of Human Evolution, 65*, 806–809.

Higham, T., Douka, K., Wood, R., Bronk Ramsey, C. B., Brock, F., Basell, L., et al. (2014). The timing and spatiotemporal patterning of Neanderthal extinction. *Nature, 512*(7514), 306–309.

Holt, B. M., Negrino, F., Formicola, V., Riel-Salvatore, J., Churchill, S. E., Vicino, G., & Del Lucchese, A. (2006). The Mousterian-Aurignacian transition at the Riparo Bombrini (Liguria, Italy) rockshelter. *PaleoAnthropology, 2006*, A92.

Holt B., Negrino F., Riel-Salvatore J., Formicola V., Vicino G., Churchill S. E., Arellano A., Arobba D., Boschian G., Di Canzio E., & Cristiani E. (n.d.). The Middle-Upper Paleolithic transition at Riparo Bombrini (Liguria, Italy). Unpublished manuscript, under review.

Hublin, J.-J. (2015). The modern human colonization of western Eurasia: When and where? *Quaternary Science Review, 118*, 194–210.

Kelly, R. L. (1995). *The foraging Spectrum: Diversity in hunter-gatherer lifeways*. Washington D. C: Smithsonian Institution Press.

Kuhn, S. L. (2002). Pioneers of Microlithization: The "proto-Aurignacian" of southern Europe. In R. G. Elston & S. L. Kuhn (Eds.), *Thinking small: Global perspectives on Microlithization* (pp. 83–94). Washington: American Anthropological Association.

Kuhn, S. L. (2004). Middle Paleolithic assemblage formation at Riparo Mochi. In A. L. Johnson (Ed.), *Processual archaeology* (pp. 31–60). Westport, CT: Praeger.

Kuhn, S. L., & Stiner, M. C. (1998). The earliest Aurignacian of Riparo Mochi (Liguria, Italy). *Current Anthropology, 38*, S175–S189.

Kuhn, S. L., & Clark, A. E. (2015). Artifact densities and assemblage formation: evidence from Tabun Cave. *Journal of Anthropological Archaeology, 38*, 8–16.

Laplace, G. (1966). *Recherches sur l'Origine des Complexes Leptolithiques*. Mélanges d'Archéologie et d'Histoire de l'École Française de Rome, no. 4. E. de Broccard, Paris.

Laplace, G. (1977). Il Riparo Mochi ai Balzi Rossi di Grimaldi (Fouilles 1938-1949). Les industries leptolithiques. *Rivista di Scienze Preistoriche, 32*, 3–131.

Lowe, J., Barton, N., Blockley, S., Ramsey, C. B., Cullen, V. L., Davies, W., Gamble, C., Grant, K., Hardiman, M., Housley, R., Lane, C. S., Lee, S., Lewis, M., MacLeod, A., Menzies, M., Müller, W., Pollard, M., Price, C., Roberts, A. P., Rohling, E. J., Satow, C., Smith, V. C., Stringer, C. B., Tomlinson, E. L., White, D., Albert, P., Arienzo, I., Barker, G., Borić, D., Carandente, A., Civetta, L., Ferrier, C., Guadelli, J. L., Karkanas, P., Koumouzelis, M., Müller, U. C., Orsi, G., Pross, J., Rosi, M., Shalamanov-Korobar, L., Sirakov, N., & Tzedakis, P. C. (2012). Volcanic ash layers illuminate the resilience of Neanderthals and early modern humans to natural hazards. *Proceedings of the National Academy of Sciences, 109*, 13532–13537.

Meese, D. A., Gow, A. J., Alley, R. B., Zielinski, G. A., Grootes, P. M., Ram, M., Taylor, K. C., Mayewski, P. A., & Bolzan, J. F. (1997). The Greenland ice sheet project 2 depth-age scale: Methods and results. *Journal of Geophysical Research, 102*, 26411–26423.

Mellars, P. (2005). The impossible coincidence. A single-species model for the origins of modern human behavior in Europe. *Evolutionary Anthropology, 14*, 12–27.

Mellars, P. (2006). Archaeology and the dispersal of modern humans in Europe: Deconstructing the "Aurignacian." *Evolutionary Anthropology, 15*, 167–182.

Negrino, F. (2002). *Modificazioni Tecno-Tipologiche ed Utilizzo delle Materie Prime nell'Appennino Tosco-Emiliano e nell'Arco Ligure tra Paleolitico Medio Recente e Paleolitico Superiore Antico*. Rome: Università di Roma "La Sapienza".

Negrino, F. (2005). Riparo Bombrini, Balzi Rossi (Ventimiglia,IM): la campagna 2005. *Ligures, 3*, 194–196.

Negrino, F., Martini, S., Ottomano, C., & Del Lucchese, A. (2006). Palaeolithic evidence for quarrying activity at "I Ciotti" (Mortola Superiore, Ventimiglia, imperia, Italy). In G. Körlin & G. Weisgerber (Eds.), *Stone age – Mining age, VIII international Flint symposium* (pp. 153–162). Bochum: Der Anschnitt 19.

Negrino, F., & Starnini, E. (2003). Patterns of lithic raw material exploitation in Liguria from the Palaeolithic to the copper age. *Préhistoire du Sud-Ouest, 5*, 235–243.

Nigst, P., Haesaerts, P., Damblon, F., Frank-Fellner, C., Mallol, C., Viola, B., Götzinger, M., Niven, L., Trnka, G., & Hublin, J. J. (2014). Early modern human settlement of Europe north of the alps occurred 43,500 years ago in a cold steppe-type environment. *Proceedings of the National Academy of Sciences, 111*(40), 14394–14399.

O'Farrell, M. (2005). Étude préliminaire des éléments d'armature lithique de l'Aurignacien ancien de Brassempouy. In F. LeBrun-Ricalens (Ed.), *Productions Lamellaires Attribuées à l'Aurignacien : Chaînes Opératoires et Perspectives Technoculturelles* (pp. 395–412). Luxembourg: Musée national d'histoire et d'art.

Palma di Cesnola, A. (1993). *Il Paleolitico Superiore in Italia: Introduzione allo Studio*. Florence: Garlatti e Razzai Editori.

Pelegrin, J., & O'Farrell, M. (2005). Les lamelles retouchées ou utilisées de Castanet. In F. LeBrun-Ricalens (Ed.), *Productions Lamellaires Attribuées à l'Aurignacien: Chaînes Opératoires et Perspectives Technoculturelles* (pp. 103–121). Luxembourg: Musée national d'histoire et d'art.

Porraz, G., Simon, P., & Pasquini, A. (2010). Identité technique et comportements économiques des groupes proto-aurignaciens à la grotte de l'Observatoire (principauté de Monaco). *Gallia Préhistoire, 52*, 33–59.

Riel-Salvatore, J. (2007). *The Uluzzian and the Middle-Upper Paleolithic Transition in Southern Italy*. Unpublished Ph.D. Dissertation, Arizona State University.

Riel-Salvatore, J. (2010). A niche construction theory perspective on the middle-upper Paleolithic transition in Italy. *Journal of Archaeological Method and Theory, 17*(4), 323–355.

Riel-Salvatore, J., & Barton, C. M. (2004). Late Pleistocene technology, economic behavior and land-use dynamics in southern Italy. *American Antiquity, 69*, 257–274.

Riel-Salvatore, J., & Barton, C. M. (2007). New quantitative perspectives on the middle– Upper Paleolithic transition: The view from the northern Mediterranean. In J. Riel-Salvatore & G. A. Clark (Eds.), *Transitions great and small: New approaches to the study of early upper Paleolithic 'Transitional' Industries in western Eurasia* (pp. 61–74). Oxford: Archaeopress.

Riel-Salvatore, J., Popescu, G., & Barton, C. M. (2008). Standing at the gates of Europe: Human behavior and biogeography in the southern Carpathians during the late Pleistocene. *Journal of Anthropological Archaeology, 27*, 399–417.

Riel-Salvatore, J., Ludeke, I. C., Negrino, F., & Holt, B. M. (2013). A spatial analysis of the late Mousterian levels of Riparo Bombrini (Balzi Rossi, Italy). *Canadian Journal of Archaeology, 37*(1), 70–92.

Riel-Salvatore, J., & Negrino, F. (2009). Early upper Paleolithic population dynamics and raw material procurement patterns in Italy. In M. Camps i Calbet & C. Szmidt (Eds.), *The Mediterranean between 50–25,000 BP: Turning points and new directions* (pp. 205–224). Oxford: Oxbow Books.

Roebroeks, W., & Corbey, D. (2000). Periodisations and double standards in the study of the Palaeolithic. In W. Roebroeks, M. Mussi, J. Svoboda, & K. Fennema (Eds.), *Hunters of the golden age* (pp. 77–86). Leiden: University Press and European Science Foundation.

Ronchitelli, A. M., Benazzi, S., Boscato, P., Douka, K., & Moroni, A. (2013). Comments on "human-climate interaction during the early upper Paleolithic: Testing the hypothesis of an adaptive shift between the proto-Aurignacian and the early Aurignacian" by William E. Banks, Francesco d'Errico, and João Zilhão. *Journal of Human Evolution, 73*, 107–111.

Slimak, L., Pesesse, D., & Giraud, Y. (2006). Reconnaissance d'une installation du Protoaurignacien en vallée du Rhône. Implications sur nos connaissances concernant les premiers hommes modernes en France méditerranéenne. *Comptes Rendus Palevol, 5*, 909–917.

Szmidt, C. C., Normand, C., Burr, G., Hodgins, G., & LaMotta, S. (2010). AMS ^{14}C dating the Protoaurignacian/early Aurignacian of Isturitz, France. Implications for Neanderthal-modern human interaction and the timing of technical and cultural innovations in Europe. *Journal of Archaeological Science, 37*, 758–768.

Tsanova, T., Zwyns, N., Eizenberg, L., Teysandier, N., Le Brun-Ricalens, F., & Otte, M. (2012). Le plus petit dénominateur commun : réflexion sur la variabilité des ensembles lamellaires du Paléolithique supérieur ancien d'Eurasie. Un bilan autour des exemples de Kozarnika (est des Balkans) et Yafteh (Zagros central). *L'Anthropologie, 116*, 469–509.

Van Andel, T. H., Davies, W., & Weninger, B. (2003). The human presence in Europe during the last glacial period I: Human migrations and changing climate. In T. H. van Andel & W. Davies (Eds.), *Neanderthals and modern humans in the European landscape during the last glaciation* (pp. 31–56). Cambridge: McDonald Institute of Archaeological Research.

Vicino, G. (1984). Scavo Paleolitico al Riparo Bombrini (Balzi Rossi di Grimaldi, Ventimiglia). *Rivista Ingauna e Intemelia, 39*, 1–10.

Chapter 9
Environmental Change and Technological Convergence in Southern Germany

Michael Jochim

Introduction

Separated by at least 5000 years of environmental change, hunter-gatherers of the Magdalenian and the Late Mesolithic of southern Germany inhabited very different worlds. The Magdalenians lived in an open landscape of steppe-tundra, populated by reindeer, horse, and arctic hare, while the people of the Late Mesolithic coped with dense, temperate forests that were home to deer, boar, and a variety of woodland furbearers. A progressive set of environmental changes over the course of the later Pleistocene and early Holocene led to the radical, progressive transformation of the landscape and to a changing set of challenges to human adaptation. Lithic technology was one of the means by which people coped with these changes. Despite the specific, directional nature of the environmental changes, however, by the Late Mesolithic, there is a convergence in some of the technological strategies with those of the Magdalenian, a convergence that appears to be an adjustment to the convergence of features of the environment. This discussion focuses solely on environmental and archaeological data from southwestern Germany, recognizing that other parts of Europe, particularly northern and northwestern Europe, differ considerably from this region.

The relationship between environmental change and lithic technology is clearly complex and is mediated by a host of factors including degree and organization of mobility, raw material quality and availability, and specific resource characteristics. Nevertheless, a reasonable assumption with which to begin investigation of this relationship is that directional changes in the environment should be accompanied by directional changes in lithic technology. In the late and postglacial periods, southwestern Germany witnessed generally directional environmental changes in

M. Jochim (✉)
Department of Anthropology, University of California, Santa Barbara, CA, USA
e-mail: jochim@anth.ucsb.edu

E. Robinson, F. Sellet (eds.), *Lithic Technological Organization and Paleoenvironmental Change*, Studies in Human Ecology and Adaptation 9,
DOI 10.1007/978-3-319-64407-3_9

terms of increasing temperatures and progressive reforestation and vegetational succession. Over the course of this period, stone tool assemblages showed a number of dramatic changes as well, but only some showed a linear development. Other changes are not linear either but do show patterning and, in some ways, result in similarities between the Magdalenian and the Late Mesolithic in their technological solutions to environmental challenges. A consideration of the nature of the environmental transformations may shed light on the specific trajectory shown by the lithic data.

Environmental Changes

Climate change usually alters the specific plants and animals in an area, with potentially profound effects on lithic technology. In addition, a dramatic change in the environment may have important implications for lithic technology by transforming the overall *structure* of the habitat. One of the potential effects of environmental change, for example, is the degree of vegetational patchiness and, with it, the degree of predictability of resource distributions in time and space. Patchiness is notoriously difficult to define, as it varies in scale and depends on both the objective distribution of species and the perception of this distribution by organisms (Wiens 1976; Winterhalder 1980). In general, there is a positive correlation between species diversity and spatial diversity, and thus an increase in spatial patchiness with succession (Odum 1969), although this is not always the case (e.g., Auclair and Goff 1971). With decreasing vegetational homogeneity and increasing patchiness, animals may show increasingly disparate distributions among microhabitats, and more predictable seasonal movements as well, in light of the differing tolerances and preferences among species. This has implications for the techniques and costs of searching for resources, and ultimately for the organization of foraging and foraging technology. Lithic organization should therefore respond in part to such structural changes in the habitat.

Southwestern Germany demonstrates these patterns during the late and postglacial periods (Jochim 1998; Fisher 2000; Harris 2006). Over the course of the time period of roughly 16,000– 6500 cal BP, this area witnessed progressive deglaciation and a succession from steppe-tundra vegetation to pine and birch forests to increasingly diverse mixed oak forests. Concomitant changes in animal species included reindeer and horse communities gradually being replaced by moose, horse, red deer, and beaver and, in turn, by red and roe deer, wild boar, and a large variety of small forest species. Culturally, this period is represented by the Magdalenian, the "Late Paleolithic" (similar to the Azilian and Federmesser of the west and north), and the Early and Late Mesolithic (Fig. 9.1).

During the Final Pleniglacial (Dryas I, ca. 16,000–13,000 cal BP), the entire region was broadly characterized by steppe-tundra (Weniger 1982: 49–52; Rösch 1990), but "…differences in substrate, drainage, and relief suggest larger-scale habitat variation between upland steppe communities and lowland tundra communities" (Fisher 2000: 82). The principal relevant distinguishing feature between these two

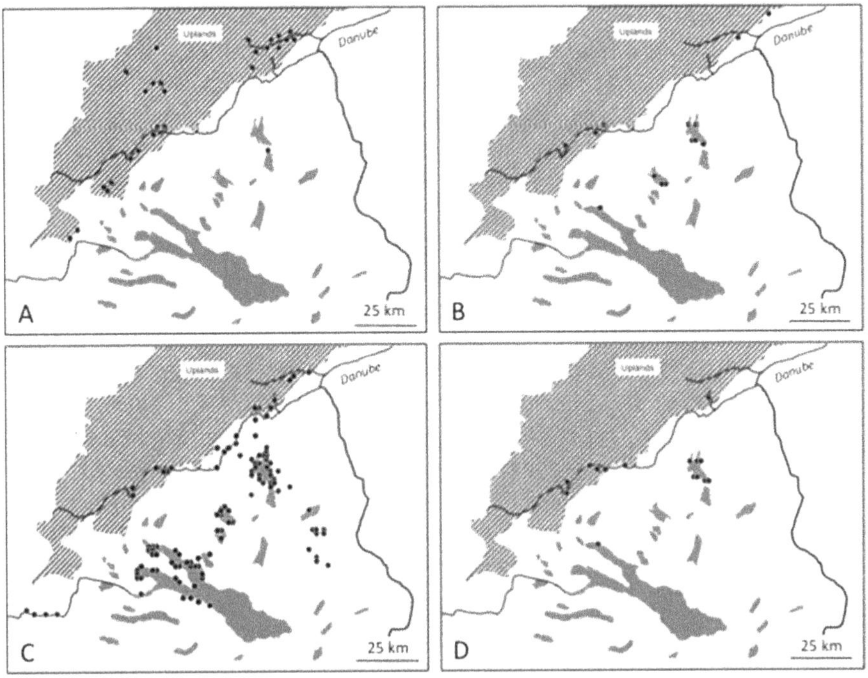

Fig. 9.1 Study area and sites (**a**) Late Magdalenian; (**b**) Late Paleolithic; (**c**) Early Mesolithic; (**d**) Late Mesolithic

regions is the available groundwater; the uplands consist of a limestone plateau with karst characteristics creating much drier conditions. These two vegetational communities contained partially contrasting animal communities. Reindeer and hare dominated the lowlands, while horse, reindeer, ibex, and grouse were found in the uplands (Eriksen 1996; Fisher 2000). Reindeer were predictably in the lowlands during parts of their concentrated fall and spring migrations as well as throughout the winter. In the late spring and summer, by contrast, the various animals were dispersed throughout the drier uplands (Weniger 1982). Because of the rather distinct habitats, as well as the migratory herd behavior of reindeer, Magdalenian hunters could predict with considerable success the location of prey in different places and different seasons, and position their settlements accordingly. Large sites for intercepting herd migrations in fall were located in the lowlands along migratory routes, while small sites in summer were dispersed in small valleys in the uplands (Weniger 1982: 199ff; Eriksen 1996: 115).

This situation changed during the Bølling-Allerød and Younger Dryas periods (ca. 13,000–11,500 cal BP). Rapid warming in the early Bølling was accompanied by an increase in juniper, followed by tree birch and then pine. During the Allerød, tree pollen, especially pine, came to dominate pollen spectra, with some fluctuations throughout this period (Rösch 1990). The appearance of pine and birch, and their spread throughout the entire region, created a more homogeneous vegetation. During the Younger Dryas, studies indicate that, despite a decrease in temperatures (varying with altitude and other factors), vegetational communities in this region were not drastically disrupted (Eicher and Siegenthaler 1976; Rösch 1990; Esterhues et al. 2002; Magny 2004; Lang 2006). The treeline descended somewhat, forests became more open, at least in places, and the proportions of pine and birch varied. In addition, it seems that the climatic deterioration had a less marked impact upon the local vegetation around lakes than on the broader regional vegetation, with greater tree cover persisting at the lakes (Peyrona 2005). Lake levels fluctuated, but lower water levels predominated, and erosion was at times quite marked.

The range of animal species available in the region changed along with these vegetational changes. Mammoth, wooly rhinoceros, reindeer, and horse, for example, were present already during the Late Pleniglacial and Meiendorf, based on both archaeological and paleontological finds (Leesch et al. 2012). By the Oldest Dryas period, the presence of a number of additional herbivore species is documented, including steppe bison, chamois, ibex, musk ox, and red deer, while moose and beaver may also have appeared at least as early as the Oldest Dryas (Weniger 1982; Fahlke 2009). During the course of the Bølling warm period, both mammoth and wooly rhinoceros disappeared from the region, reindeer and horse were still present, moose and beaver increased in numbers, and aurochs first appeared (Eriksen 1991: 36). By the beginning of the Allerød period, musk ox and bison disappeared; reindeer, ibex, and, to some extent, horse decreased in numbers; moose, red deer, and beaver continued to increase; and roe deer appeared.

A major transformation of the carnivore species present occurred during the Bølling/Older Dryas periods, when cave lion, cave bear, cave hyena, wolverine, and (somewhat later) arctic fox disappeared and badger, otter, wildcat, red fox, and pine marten joined the ongoing communities of bear, wolf, and lynx (Eriksen 1991: 37). In addition to a number of important land birds, such as grouse, capercaillie, and partridge, waterfowl proliferated, beginning mainly during the Bølling – including swan, mallard, teal, pintail, and tufted duck (Eriksen 1991: 38). Fish species increased in diversity over the course of these periods, with many appearing first during the Bølling (Eriksen 1991: 39).

The gradual disappearance of reindeer herds, and growing dominance of moose, red deer, and horse (none of which shows the large aggregations and organized migrations of reindeer), created a new landscape in which visibility and the spatial predictability of prey were considerably less. Such boreal forests, of course, are not truly homogeneous but are much more so than many other habitats. The rather monotonous pine–birch forest, with low species diversity, was broken largely by the various lakes and streams of the region. The implications of boreal forest patterns for hunters in general are noted by Winterhalder (1983: 209):

> The prey available to the Cree are solitary (or in small 'packets') and dispersed. Some aggregation occurs for reasons of habitat, but patch-types are themselves small and irregularly dispersed. As a consequence Cree foragers should spend considerable time searching, and, as a result, have a fairly broad diet breadth.

In this region, the only archaeological entity between the Magdalenian and the Mesolithic is the Late Paleolithic, which is similar to the Azilian of both France and Switzerland, and the Federmesser of areas farther north. Late Paleolithic sites in this region are uniformly quite small and, despite wide-ranging surveys (Jochim et al. 1998), were located largely around lakes in the lowlands or along streams at the edge of the uplands – in other words, in the most visible and salient breaks in the forest – and contained a diverse array of prey species, including some fish.

During the Early Mesolithic of the Preboreal and Boreal periods (ca. 11,500–8000 cal BP), oak and hazel spread, displacing the pine and birch forests in some areas and, according to the differing requirements and tolerance of various species, creating a somewhat more varied vegetational mosaic of deciduous and coniferous trees. Deer, boar, and fur-bearing animals certainly became increasingly separated by habitat and therefore were likely to have been somewhat more predictable in their location. Moose, for example, were closely associated with lakes, ponds, and marshes (Phillips et al. 1973). The aurochs were predominantly grazers, feeding on grasses available in open areas, but grasslands were becoming scarce, and relatively open sedge marshes along rivers and ponds probably became a favored habitat (Van Vuure 2005). Red deer were found in a wide variety of habitats, and in regions of considerable topographic relief the animals undertook seasonal movements up- and downslope (Gebert and Verheyden-Tixier 2001). Horses were almost exclusively grazers and would have been increasingly restricted in distribution to higher elevations and lowland clearings. Many smaller mammals showed some preference for the edges of streams and lakes, especially because of the diverse vegetation. Fish productivity and diversity increased (Torke 1981) and being confined to lakes and rivers were certainly spatially relatively predictable. Early Mesolithic sites are much more abundant than earlier sites and are situated in a greater variety of locations, including around large and small lakes and along rivers and streams in the lowlands as well as in river valleys and on hilltops in the uplands (Jochim et al. 1998).

By the Late Mesolithic (ca. 8000–6500 cal BP) of the Early Atlantic period, forest succession had progressed with the influx and spread of maple, elm, linden, and other species, creating an even more diverse habitat mosaic, a common pattern of temperate deciduous forests (Röhrig and Ulrich 1991: 40). In this mosaic, the forest species of animals, with their differing preferences and tolerances, would have likely become even more differentiated in their distributions with a greater spatial predictability at different times of the year. Appropriate locations for seasonal sites should have been relatively predictable. Late Mesolithic sites are relatively few in number and are concentrated around a few large lakes as well as in narrow portions of the Danube Valley within the uplands.

Directional Changes in Lithic Technology

Magdalenian assemblages are characterized by backed bladelets and backed points used as components of hunting projectiles, as well as relatively large burins, scrapers, and other manufacturing tools. One of the hallmarks of Late Paleolithic technology is its simplicity in relation to that of the Magdalenian, with fewer well-defined tool types and generally smaller artifacts. Characteristics of hunting equipment in these assemblages include smaller backed points and fewer backed bladelets. Manufacturing tools are also smaller and include thumbnail scrapers. Major components of Early Mesolithic technology are geometric microlithics, used as elements of composite projectiles. In the Late Mesolithic, by contrast, microliths become much scarcer, suggesting simpler projectiles with single trapezoidal points.

A number of the technological changes can be interpreted as attempts to cope with the progressive environmental changes. In the Magdalenian, seasonal movements of reindeer and horse may have been relatively pronounced and fairly predictable, so that the hunters could focus on intercepting herds in fall and spring at particular points on the landscape. With reforestation, however, the herds of reindeer and horse were decreasing in abundance. Animals favored by the vegetational changes are more solitary than the gregarious reindeer and horse, and likely to have been more dispersed throughout the environment; their movements were shorter and less predictable, and prey visibility was poorer as well. As a result, an intercept strategy was increasingly less likely to be productive. Walking through the forest to encounter prey probably became more important as a hunting technique. With a huge increase in search costs (and probably pursuit costs as well), the overall costs of hunting must have increased, leading to a broadening of the diet to include smaller mammals and more fish and birds (Jochim 1998).

Although hare, arctic fox, and grouse were frequent prey of Magdalenian groups, it is apparent that reindeer and horse were consistently the most important food resources in this area (Weniger 1982: 213). Fish and waterfowl appear to have played a very minor role in Magdalenian subsistence. In the well-excavated, small, warm-weather site of Felsställe, for example, there were only 16 bones of fish and waterfowl among the total of 2418 bones in the Magdalenian levels (Kind 1987). The one known lakeshore site of Schussenquelle reflects this economic orientation. Although much of the faunal collection has been lost since its primary excavation in 1866, it is clear that reindeer were the focus of activities. Very few fish and waterfowl were found. A recent reanalysis of this site concludes that the location was chosen for its hunting potential: the narrow ridge of land between lakes channeled animal movements, allowing easier intercept hunting (Schuler 1994). It should be noted that two other well-excavated Magdalenian lakeshore sites discovered in Switzerland, Monruz and Hauterive-Champreveyres, show a similar pattern (Leesch 1997; Morel and Müller 1997). Although both are situated directly on the shore of Lake Neuchatel, both are dominated by big game, with very few fish and birds.

In the Late Paleolithic level of the site of Henauhof NW on the Federsee, fish and birds show a somewhat greater importance (Jochim 1998). It is also clear that the number of fish bones increased significantly from the Magdalenian to the Late Paleolithic in both the rockshelter site of Zigeunerfels and the cave of Dietfurt, both on the Danube River, although detailed faunal analyses are not available (Torke 1981). During the succeeding Early and Late Mesolithic, the role of fish, birds, and now plant foods continued to increase (Jochim 1998: 209–211).

Nevertheless, in all Late Paleolithic and Mesolithic sites with faunal assemblages, large game animals made up the majority of identified finds and constituted the major components of the diet. In light of the increasing costs and difficulty of hunting, a number of technological changes, over the course of this time period, appear to represent attempts to increase the killing power of projectiles, and thus the productivity of hunting. The decreasing size of projectile points from the Magdalenian to the Late Paleolithic, for example, may in part reflect the need for greater portability, in association with changes in mobility, but it may also reflect steps to increase the penetration of projectiles into prey targets, as well as to decrease the weight of spears or arrows and allow larger striking distances. The development of arrows with multiple microlithic barbs in the Early Mesolithic may have been one way of trying to increase the lethalness of projectiles. It has also been suggested that the prevalence of intentional heat-tempering of stone raw material, primarily during the Early Mesolithic, may have increased not only the ease of working the material but also its brittleness, allowing more frequent fragmentation upon impact and thus more internal bleeding (Harris 2006: 300–306). Finally, the use of arrows with single trapezoidal points in the Late Mesolithic, with the broad cutting edge in the front, may have further increased the internal bleeding and led to increased success of prey kills.

Another directional trend in technological changes is increasing attempts to compensate for what may have been decreasing visibility of, and access to, stone raw material. The highest-quality raw material in the region is Jurassic chert, occurring in the limestone formation of the Swabian Alb north of the Danube as well as in gravels of the Danube and its tributaries. Compared to northern flints, however, this material is relatively coarse-grained and variable within single nodules (Eriksen 1997: 327). Other raw materials used during these periods include radiolarites and other materials from the Alps, available in the gravels south of the Danube and other cherts and jasper from more distant locations. The growing density of vegetation, including forest underbrush, during the period under consideration may have increased the difficulty of locating high-quality stone raw materials, which would have encouraged attempts to maximize the utility of such materials (Eriksen 1997: 328). As mentioned above, intentional heat-tempering has been primarily interpreted as one means of increasing the ease of working stone materials (Eriksen 1997: 328; Harris 2006: 300–306). In southern Germany, the use of this technique shows a rise in frequency from the Magdalenian to the Early Mesolithic, with general percentages of occurrence in assemblages of 2–8% in the Magdalenian, 3–15% in the Late Paleolithic, and 19–66% in the Early Mesolithic (Harris 2006: 308; Kind 1997: 71; Fig. 9.2). In the Late Mesolithic, however, the frequency drops to 5–13%, which may be explained by the adoption of a different means to increase control over raw material use: the punch technique for removing blanks (Kind 1997: 58, 2012: 82).

Environmental and Technological Convergence

Despite the specific and progressive changes in vegetation and animal species creating very different environments through this period of cultural change, the structure of the habitat appears to show a nonlinear alternation from periods of relatively high

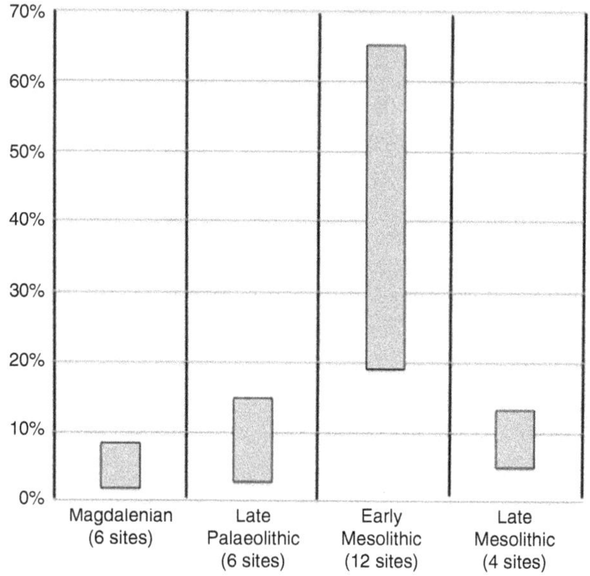

Fig. 9.2 Average percentage of heat-tempered artifacts in assemblages by period

spatial predictability (Magdalenian) to low spatial predictability (Late Paleolithic) to low–moderate predictability (Early Mesolithic) and finally to relatively high predictability again (Late Mesolithic). If so, it may be hypothesized that, despite specific differences in stone tool typology, elements of technological organization that are responsive to spatial and temporal predictability would be similar in the Magdalenian and Late Mesolithic and would differ from the Late Paleolithic and to some extent from the Early Mesolithic as well.

Gearing Up

If resources are relatively predictable, tools can be made in advance to suit anticipated needs. Given the relatively high spatial and temporal predictability of reindeer, the major resource for Magdalenian hunters during the fall and winter, lithic tools could, indeed, be made in advance. As Fisher (2000: 296) notes, "Tool-makers producing implements in advance of targeted, time-stressed subsistence activities can be expected to carefully produce tool forms according to design criteria chosen for their effectiveness, minimizing deviation from a desired shape or size." Among criteria for identifying such gearing up in lithic assemblages, she proposes that "a high ratio of laminar to flake debitage, consistent control over platform angle and shape, and consistent width and thickness of blades…[would be] aspects of debitage assemblages reflecting 'gearing up' for targeted subsistence activities" (Fisher 2000: 322). In her study of south German sites, Fisher (2000:325) documents a decrease in laminar debitage as a steady decline in the average proportion of blades among all blades and flakes, from 50% in the Magdalenian, to 32% in the Late Paleolithic, to 25% in the Early Mesolithic. In a separate study, Harris (2006:278) finds that the percentage of blades increases again to 53% in the Late Mesolithic (Fig. 9.3). Control over the preparation of cores appears to show a similar pattern. For example, there is a significant decrease in the percentage of faceted platforms from the Magdalenian to the Late Paleolithic, followed by an increase from the Late Paleolithic to the Early Mesolithic and particularly to the Late Mesolithic (Fisher 2000: 331; Harris 2006: 313; Fig. 9.4). Fisher's work (2000: 331) also documents an increasing coefficient of variation of core length from the Magdalenian to the Early Mesolithic.

Another indication of gearing up is the differential distribution among sites of intensive lithic reduction and tool manufacture. Anticipatory production for expected tool needs should lead to the concentration of bouts of production in certain sites and much less so in others. As Fisher (2000: 273) notes, "During the Magdalenian, backed bladelets are discarded in highly varying numbers, suggesting that their manufacture and use is not evenly or continuously distributed through time and space, but clusters around anticipated food-getting events at particular kinds of locations." In the Late Paleolithic, by contrast, the evidence of tool manufacture and repair is widespread, occurring at most sites, and is characterized by diverse core types with little preshaping and relatively more flakes than more carefully made blades. Similarly, tool manufacture is a common component of many large and small Early Mesolithic sites as well. Kind (1996), however, does point out contrasts among sites in the intensity of primary reduction as indicated by the

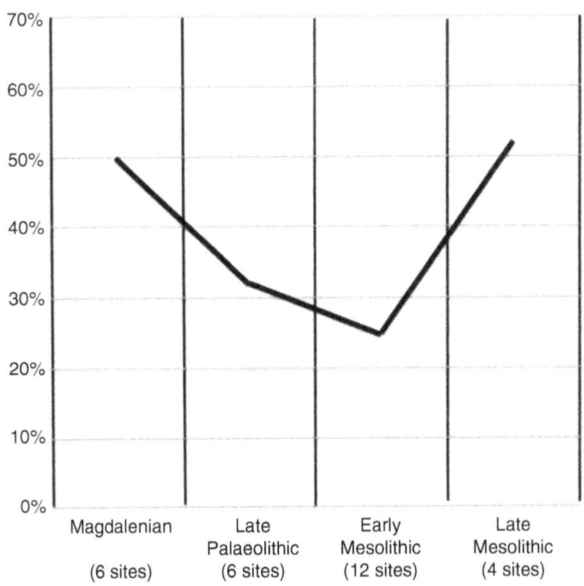

Fig. 9.3 Average percentage of blades in assemblages by period

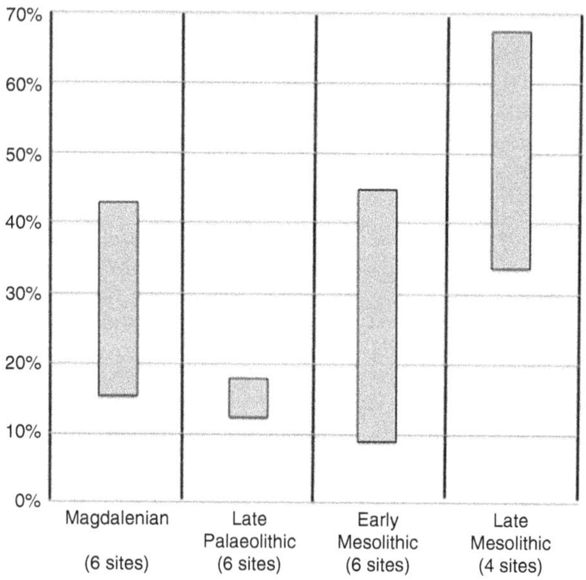

Fig. 9.4 Average percentage of faceted core platforms in assemblages by period

number of cores and uses this to suggest a distinction between residential and logistical camps. A different pattern of Early Mesolithic variation in the lowlands is shown by a contrast between large sites on the shores of large lakes, and those elsewhere. All show evidence of lithic reduction, but the former contain more evidence of lithic reduction and manufacture in the form of percentage of cores (11%–18% vs. 8%–13%) and artifacts with cortex (37%–50% vs. 29%–38%) (Jochim 2006). All of these sites appear to be residential or base camps, to judge from a high diversity of tool types as well as still significant evidence of lithic reduction. By the Late Mesolithic, on the other hand, a number of sites such as Henauhof Nord II show very little evidence of lithic manufacture. This site contains only one core and little primary debitage, but numerous, relatively standardized blades and blade segments of nonlocal chert, apparently manufactured in advance elsewhere and transported to this location.

Furthermore, in situations of relatively high spatial patchiness and predictability, individuals know with some certainty, and in advance, where resources are likely to be found, as well as what implements will be needed. These necessary items can be manufactured ahead of time and carried to the selected locations. The implements may be made from raw material available at the site of manufacture or from materials, usually of high quality, that have been brought to it. The locus of manufacturing should contain abundant evidence for the manufacture in terms of cores and debitage. The destination site should show evidence of imported tools made from material not necessarily found near that location and whatever products of manufacture that do occur, such as cores and debitage, could largely reflect raw materials local to that site. Consequently, there should be significant variability among sites in the intensity of manufacture and in the kinds and amounts of different raw materials represented in cores versus finished tools and blades. The destination sites might be either logistical camps targeting particular areas and resources or residential camps in areas of low-quality raw material.

Therefore, another source of evidence of the degree of anticipatory lithic production is the stone raw material represented by cores compared to that of tools and blades. In the case of the Late Mesolithic site of Henauhof Nord II, for example, the single core is of local radiolarite, whereas the bulk of the blades and tools are of nonlocal Jurassic chert and were evidently made elsewhere. Although the sample is small and available for only the later three time periods, an examination of the importance of nonlocal material in cores vs. blades and tools at different sites reveals some differences through time (Fig. 9.5). Both the Late Paleolithic and Early Mesolithic samples show remarkable similarities across sites between these two measures: when nonlocal material dominates the tools and blades, so too does it dominate the cores, and vice versa, suggesting ongoing manufacture with little tool transport. In the small Late Mesolithic sample, on the other hand, the percentage of nonlocal material utilized in tools and blades is much higher than that of the cores. This is true of the apparently specialized, logistical camp of Henauhof Nord II mentioned above, but also of the residential camp of Sieben Linden 3–5 (Kind et al. 2012: 77).

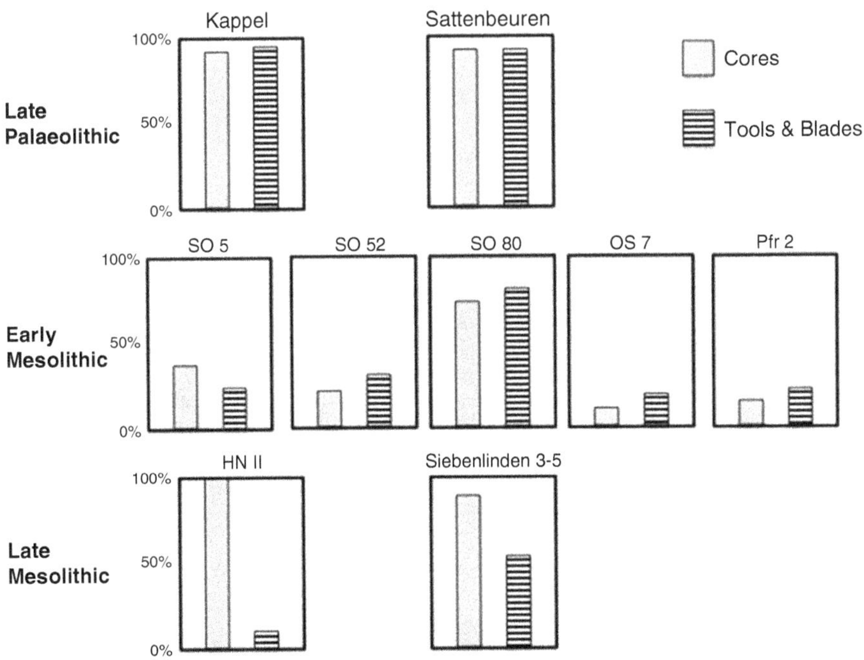

Fig. 9.5 Percentage of local (<10 km) raw material in cores and in blades in selected sites of the Late Paleolithic, Early Mesolithic, and Late Mesolithic

Summary and Conclusions

Despite considerable differences in typology, the lithic assemblages of the Magdalenian and Late Mesolithic of southwestern Germany share a number of characteristics that differ from those of the intervening periods of the Late Paleolithic and Early Mesolithic. These characteristics include the percentage of carefully made blades, the frequency of core preparation, the degree of variation among cores, the distribution of lithic manufacturing debris among sites, and the proportion of obviously imported tools and blades. All of these characteristics reflect the practice of anticipatory manufacture and, in turn, appear to be attuned to structural aspects of the environment. In both cases, the spatial predictability of the resource location is relatively high, allowing for advance planning of tool needs. For the Magdalenian, this derives from the differential distribution of steppe and tundra elements between uplands and lowlands, as well as the relatively predictable behavior of reindeer. For the Late Mesolithic, the diverse forest vegetation created a heterogeneous mosaic of richer and poorer patches, some of which could be seasonally targeted. The intervening periods, by contrast, were characterized by greater habitat homogeneity and less resource predictability. In some ways, the environmental changes of the late and postglacial periods came full circle in terms of general habitat structure and the technology was adjusted accordingly.

Acknowledgments I have relied heavily on the research by Lynn Fisher and Susan Harris in this chapter and want to express my thanks and appreciation for their work and their friendship over the years.

References

Auclair, A., & Goff, G. (1971). Diversity relations of upland forests in the western Great Lakes area. *The American Naturalist, 105*(946), 499–528.

Eicher, U., & Siegenthaler, U. (1976). Palynological and oxygen isotope investigations on late-glacial sediment cores from Swiss lakes. *Boreas, 5*, 109–117.

Eriksen, B. (1991). *Change and continuity in a prehistoric hunter-gatherer society, Archaeologica Venatoria 12*. Tübingen: Archaeologica Venatoria.

Eriksen, B. V. (1996). Resource exploitation, subsistence strategies, and adaptiveness in late Pleistocene-early Holocene northwest Europe. In L. Straus, B. V. Eriksen, J. M. Erlandson, & D. R. Yesner (Eds.), *Humans at the end of the ice age: The archaeology of the Pleistocene-Holocene transition* (pp. 101–128). New York: Plenum.

Eriksen, B. V. (1997). Implications of thermal pretreatment of chert in the German Mesolithic. In R. Schild & Z. Sulgostowska (Eds.), *Man and Flint: Proceedings of the VIIth International Flint Symposium* (pp. 325–329). Warsaw: Institute of Archaeology and Ethnology, Polish Academy of Sciences.

Esterhues, K., Lechterbeck, J., Schneider, J., & Wolf-Brozio, U. (2002). Late- and post-glacial evolution of Lake Steisslingen (I): Sedimentary, history, palynological record and inorganic geochemical indicators. *Palaeogeography, Palaeoclimatology, Palaeoecology, 187*, 341–371.

Fahlke, J. (2009). Der Austausch der terrestrischen Säugetierfauna and der Pleistozän/Holozän-Grenze in Mitteleuropa (ungedr. Diss. Universität Bonn 2009). http://hss.ulb.uni-bonn.de/2009/1836-1.pdf

Fisher, L. (2000). *Land use and technology from Magdalenian to Early Mesolithic in Southwestern Germany.* Unpublished Ph.D. dissertation, University of Michigan, Ann Arbor.

Gebert, C., & Verheyden-Tixier, H. (2001). Variations of diet composition of red deer in Europe. *Mammal Review, 21*, 189–201.

Harris, S. (2006). *Settlement pattern and lithic tool technology changes during the Mesolithic of Southwest Germany.* Unpublished Ph.D. dissertation, University of California, Santa Barbara.

Jochim, M. (1998). *A hunter-gatherer landscape, southwest Germany in the late Palaeolithic and Mesolithic*. New York: Plenum Press.

Jochim, M. (2006). Settlement variability in the early Mesolithic of southwestern Germany. In C.-J. Kind (Ed.), *After the ice age: Settlements, subsistence and social development in the Mesolithic of Central Europe, Materialhefte zur Archäologie 78* (pp. 175–180). Stuttgart: Konrad Theiss Verlag.

Jochim, M., Glass, M., Fisher, L., & McCartney, P. (1998). Mapping the stone age: An interim report on the south German survey project. In N. Conard (Ed.), *Aktuelle Forschungen zum Mesolithikum/Current Mesolithic Research, Urgeschichtliche Materialhefte 12* (pp. 121–132). Tübingen: Mo Vince Verlag.

Kind, C.-J. (1987). *Das Felsställe, Forschungen und Berichte zur Vor- und Frühgeschichte in Baden-Württemberg 23*. Stuttgart: Konrad Theiss Verlag.

Kind, C.-J. (1996). Bemerkungen zur Diversität des südwestdeutschen Frühmesolithikums. In I. Campen, J. Hahn, & M. Uerpmann (Eds.), *Spuren der Jagd – Die Jagd nach Spuren, Tübinger Monographien zur Urgeschichte 11* (pp. 325–330). Tübingen: Mo Vince Verlag.

Kind, C.-J. (1997). *Die letzten Wildbeuter, Materialhefte zur Archäologie*. Stuttgart: Konrad Theiss Verlag.

Kind, C.-J., Beutelspacher, T., David, E., & Stephan, E. (2012). *Das Mesolithikum in der Talaue des Neckars*. Stuttgart: Konrad Theiss Verlag.

Lang, G. (2006). Late-glacial fluctuations of timberline in the Black Forest (SW Germany). *Vegetational History and Archaeobotany, 15*, 373–375.

Leesch, D. (1997). *Un Campement Magdalenien au Bord du Lac de Neuchatel: Cadre Chronologique et Culturel, Mobilier et Structures, Analyse Spatiale (Secteur 1)*. Neuchatel: Musee Cantonal d'Archeologie.

Leesch, D., Müller, W., Nielsen, E., & Bullinger, J. (2012). The Magdalenian in Switzerland: Re-colonization of a newly accessible landscape. *Quaternary International, 30*, 1–18.

Magny, M. (2004). Holocene climate variability as reflected by mid-European lake-level fluctuations and its probable impact on prehistoric human settlements. *Quaternary International, 113*, 65–79.

Morel, P., & Müller, W. (1997). *Un Campement Magdalenien au Bord du Lac de Neuchatel: Etude Archeozoologique (Secteur 1)*. Neuchatel: Musee Cantonal d'Archeologie.

Odum, E. (1969). The strategy of ecosystem development. *Science, 164*(3877), 262–270.

Peyrona, O. (2005). Late-glacial climatic changes in eastern France (Lake Lautrey) from pollen, lake-levels, and chironomids. *Quaternary Research, 62*, 197–211.

Phillips, R., Berg, W. E., & Siniff, D. B. (1973). Moose movement patterns and range use in northwestern Minnesota. *Journal of Wildlife Management, 37*(3), 266–278.

Röhrig, E., & Ulrich, B. (1991). *Temperate deciduous forests*. Amsterdam: Elsevier.

Rösch, M. (1990). Vegetationsgeschichtliche Untersuchungen im Durchenbergried. *Sidelungsarchäologie im Alpenvorland, 2*, 9–64.

Schuler, A. (1994). *Die Schussenquelle, Materialhefte zur Archäologie*. Stuttgart: Konrad Theiss Verlag.

Torke, W. (1981). *Fischreste als Quellen der Ökologie und Ökonomie in der Steinzeit Südwest-Deutschlands, Urgeschichtliche Materialhefte 4*. Tübingen: Verlag Archaeologica Venatoria.

Van Vuure, C. (2005). *Retracing the aurochs: History, morphology and ecology of an extinct wild ox*. Sofia: Pensoft Publishers.

Weniger, G. (1982). *Wildbeuter und ihre Umwelt: Ein Beitrag zum Magdalenien Südwestdeutschlands aus ökologischer und ethnoarchäologischer Sicht*. Tübingen: Archaeologica Venatoria.

Wiens, J. (1976). Population responses to patchy environments. *Annual Review of Ecology and Systematics, 7*, 81–120.

Winterhalder, B. (1980). Environmental analysis in human evolution and adaptation research. *Human Ecology, 8*(2), 136–170.

Winterhalder, B. (1983). Boreal foraging strategies. In A. T. Steegmann (Ed.), *Boreal forest adaptations* (pp. 201–242). New York: Plenum.

Chapter 10
Technology and Human Response to Environmental Change at the Pleistocene-Holocene Boundary in Eastern Beringia: A View from Owl Ridge, Central Alaska

Angela K. Gore and Kelly E. Graf

Introduction

Archaeological investigations in Alaska are significant in providing information about initial human occupation of Beringia, the entry point from an Asian homeland for first Americans (Meltzer 2004; Goebel et al. 2008). Recent research in eastern Beringia has revealed a complex record of terminal Pleistocene-aged sites important to understanding how the Americans were settled. Shortly after initial colonization of eastern Beringia, so far identified at the Swan Point site and dated to ~14,100 calendar years before present (cal BP) (Potter et al. 2014a), the Beringian record became highly variable. One case of this variability comes from central Alaska and is represented by two technological complexes, Nenana and Denali (Powers and Hoffecker 1989; Hoffecker 2001; Graf and Bigelow 2011; Graf et al. 2015).

We explore this variability in central Alaska by examining how early and later inhabitants of the Owl Ridge site organized their technologies in response to Late Pleistocene and early Holocene environmental fluctuations. We use the established terms, Nenana complex and Denali complex, heuristically, not in an attempt to define human groups or archaeological traditions but to classify observed technologies that represent technological strategies humans adopted while responding to past environmental change. We focus specifically on lithic raw material (or toolstone) procurement and selection behaviors to explain how humans responded to climate change during this interval while arriving in central Alaska and subsequently settling in the region.

A.K. Gore (✉) • K.E. Graf
Center for the Study of the First Americans, Department of Anthropology, Texas A&M University, College Station, TX, USA
e-mail: angela_gore@tamu.edu; kgraf@tamu.edu

E. Robinson, F. Sellet (eds.), *Lithic Technological Organization and Paleoenvironmental Change*, Studies in Human Ecology and Adaptation 9,
DOI 10.1007/978-3-319-64407-3_10

Background

Archaeological Context

As mentioned above, the earliest unequivocal evidence of humans in eastern Beringia comes from Swan Point, located in the middle Tanana Valley 100 km southeast of Fairbanks, Alaska, dating to 14,100 cal BP, and containing a Siberian late Upper Paleolithic technology based on wedge-shaped microblade-core production (Gomez Coutouly 2011, 2012; Holmes 2011). Following this, humans continued occupying central Alaska through the Late Pleistocene and early Holocene (Potter 2008; Graf and Bigelow 2011), but toolkits changed. The regional pattern of technological variability that emerged after initial exploration has led some to recognize a Nenana complex chronologically and technologically distinct from the Denali complex first identified by West (1967). In this view, Nenana complex assemblages, found at several multicomponent sites in the Nenana and Tanana valleys, date to 13,500–13,000 cal BP and contain unifacial tools (end scrapers, retouched blades and flakes, gravers, and wedges), diagnostic Chindadn-type bifacial points that are teardrop-shaped or triangular-shaped and sometimes only marginally retouched, other bifaces, and cobble tools (Powers and Hoffecker 1989; Hoffecker et al. 1993; Goebel et al. 1991; Yesner 1996, 2001; Hoffecker 2001; Graf and Goebel 2009; Graf and Bigelow 2011; Graf et al. 2015). Similar technologies dating to the same period of time have even been reported from the Ushki Lake and Berelekh sites in western Beringia (Dikov 1977; Mochanov 1977; Goebel et al. 2003, 2010; Pitulko 2011).

In contrast, Denali complex assemblages, many from the same multicomponent sites with temporally and stratigraphically distinct lower Nenana complex components, date to 12,600–10,000 cal BP and contain toolkits with lanceolate and concave-based bifacial points, unifacial tools (side scrapers and retouched flakes), as well as burin and microblade technologies. In Nenana Valley sites, lanceolate and concave-based points, burins, and microblade technologies are absent from older Nenana complex components (i.e., Owl Ridge, Dry Creek, Walker Road, Moose Creek) (Powers and Hoffecker 1989; Pearson 1999; Hoffecker 2001; Graf and Bigelow 2011; Graf et al. 2015). During recent investigations of the Teklanika West site, however, a lanceolate point was found in what appears to be a compressed stratigraphic context and palimpsest situation, where two horizontally overlapping artifact zones (components 1 and 2) were found in the same sedimentological unit unseparated by sterile deposits and associated with faunal remains dating to 13,100–9700 cal BP. Coffman (2011: 106) concluded that the lanceolate point could be associated with component 1 but acknowledged it could be intrusive from component 2.

Mostly because a very early microblade-bearing component at Swan Point was found to predate 14,000 cal BP, but also because two sites in the Tanana valley continue through the terminal Pleistocene to have bifacial points resembling Chindadn points from Nenana complex sites in the Nenana Valley, some archaeologists argue Nenana complex and Denali complex variability represents different behavioral facies of a pan-Beringian archaeological tradition lasting >4000 years (Holmes 2011; Potter et al. 2014a) and presumably reflects no significant adaptive change to major climatic fluctuation over this timeframe. Thereby, depending on the situation, people selected different technological strategies, bifacial versus composite osseous-microblade hunting weapons, for different immediate needs such as hunting different animals during different seasons, extracting resources in uplands versus lowlands, or proximity to toolstone sources (Holmes 2001; Gal 2002; Potter 2005; Wygal 2009, 2011; Graf and Bigelow 2011). A major issue with this reasoning is that we should expect to find Nenana and Denali complex artifacts together at some sites, but we do not observe this pattern. The only exception is the stratigraphically problematic Healy Lake site, where multiple components may have been excavated together as one (Erlandson et al. 1991; Cook 1996; Hamilton and Goebel 1999). Additionally, faunal data do not support expectations of the related different-animal-during-different-seasons hypothesis. From the Dry Creek site, fauna found in both the Nenana and Denali components indicates hunting activities during the same season (late fall/winter) as well as hunting of the same animal type (Dall sheep) with the different weapon-system technologies (first Chindadn points, then osseous-microblade composite and lanceolate points). At Broken Mammoth, hunters used the same weapon system (Chindadn points) to dispatch different animal types during different seasons. Clearly, we cannot simply claim that microblade technology was selected only during a specific season and for a specific animal type compared with bifacial technologies. We argue the use of a broad-sweeping "Beringian Tradition" oversimplifies complex patterns observed in the early Beringian record and lumping together varied technological strategies found in stratigraphically and temporally discrete contexts obscures evident variability that needs to be explained.

At least three sites in the Nenana Valley contain both Nenana and Denali assemblages in stratigraphically and chronologically separate geological contexts: Owl Ridge, Dry Creek, and Moose Creek (Pearson 1999; Graf and Bigelow 2011; Graf et al. 2015). Historically, proponents of separating Nenana and Denali complexes have argued this variability resulted from two different populations settling central Alaska from Northeast Asia (Goebel et al. 1991; Hoffecker et al. 1993; Hoffecker and Elias 2007). This interpretation certainly fits well with the recently proposed Beringian standstill model for development of Native American genetic population differentiation hypothetically staged in Beringia or far Northeast Asia (Tamm et al. 2007; Mulligan and Kitchen 2014; Raghavan et al. 2015). The hypothesis of different Beringian populations with different toolkits is difficult to test without abundant human skeletal remains preserving ancient DNA that would provide population-level genetic information. Recent skeletal finds associated with Denali complex technology at the Upward Sun River site in the Tanana Valley (Potter et al. 2011, 2014b) evidence at least two mtDNA clades present in the same population, giving us important clues about social organization at this time and genetic relatedness of early Holocene Alaskans with other Native Americans (Tackney et al. 2015); however, we need preserved human DNA from earlier Beringian sites with Nenana complex technology to begin to test the different populations hypothesis.

What does the chronological patterning of Beringian archaeological variability mean? We are more interested in understanding whether the patterns of variability can be explained as human response to variation in resource distribution resulting from climate change (following Mason et al. 2001; Graf and Goebel 2009; Graf and Bigelow 2011; Wygal 2011). We contend humans will select necessary tool-provisioning strategies to be successful in a given environmental situation and perceived landscape. In this paper we consider the observed differences between Nenana and Denali complexes the result of humans selecting different hunting strategies as they became increasingly familiar with the local landscape and responded to climate change and shifts in habitat and resource availability. Before delving into the details of our lithics study, we first review the central Alaskan paleoecological record to establish ecological parameters humans faced at the Pleistocene-Holocene boundary.

Paleoenvironmental Context

Paleoecologists working in Alaska have long been interested in identifying climatic fluctuations between about 15,000 and 10,000 cal BP. Therefore, the paleoenvironmental record for the region is reasonably robust and can be used to infer major climatic events and changes in biome composition. Such data allow us to predict resource distributions for humans inhabiting the region and provide a means to evaluate paleoecological constraints faced by the region's earliest inhabitants. In particular we are interested in the effects of Northern Hemispheric climatic events such as the Older Dryas, Allerød, Younger Dryas, and Holocene Thermal Maximum on central Alaskans (Bigelow and Edwards 2001; Bigelow and Powers 2001; Kaufman et al. 2004; Kokorowski et al. 2008; Graf and Bigelow 2011). These specific climatic events characterize the time before, during, and after hunter-gatherers inhabited Owl Ridge.

Regional late glacial pollen records, predating 14,000 cal BP, indicate herb-tundra vegetation. The landscape would have been open with few trees and dominated by an herbaceous plant combination of short grasses, sedges, and *Artemisia* sp. (Bigelow and Powers 2001; Anderson et al. 2004). Animals would have included woolly mammoth, horse, bison, wapiti, and moose as well as other smaller species (Guthrie 2017, 2006; Meiri et al. 2014). By about 14,000 cal BP, a birch and willow shrub-tundra vegetation community came to dominate the region (Bigelow and Powers 2001; Anderson et al. 2004; Brubaker et al. 2005). Rises in lake levels through the Allerød (14,000–13,000 cal BP) indicate relatively warmer temperatures and higher humidity than immediately before or after this time (Abbott et al. 2000; Bigelow and Edwards 2001). As a result, obligate grazers such as horse and mammoth went extinct by 13,500 cal BP, while bison and wapiti (grazers who also browse) and moose (an obligate browser) populations were maintained (Guthrie 2006), and abundance of waterfowl in the Broken Mammoth faunal assemblage indicates the presence and use by humans of more mesophilic species (Yesner 2007).

In some regions of the Northern Hemisphere, the Younger Dryas was not significantly felt, but in northern latitudes its effects were more pronounced (Kokorowski et al. 2008). In fact, central Alaskan paleoecological records suggest much drier conditions, especially north of the Alaska Range due to an interior Alaskan rain-shadow effect, reflected by a significant increase in *Artemisia* sp. pollen, lowered lake levels, and deposition of eolian sand layers (Hu et al. 1993; Bigelow et al. 1990; Abbott et al. 2000; Bigelow and Edwards 2001; Bigelow and Powers 2001). Bison and wapiti populations were maintained during this arid interval; however, moose became far less prevalent (Guthrie 2006). Archaeological sites in the region also indicate the presence of caribou (Yesner 2001, 2007; Bowers and Reuther 2008).

Within a few centuries following the Younger Dryas and by 11,000 cal. BP, the onset of the Holocene Thermal Maximum had begun with expansion of *Populus*, representing the first trees to inhabit the Alaskan interior since marine oxygen isotope stage (MIS) 3 (~35,000–26,000 cal. BP). *Populus* is known to be cold-tolerant yet thrives in warm summer conditions. Regional lake levels were lower than today, indicating an early Holocene climate warmer and drier, especially during summer months. Following 10,000 cal BP, *Picea* spread to the region and lake levels increased, signaling a shift from an open-forest parkland to boreal-forest biome and the relatively warm, moist conditions of today (Abbot et al. 2000; Barber and Finney 2000; Bigelow and Powers 2001; Lloyd et al. 2006). Faunal compositions during the early Holocene also mimic the later Holocene pattern with wapiti extinct, but populations of moose and bison maintained (Guthrie 2006).

The paleoecological record of central Alaska indicates initial migrants from Siberia were faced with a frigid, dry landscape with little woody vegetation for fire production and maintenance at 14,100 cal BP, though large mammal populations of the herb tundra would have provided high-protein resources and a source of slow-burning fuel once a fire could be established with wood (Crass et al. 2011). A fire fueled with bones, however, burns with a high flame and does not carry embers, so it is good for lighting, drying, and curing, but not necessarily for more thorough cooking (Théry-Parisot et al. 2002). Perhaps this is why only one interior Alaskan archaeological site to date has been recorded for the period just prior to the Allerød (Hoffecker and Elias 2007). During the Allerød, wetter conditions resulted in spread of shrub-tundra vegetation increasing burning opportunities for people so they could maintain fires for both cooking and curing as well as drying and warmth. Bison, wapiti, and moose were available for hunting and so were smaller wetland resources, such as waterfowl. During the Younger Dryas, a brief reversal to drier conditions meant that more mesophilic taxa, such as moose, were less available for human use (Yesner 2007). Following the Pleistocene, warmer and eventually more humid conditions returned and persisted, altering the biome of central Alaska. The eventual emergence of the boreal forest led to lower numbers and more dispersed large fauna with bison relegated to lowland settings, wapiti eventually becoming extinct locally, and solitary moose widely dispersed across the landscape.

Below we use the archaeological record from Owl Ridge to test the hypothesis that technological changes during the terminal Pleistocene resulted from human response to climate change and associated changes in fuel and food resource distributions. We expect that human decisions to select specific adaptive strategies are reflected in the technologies they used and that these decisions were made in response to environmental change, such as change in composition, proportion, and distribution of natural resources around them (Nelson 1991; Kuhn 1995; Elston and Brantingham 2002; Andrefsky 2009; Graf 2010; Graf and Bigelow 2011).

Materials and Methods

Owl Ridge Basics

Owl Ridge is located in the northern foothills of the Alaska Range along the Teklanika River, a glacially fed tributary to the Nenana River (Fig. 10.1). The site is situated in interbedded loess, cliff-head sand, and colluvial deposits capping a glacial outwash terrace of the Teklanika River and resting approximately 61 m above the confluence of the river and First Creek, a small clear stream draining the immediate foothills. Given conditions of the herb-tundra and shrub-tundra landscape, this location would have provided hunter-gatherers of the terminal Pleistocene an advantageous, unobstructed view of game and lithic resources located in the surrounding area as well as a source of clear water. The Owl Ridge site was initially discovered

Fig. 10.1 Map of the Nenana and Teklanika River Valleys with the location of the Owl Ridge site (**a**). Picture of two rock samples from the glacial outwash terrace that the site rests on (**b**). Picture of location of Owl Ridge relative to the Teklanika River and First Creek floodplains (**c**)

in 1976 during a backcountry survey of the Teklanika River (Plaskett 1976), and it was tested in 1977–1979 and 1982–1984 by the University of Alaska Fairbanks archaeologists. Following the 1980s testing project, three cultural components were identified. Based primarily on stratigraphy and several conventional radiocarbon (^{14}C) dates, Phippen (1988) assigned the lowermost component to the then recently defined Nenana complex and the upper two components to the Denali complex, one dating to the Younger Dryas and the other dating to the middle-late Holocene. In 2007, 2009, and 2010, we returned to Owl Ridge to conduct full-scale excavations, opening an additional 54 m^2. We found site deposits to be approximately 125 cm thick, consisting of three sandy loams, separated by two sand layers (Fig. 10.2). The sandy loams represent three loess-deposition events: loess 1, loess 2, and loess 3. The lowermost sand, sand 1, is a relatively thin eolian deposit, most likely resulting from cliff-head sand deposition, and the upper sand, sand 2, is a thick set of colluvial deposits.

Three cultural components were found in three stratigraphically separated strata. The earliest, component 1, was found in the upper 5 cm of loess 1. One conventional ^{14}C date obtained by Phippen (1988) on a bulk charcoal sample provided an age of 11,340 ± 150 (Beta-11,209) ^{14}C BP, and an additional AMS

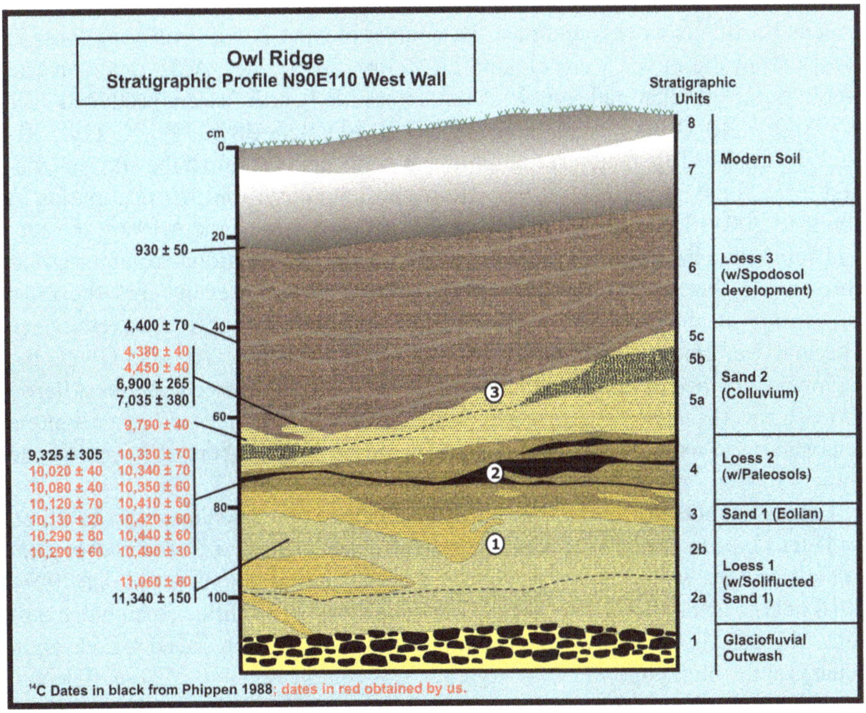

Fig. 10.2 Representative stratigraphic profile of the Owl Ridge site, showing stratigraphic locations of radiocarbon dates obtained and cultural components (*1*, *2*, and *3*) identified during site investigations

date obtained by our team on a single piece of naturally occurring wood (*Salix* sp.) charcoal from loess 1 within a component 1 artifact cluster provided the age of 11,056 ± 59 (AA-86969) ^{14}C BP (Graf and Bigelow 2011). Together these dates indicate a range of occupation of about 13,300–13,000 cal BP (all ^{14}C dates in this chapter were calibrated using the Intcal13 curve in the Calib 7.0.2 downloadable program for MS Windows [Reimer et al. 2013]). Component 1, therefore, dates to the end of the Allerød and immediately prior to the Younger Dryas (Graf and Bigelow 2011). Component 2 artifacts were consistently found associated with a paleosol (buried A/B horizon) in loess 2. In our excavations, we obtained 13 radiocarbon samples of naturally occurring wood (*Salix* sp.) charcoal isolated in the paleosol and found within component 2 artifact concentrations. These dates overlap at 2-sigma standard deviation and range from 10,485 ± 25 (UCIAMS-71261) to 10,020 ± 40 (Beta-289,378) ^{14}C (12,550–11,315 cal) BP, dating the paleosol and deposition of artifacts to the Younger Dryas (Graf et al. 2010; Graf and Bigelow 2011). Given that dated materials and artifacts from component 2 were found in a paleosol of loess 2, signaling a stable surface and relatively mild climate, plus they are directly overlying cliff-head sand deposits signaling a relatively windy, dry period, we argue that locally the Younger Dryas climatic reversal was brief and can be dated to the intervening 450 years between component 1 and component 2 site visits. Finally, component 3 artifacts were found near the contact of sand 2 with overlying loess 3, most within the upper 5 cm of sand 2 (Melton 2015). Two AMS dates on two wood (*Salix* sp.) charcoal samples from a possible hearth feature produced ages of 9880 ± 40 (Beta-330,127) and 9790 ± 40 (Beta-289,379) ^{14}C (11,390–11,170 cal) BP. Together, stratigraphic and AMS data establish the site was visited three times at the Pleistocene-Holocene boundary: the first occupation at about 13,300–13,000 cal BP or during the last centuries of the Allerød; the second some time between 12,550–11,320 cal BP during the global Younger Dryas chronozone, but after the local Younger Dryas climatic event; and the third occupation at about 11,390–11,170 cal BP, immediately before the Holocene Thermal Maximum as forests were emerging in central Alaska. Given the regional sequence of climatic and biome changes that occurred from the Allerød through the Holocene Thermal Maximum, with Owl Ridge, we have a unique opportunity to examine human adaptive response by members of a small-scale society to fairly rapid shifts in local climate.

Lithic assemblages analyzed for this paper include excavated materials collected by Peter Phippen currently housed at the University of Alaska Fairbanks Museum of the North, as well as materials collected from our excavations during the 2007–2010 field seasons. Taken together, the analyzed Owl Ridge lithic assemblage presented here totals 4104 artifacts. An additional 223 artifacts were found in excavation squares at the bluff edge where stratigraphy was compressed into <50 cm of deposits, and assignment of these pieces to specific stratigraphic units and cultural components could not be confidently undertaken, and therefore are omitted from our analysis presented here.

Technological Organization and Human Response to Environmental Change

One way to gain a clearer picture of how people responded to environmental change is to explore how they organized their technologies, subsistence, and land-use strategies. The terminal Pleistocene archaeological record in central Alaska, however, is largely a lithic record. Faunal preservation is almost nonexistent with only a handful of sites preserving identifiable specimens (e.g., Dry Creek, Broken Mammoth, Carlo Creek, Swan Point, Upward Sun River, and Gerstle River Quarry) (Bowers 1980; Yesner 2001; Potter 2007; Graf and Bigelow 2011; Graf et al. 2015), and of these only the Gerstle River Quarry and Broken Mammoth assemblages have been analyzed beyond number of identified specimens (Potter 2007; Yesner 2007). Therefore, with the data at hand, little about subsistence organization can be directly garnered from the record. We are left to rely mostly on the lithic record to reconstruct how people organized themselves on the landscape, how they made a living, and why. Owl Ridge is no exception to this pattern. Here we analyze the lithic assemblages from the site's three terminal Pleistocene components to explore changes in technological organization, provisioning, and use of the lithic landscape.

Because climate in central Alaska was variable during the terminal Pleistocene and resource distributions changed as a result of this variability, we expect humans to have altered their mobility and technological strategies in response. We approach this problem from a human ecology, resilience theory perspective (Redman 2005; Cooper and Sheets 2012; Birks et al. 2015). Humans organize mobility, subsistence, and technological strategies around solving the problem of procuring food (Binford 1980; Bleed and Bleed 1987; Nelson 1991; Kuhn 1995; Morgan 2009). Human interaction with the environment guides technological, subsistence, and land-use decisions. In response to changing climate and resource availability and distribution, humans may show resilience by staying in the changing environment but making necessary alterations to behavioral strategies and adapting to the changing ecosystem. In contrast, however, they may decide to migrate or even resist change and be driven to extinction (Redman 2005; Fitzhugh 2012; Birks et al. 2015). Decisions to alter technological organization or the selection of specific strategies for making, curating, transporting, and discarding tools happen in response to resource distribution, productivity, and predictability (Binford 1979; Shott 1986; Bamforth 1991; Nelson 1991; Andrefsky 2009). To explain technological behavioral patterns reflected in the archaeological record in central Alaska and how these patterns may represent human response to climate and environmental change, this paper will examine toolstone procurement and selection behaviors represented in components 1, 2, and 3 at Owl Ridge. By analyzing variables that inform on ways toolstones were procured then selected for tool manufacture, we can make inferences regarding how site occupants used their landscape. By comparing the cultural occupations, we will detect behavioral responses to environmental change through time.

As central Alaskan climate, biomes, and landscapes changed throughout human occupation (13,300–11,200 cal BP) at Owl Ridge, we expect to see changes in the technological strategies and ways people used the site and surrounding landscape as they responded to these shifts. This will help us document and consider the resilience of hunter-gatherers in the region during the last major global warming event.

We examine lithic raw material availability (lithic landscape), variability, and transport to explain toolstone procurement at the site. The availability and distribution of potential toolstones affect decisions to procure those materials (Kuhn 1995; Andrefsky 2009; Graf 2010). Below, we discuss current knowledge of the lithic landscape local to the site and within the greater Nenana and Teklanika Valleys. We consider frequency of raw material classes, such as cryptocrystalline silicate (CCS) or the fine-grained cherts and chalcedonies, microcrystalline silicate (MCS) or coarse-grained cherts, andesite, basalt, rhyolite, and other less common raw materials or quartzites, granodiorites, and greywackes, by archaeological component to understand toolstone variability. This allows us to assess which available resources in the lithic landscape were economically significant to the inhabitants of Owl Ridge and whether these procurement patterns changed through time. In our analyses we identified toolstones through visual inspection. Geochemical characterization and sourcing studies have been successfully accomplished on Alaskan obsidians (Reuther et al. 2011); however, because obsidian is lacking from the Owl Ridge assemblage and little is known about specific basalt and rhyolite sources in central Alaska (but see Coffman and Rasic 2015 for preliminary investigation of rhyolite use), we did not use geochemical characterization to identify specific raw material source locations. One of us (Gore) is currently working on geochemical characterization of all local basalts and andesites from the Nenana Valley. In this study, we identified the presence of cortex on toolstones to explore relative degrees of transport. We assume specific toolstone types always found without cortex originated offsite and were not locally procured. Toolstone types expressing cortex, especially alluvial cobble cortex, were locally procured on-site in the glacial outwash or nearby in the creek and river alluvium. Variables we used to highlight toolstone transport behaviors include number of toolstone types expressing cortex and, therefore, representing locally procured raw materials, frequency of nonlocal toolstone types in the site assemblage, and frequency of local versus nonlocal toolstones by component.

We used three integrative lithic variables to explore technological activities: primary versus secondary reduction activities, formal versus informal technologies, and bifacial versus unifacial technologies (Graf and Goebel 2009). To understand how Owl Ridge foragers selected toolstones for these activities, we considered each variable first by toolstone type and second by nonlocal/local toolstone. Primary reduction artifacts related to core reduction and tool-blank production include cores, cortical spalls, flakes (>1 cm^2 in total dimension), blade-like flakes, bladelets, microblades, technical spalls (diagnostic of blade or microblade-core production), and angular shatter (Graf and Goebel 2009).

Secondary reduction artifacts related to tool manufacture and rejuvenation include tool trimming flakes or "retouch chips" (<1 m^2 in total dimension), biface-thinning flakes, burin spalls, and tools (Graf 2008; Graf and Goebel 2009). Formal technologies include prepared cores (blade and microblade cores) and tools manufactured to have long use-life histories (bifaces, side scrapers, end scrapers, and combination tools). Informal or expediently produced cores and tools include flake cores, tested cobbles, retouched flakes and blades, gravers, burins, and cobble tools. These tools evidence little retouch, shaping, preparation, and short use-life histories (Kuhn 1995; Graf 2008, 2010; Andrefsky 2009; Graf and Goebel 2009). Bifacial technologies include all bifaces and bifacial thinning flakes (Graf 2008; Graf and Goebel 2009). Unifacial technologies include all unifacial tools and retouch chips with smooth platforms, representing debitage removed from unifacial edges (Graf 2008; Graf and Goebel 2009).

Results

Character of Owl Ridge Lithic Assemblages

The analyzed Owl Ridge assemblage totaled 4104 artifacts (Table 10.1), 894 from component 1, 1343 from component 2, and 1867 from component 3. Within component 1 there was 1 tested cobble, 870 debitage pieces, and 23 tools. Debitage includes cortical spalls, flakes and flake fragments, blade-like flakes, bladelets, a blade core tablet, angular shatter, retouch chips, biface-thinning flakes, and burin spalls. Four triangular-shaped bifacial points manufactured on flake blanks were identified in the tool assemblage, but only one of these was found in a nearly complete condition, only missing its tip (Fig. 10.3a). Other tools included bifaces, retouched flakes, and an anvil stone. In component 2 there were 9 cores, 1300 debitage pieces, and 34 tools. Cores included tested cobbles and unidirectional flake cores. Debitage consisted of cortical spalls, flakes and flake fragments, blade-like flakes, one proximal blade, microblades, microblade-reduction technical spalls, angular shatter, retouch chips, biface-thinning flakes, and one burin spall. Three lanceolate-shaped bifacial points made on biface tool blanks were identified in the tool assemblage. The rest of the tools included bifaces, a scraper-biface combination tool, side scrapers, end scrapers, a dihedral burin, retouched flakes, and cobble tools (scraper planes, hammerstones, and an abrader). In component 3, there were a total of 9 cores, 1835 debitage pieces, and 23 tools. Cores included tested cobbles, bidirectional flake cores, and a multidirectional flake core. Debitage included cortical spalls, flakes and flake fragments, a blade-like flake, a blade midsection, a microblade, angular shatter, retouch chips, and biface-thinning flakes. Tools consisted of bifaces, a scraper-biface combination tool, side scrapers, an end scraper, retouched flakes, a cobble-spall scraper, a cobble tool, and hammerstones.

Table 10.1 Presentation of artifact types by component

Artifact class	Component 1	Component 2	Component 3
Cores			
Tested cobbles	1 (0.1%)	6 (0.5%)	6 (0.3%)
Unidirectional flake cores	0 (0.0%)	1 (0.1%)	0 (0.0%)
Bidirectional flake cores	0 (0.0%)	0 (0.0%)	2 (0.1%)
Multidirectional flake cores	0 (0.0%)	2 (0.1%)	1 (0.1%)
Subtotal	*1 (0.1%)*	*9 (0.7%)*	*9 (0.5%)*
Debitage			
Cortical spalls	87 (9.7%)	83 (6.2%)	367 (19.7%)
Flakes and flake fragments	527 (58.9%)	694 (51.7%)	1141 (61.1%)
Blade-like flakes	7 (0.8%)	11 (0.8%)	2 (0.1%)
Blades	3 (0.3%)	1 (0.1%)	2 (0.1%)
Microblades	0 (0.0%)	3 (0.2%)	1 (0.1%)
Technical spalls	1 (0.1%)	2 (0.1%)	1 (0.1%)
Angular shatter	9 (1.0%)	65 (4.9%)	78 (4.2%)
Resharpening chips	158 (17.8%)	342 (25.5%)	162 (8.6%)
Biface thinning flakes	77 (8.6%)	95 (7.1%)	81 (4.3%)
Burin spalls	1 (0.1%)	4 (0.3%)	0 (0.0%)
Subtotal	*870 (97.3%)*	*1300 (96.8%)*	*1835 (98.3%)*
Tools			
Bifaces	15 (1.7%)	10 (0.7%)	4 (0.2%)
Side scrapers	0 (0.0%)	4 (0.3%)	3 (0.1%)
End scrapers	0 (0.0%)	2 (0.1%)	1 (0.1%)
Combination tools	0 (0.0%)	1 (0.1%)	1 (0.1%)
Burins	0 (0.0%)	1 (0.1%)	0 (0.0%)
Retouched flakes	7 (0.8%)	5 (0.4%)	7 (0.3%)
Scraper on cobble	0 (0.0%)	0 (0.0%)	1 (0.1%)
Planes	0 (0.0%)	3 (0.2%)	0 (0.0%)
Hammerstones	0 (0.0%)	6 (0.4%)	6 (0.3%)
Anvil	1 (0.1%)	0 (0.0%)	0 (0.0%)
Abraders	0 (0.0%)	1 (0.1%)	0 (0.0%)
Flaked pebble	0 (0.0 %)	1 (0.1%)	0 (0.0%)
Subtotal	*23 (2.6%)*	*34 (2.5%)*	*23 (1.2%)*
Component totals	*894 (100%)*	*1343 (100%)*	*1867 (100%)*

Raw Material Procurement

Lithic Landscape

Today, the local lithic landscape within 5 kilometers surrounding the Owl Ridge site is characterized by glaciofluvial outwash terraces, alluvium and floodplain deposits of the Teklanika River and First Creek, and exposures of adjacent bedrock formations and associated colluvium. Bedrock formations include the Nenana Gravel formation, a Tertiary-aged conglomerate of ancient northern

Fig. 10.3 Representative sample of tools in each component at Owl Ridge. *Component 1* artifacts shown include triangular-shaped Chindadn point (**a**), bifaces (**b–e**), scraper fragment (**f**), and retouched flake (**g**). *Component 2* artifacts include concave-based point (**h**), lanceolate point (**j**), bifaces (**k**, **m**), concave-based point (**h**), double-ended scraper (**i**), and retouched flake fragment (**l**). *Component 3* artifacts include bifaces (**n**, **q**), an end scraper (**r**), a retouched flake (**p**), and a cobble-spall scraper (**o**)

Alaska Range alluvium (Wahrhaftig 1958, 1970a), and the Metamorphic Rocks North of Fish and Panguingue Creeks (MRNFPC) formation complex primarily composed of schist and slate and presumed to date to the Paleozoic/Precambrian (Wahrhaftig 1970a). The site rests directly on the Healy glacial outwash terrace of the Teklanika River, presumed to date to MIS 3 or before (Wahrhaftig 1958; Thorson 1986; Dortch et al. 2010). Glacial outwash in this area contains gravels reworked from the Birch Creek formation in the Alaska Range and from both

Nenana Gravel and MRNFPC formations in the foothills immediately nearby the site. Together the common rock types include gneiss, gabbro, diabase, andesite, basalt, quartz-sericite schist, quartzite, slate, and metachert (Wahrhaftig 1958, 1970a; Wahrhaftig and Black 1958).

A few dispersed basalt and rhyolite dikes, presumed to have formed during the early Tertiary, are mapped in the Birch Creek formation far upslope in the Alaska Range. Today, the nearest of these include several basalt dikes located about 30 km south of the site along the divide (western slope of Mt. Healy) between the Nenana and Teklanika Rivers (Wahrhaftig 1970a). The nearest rhyolite dikes are mapped about 31 km east of Owl Ridge at the headwaters of Eva Creek and 43 km southeast on Sugarloaf Mountain, both locations lie on the east side of the Nenana River (Wahrhaftig 1970b, c). A raw material survey in the immediate vicinity of Owl Ridge during the 2007, 2009, and 2015 field seasons confirmed that all raw material classes discussed above are present in both the glacial outwash on-site and in the creek and river floodplain deposits near the site. These locally available stone clasts come in the form of well-rounded to sub-rounded small boulders, cobbles, and pebbles of more brittle stones (e.g., schist, slate, and metachert) found mostly in the small cobble to pebble sizes (Fig. 10.1b).

Raw Material Variability

Raw material classes present in the Owl Ridge lithic assemblage in order of prevalence included andesite, CCS, MCS, basalt, rhyolite, and other toolstones such as quartzite, granodiorite, and greywacke (Table 10.2). Examining toolstone variability, two general patterns emerged. First, more artifacts were manufactured on andesite than all other raw materials combined, and its use dramatically increased through time. In contrast, the pattern is reversed for the next economically important raw materials. CCS and to a lesser extent MCS decreased in importance through time. Basalt and rhyolite show a similar relationship, where the use of basalt increased through time in tandem with andesite, but rhyolite use decreased through time, similar to CCS and MCS (Table 10.2).

Table 10.2 Toolstone variability by component

Raw material classes	Component 1	Component 2	Component 3	Total
CCS	308 (7.5%)	255 (6.2%)	91 (2.3%)	654 (16.0%)
MCS	209 (5.1%)	190 (4.6%)	95 (2.3%)	494 (12.0%)
Andesite	248 (6.0%)	734 (17.9%)	1241 (30.2%)	2223 (54.1%)
Basalt	37 (0.9%)	48 (1.2%)	256 (6.2%)	341 (8.3%)
Rhyolite	74 (1.9%)	3 (<0.1%)	46 (1.1%)	123 (3.0%)
Others	18 (0.4%)	113 (2.8%)	138 (3.4%)	269 (6.6%)
Total	*894 (21.8%)*	*1343 (32.7%)*	*1867 (45.5%)*	*4104 (100%)*

Raw Material Transport

We expect the presence of cobble cortex on specific raw material types to indicate these as local toolstones, whereas complete absence of cortex on specific raw material types establishes these as nonlocal toolstones. Table 10.3 illustrates the number of individual toolstones never expressing cortex by component and, therefore, the frequency of nonlocal toolstone types by component. Sixty percent of the toolstone types in component 1 are nonlocal varieties, 45% are nonlocal in component 2, and 34% are nonlocal in component 3. The number of nonlocal toolstones transported to the site decreased after initial occupation of the site.

Further examination of which of these toolstone types appear to be local and nonlocal shows some varieties of CCS, MCS, and nearly all rhyolites were nonlocal (Fig. 10.4), whereas all andesites and basalts were procured locally. Examining transport more closely, local toolstones dominated the Owl Ridge assemblage; however, there were

Table 10.3 Frequency of toolstone types never expressing cortex

	Total number of toolstone types	Number of toolstone types without cortex
Component 1	35	21 (60%)
Component 2	31	14 (45%)
Component 3	38	13 (34%)

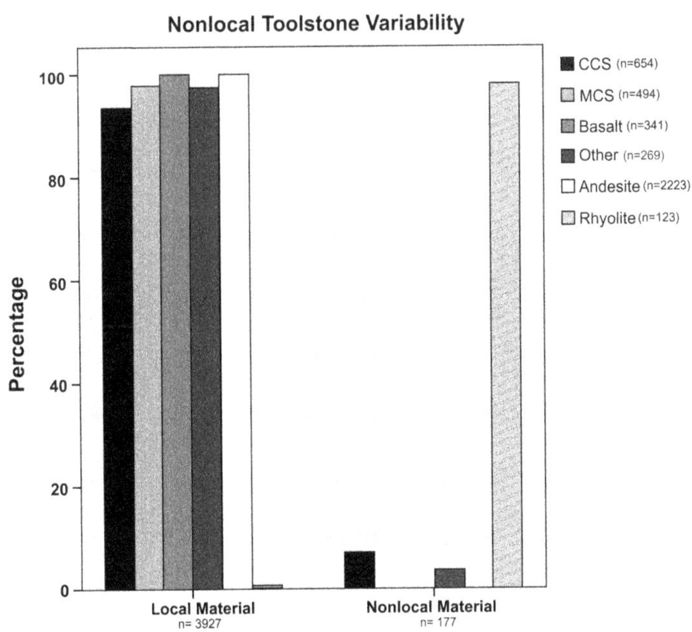

Fig. 10.4 Bar chart expressing which toolstone types are local versus nonlocal

differences through time (Fig. 10.5). Though frequencies of nonlocal toolstones were relatively low in all three components (12–1%), there were significantly more-than-expected nonlocal materials transported to the site by component 1 inhabitants and significantly less than expected procured by both component 2 and component 3 inhabitants of the site. Together, raw material transport variables indicate that through time site occupants became increasingly reliant on the procurement of local toolstones.

Raw Material Selection

Primary and Secondary Reduction Activities

Primary reduction activities dominated all three components with 71% of the component 1, 63% of the component 2, and 85% of the component 3 assemblages comprised of primary reduction pieces; however, there was more-than-expected secondary reduction during the component 1 and component 2 occupation episode but more-than-expected primary reduction during the component 3 occupation (Table 10.4). Technological activities during the occupation events reflected by components 1 and 2 centered more on tool shaping and maintenance, while activities during component 3 occupation centered more on initial steps of tool-blank production.

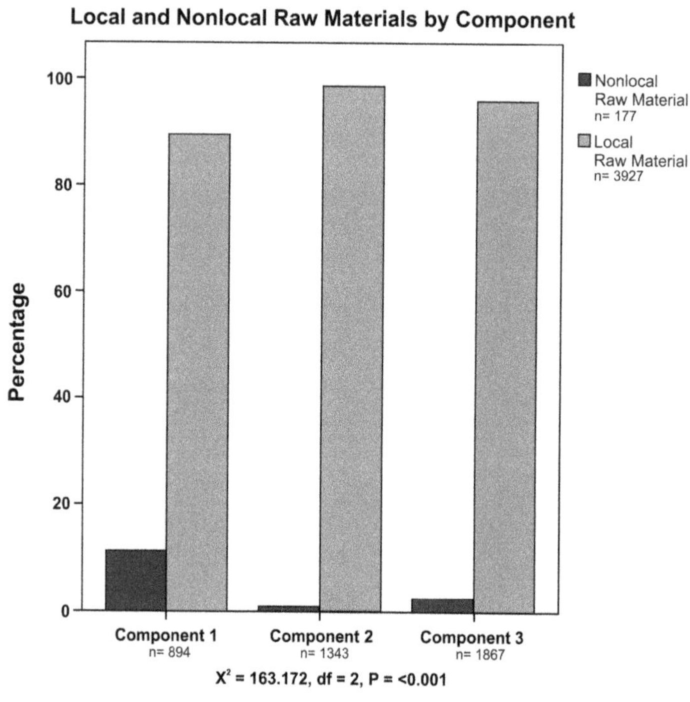

Fig. 10.5 Bar chart showing local and nonlocal toolstones by component

Table 10.4 Primary and secondary reduction activities by component

		Primary	Secondary	Total
Component 1	Count	627	258	885
	Expected count	662.1	222.9	885
	% total (within component)	(70.8%)	(29.2%)	(100.0%)
Component 2	Count	806	472	1278
	Expected count	956.2	321.8	1278
	% total (within component)	(63.1%)	(37.0%)	(100.0%)
Component 3	Count	1523	265	1788
	Expected count	1337.7	450.3	1788
	% total (within component)	(85.2%)	(14.8%)	(100.0%)
Total	Count	2956	995	3951
	Expected count	2956	995	3951
	% of total	(74.8%)	(25.2%)	(100.0%)

$\chi^2 = 202.935$; df = 2; $P < 0.001$. Note no (0.0%) cells have expected counts less than 5. The minimum expected count is 222.87

Generally speaking, chert (CCS and MCS) and/or fine-grained igneous (FGI) toolstones (basalt, andesite, and rhyolite) dominated both primary and secondary reduction activities in all three components (Fig. 10.6), meaning higher-quality, fine-grained toolstones were selected over lower-quality, coarse-grained alternatives. Examining toolstone selection by component for primary versus secondary reduction, we found some interesting patterns. For primary reduction activities, cherts were selected more than expected compared with other toolstones and FGI in component 1, but during the component 2 occupation, chert and other toolstones were selected more than expected compared with FGI, and in component 3 other toolstones and FGI were selected more than expected compared with chert. For secondary reduction activities, component 1 occupants again preferred chert over the other toolstones, component 2 occupants preferred other toolstones and FGI over chert, and component 3 inhabitants selected FGI over the others. Through time, the importance of chert as a toolstone decreased and was eventually replaced by FGI.

When examining reduction activities by local versus nonlocal toolstones, component 1 exhibited greater-than-expected selection of nonlocal toolstones for both primary and secondary reduction activities, whereas both components 2 and 3 evidenced greater-than-expected selection of local raw materials for both primary and secondary reduction (Fig. 10.6b), indicating the use of more nonlocal toolstones during the initial site visit, especially for secondary reduction activities, compared with later visits to the site.

Formal and Informal Technologies

Comparing the frequencies of formal versus informal technologies, both components 1 and 2 had more-than-expected formal technologies, whereas component 3 had less-than-expected formal technologies (Table 10.5). Through time, more effort was spent on production and maintenance of informal technologies at Owl Ridge.

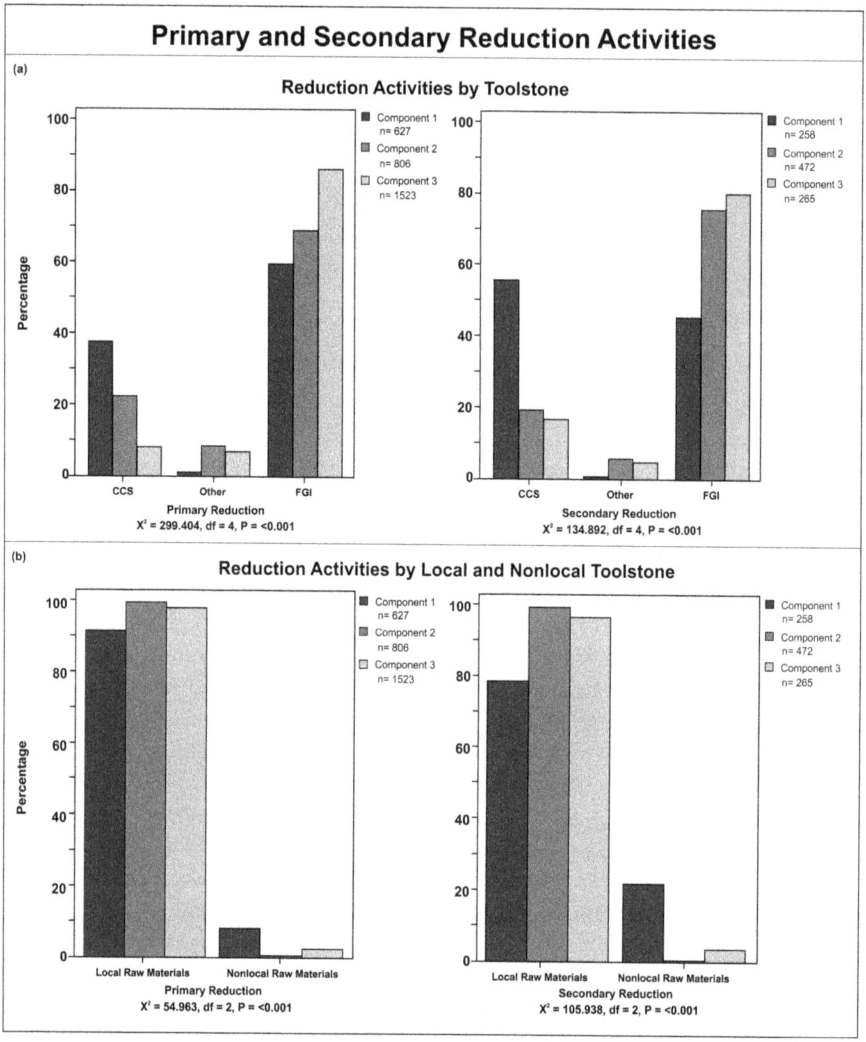

Fig. 10.6 Primary and secondary reduction activities by (**a**) toolstone types and (**b**) local and nonlocal toolstones

Similar to reduction activities, formal and informal technologies were patterned in toolstone selection (Fig. 10.7). For formal technologies, chert was selected at the expense of the other toolstones in component 1, but during the component 2 and component 3 occupations, other toolstones were selected more than chert. For manufacturing informal activities, component 1 occupants again preferred chert over the other toolstones, but component 2 occupants selected both chert and other toolstones over FGI, and component 3 inhabitants preferred both FGI and other toolstones over the chert. Through time, Owl Ridge foragers came to prefer cherts less and FGI more. This is especially true for the production and maintenance of more

Table 10.5 Formal and informal technologies by component

		Formal	Informal	Total
Component 1	Count	262	632	894
	Expected count	221.3	672.7	894
	% total (within component)	(25.8%)	(70.7%)	(100.0%)
Component 2	Count	481	862	1343
	Expected count	332.5	1010.5	1343
	% total (within component)	(35.8%)	(64.2%)	(100.0%)
Component 3	Count	273	1594	1867
	Expected count	462.2	1404.8	1867
	% total (within component)	(14.6%)	(85.4%)	(100.0%)
Total	Count	1016	3088	4104
	Expected count	1016.0	3088.0	4104
	% of total	(24.7%)	(75.3%)	(100.0%)

$\chi^2 = 201.044$; df = 2; $P < 0.001$. Note no (0.0%) cells have expected counts less than 5. The minimum expected count is 221.32

formal technologies. Examining local versus nonlocal selection for production and maintenance of formal versus informal technologies, the main difference between components is nonlocal toolstones were preferred more for formal technologies by component 1 flintknappers, whereas for both components 2 and component 3 occupants preferentially selected local toolstones for both formal and informal technologies (Fig. 10.7b).

Unifacial and Bifacial Technologies

There was no significant difference between the components in the production and maintenance of unifacial versus bifacial industries; however, component 1 had more bifacial and less unifacial technologies present compared with the other two components (Table 10.6). This pattern was upheld when looking at the number of bifacial tools relative to unifacial tools in Table 10.1. Toolstone selection for bifacial versus unifacial reduction was patterned (Fig. 10.8). For bifacial technologies, both components 1 and 2 evidenced greater-than-expected selection of chert over other toolstones, whereas component 3 evidenced greater-than-expected selection of FGI over the others. For unifacial reduction, component 1 had more-than-expected chert, component 2 had more-than-expected other toolstones, and component 3 preference was for FGI (Fig. 10.8a). Similar to the other variables, these data indicate preference for chert in component 1 for both bifacial and unifacial reduction with an increased reliance on volcanic raw materials and other toolstones for production of all tool technologies in both components 2 and 3.

Exploring local versus nonlocal toolstone selection for bifacial and unifacial technologies, again we found similar patterning. For bifacial reduction, component 1 had more-than-expected nonlocal toolstone and components 2 and 3 had

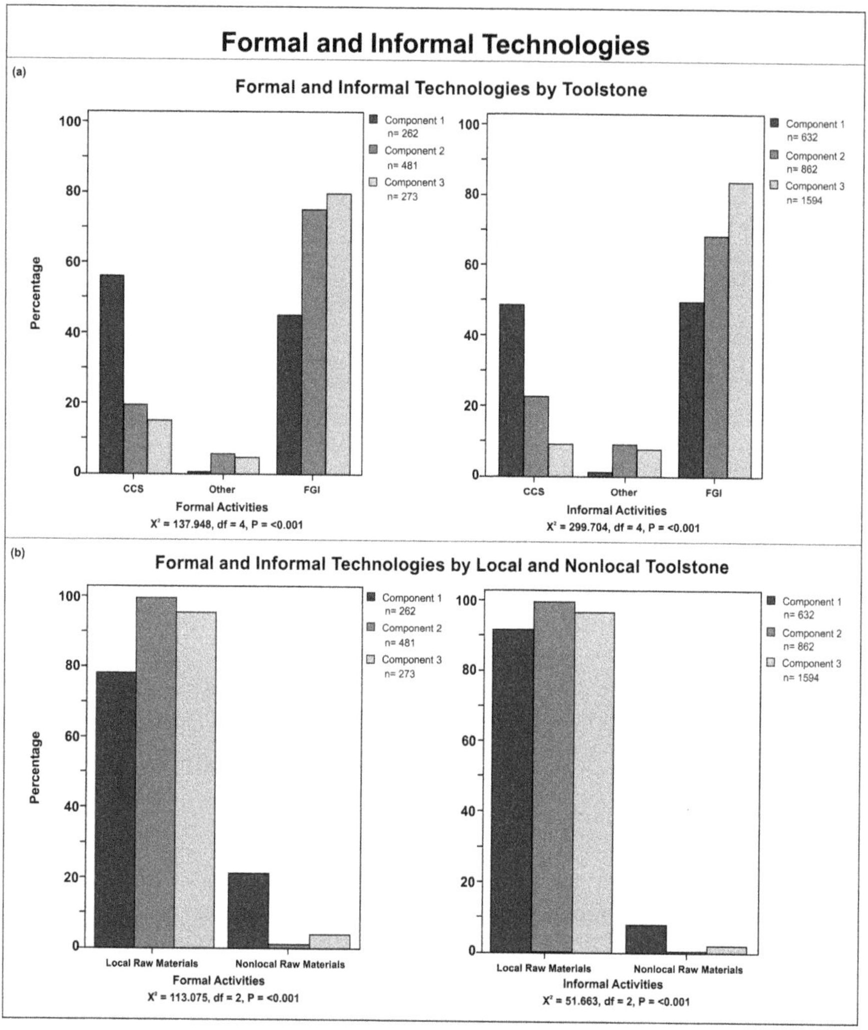

Fig. 10.7 Informal and formal technologies by (**a**) toolstone types and (**b**) local and nonlocal toolstones

more-than-expected local toolstones. For unifacial reduction, component 1 had more-than-expected nonlocal toolstones, component 2 had more-than-expected local toolstones, and component 3 evidenced no selective differences between nonlocal and local toolstones (Fig. 10.8b). During the component 1 occupation, there was clear preference for nonlocal toolstones for both bifacial and unifacial activities. For component 2, the preference was for local toolstones, and for component 3 there was a preference for local toolstones for bifacial reduction, but no clear preference in unifacial reduction.

Table 10.6 Unifacial and bifacial tool production by component

		Unifacial	Bifacial	Total
Component 1	Count	119	92	210
	Expected count	128.3	82.7	210.0
	% total (within component)	(56.4%)	(43.6%)	(100.0%)
Component 2	Count	168	106	274
	Expected count	166.7	107.3	274.0
	% total (within component)	(61.3%)	(38.7%)	(100.0%)
Component 3	Count	154	86	240
	Expected count	146	94.0	240
	% total (within component)	(64.2%)	(35.8%)	(100.0%)
Total	Count	441	284	725
	Expected count	441.0	284.0	725.0
	% of total	(60.8%)	(39.2%)	(100.0%)

$\chi^2 = 2.888$; df = 2; $P = .236$. Note no (0.0%) cells have expected counts less than 5. The minimum expected count is 82.65

Discussion

The goals of this study were threefold. We wanted to detect differences in toolstone procurement and selection behaviors between three temporally distinct cultural components at the Owl Ridge site. We also aimed to explore how these differences inform on lithic variability in Late Pleistocene-early Holocene archaeological sites in central Alaska. Finally, we wanted to understand how humans responded to global warming at the Pleistocene-Holocene boundary. Below we discuss findings of our study of the Owl Ridge lithic industries in the context of these goals.

Do Lithic Raw Material Procurement and Selection Behaviors Change Through Time at Owl Ridge?

The three cultural components at Owl Ridge have small artifact assemblages with low tool counts and diversity and the landform on which the site rests is very narrow (45 m wide). These factors, combined with lithic refit analysis, indicated the site was a repeatedly used logistical camp (Melton 2015). The site was used for special tasks and never as a long-term base camp location. We did not excavate the entire surface area; however, of the nearly 80 m^2 excavated, only 4327 artifacts were found in total. Despite the fact that the site served a similar purpose through time, data presented in this paper establish clear differences in the specific ways the site was used.

Beginning with component 1, we found the tools left behind were few but dominated by bifaces, including four triangular Chindadn points, and retouched flakes, indicating the site served as a hunting camp where hunted resources were procured

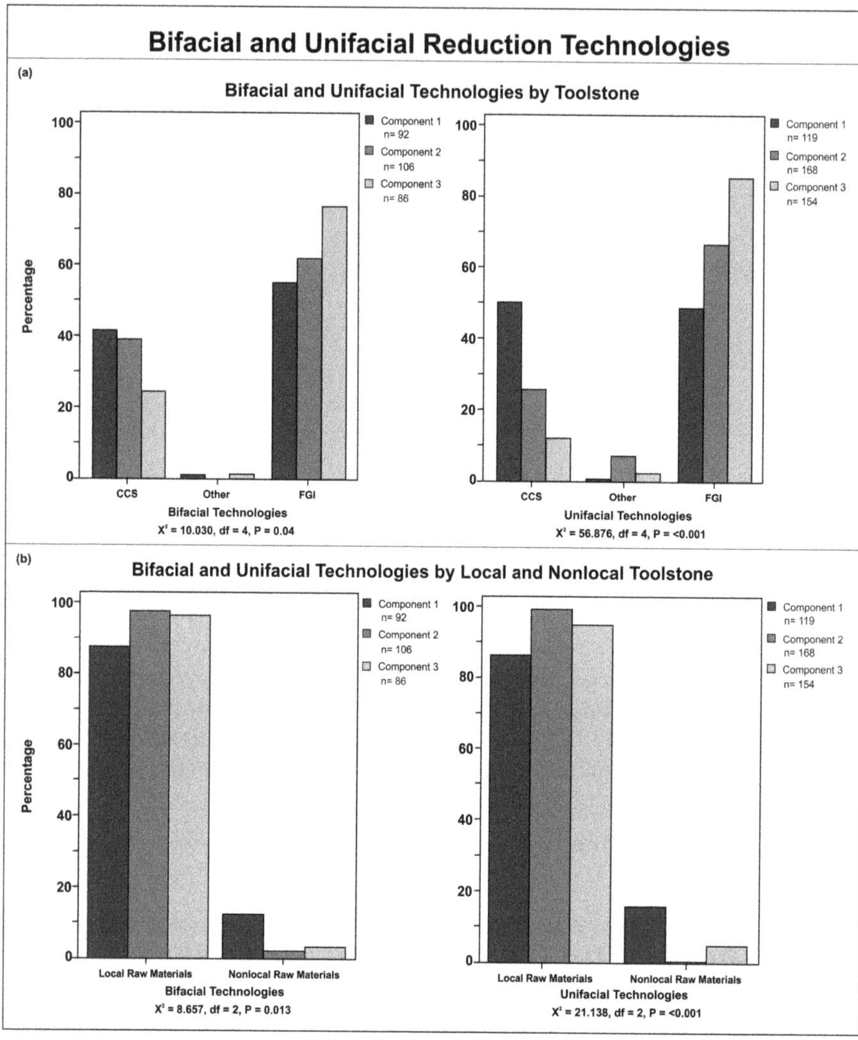

Fig. 10.8 Bifacial and unifacial reduction technologies by (**a**) toolstone types and (**b**) local and nonlocal toolstones

and initially processed presumably for transport elsewhere. No extensive processing occurred at this time because few formal processing activities were represented. Technological activities centered on both primary and secondary reduction with greater focus on informal, expedient core reduction and both bifacial and unifacial tool production and maintenance. Component 1 hunters selected both nonlocal and local toolstones for all reduction activities but preferred chert, especially the nonlocal variety. They brought nonlocal toolstones with them, mostly as finished and formal tools that they refurbished, but they also procured some of the local toolstones found in the glacial outwash at the site or in floodplain deposits nearby.

These toolstones were also used to manufacture tools transported away from the site, suggesting component 1 inhabitants retooled while visiting Owl Ridge.

The content of tools discarded during the component 2 occupation signals manufacture and maintenance of lanceolate bifacial points, scrapers, and other processing tools, suggesting component 2 occupants produced and maintained both a hunting and processing toolkit at the site. Very few nonlocal toolstones were carried to the site at this time. Mostly foragers procured locally available stones during their visit, arriving to the site nearly empty handed. Presumably, they took tools made on the local raw materials with them when they abandoned the site.

The artifact assemblage from component 3 indicates mostly primary reduction activities coupled with the manufacture and maintenance of unifacial tools. Therefore, site activities seem to be centered on processing behaviors. Different from component 1 but similar to component 2, toolstone procurement by component 3 hunter-gatherers was mostly local; however, slightly more nonlocal toolstones make up the component 3 assemblage compared with component 2. Toolstone-selection variables indicate these hunter-gatherers focused on local FGI for all reduction activities, but again these activities concentrated on primary reduction and expedient tool production, behaviors differing from earlier visits to the site.

How Can We Explain Lithic Variability at Owl Ridge and in Central Alaska During the Late Pleistocene-Early Holocene?

Our results indicate toolstone procurement and selection changed through time at Owl Ridge. As the site was first visited during the late Allerød, just a few hunters carried with them lightweight, Chindadn-type projectile points and camped at this spot for a short period of time, given that only 894 artifacts make up the component 1 assemblage. Perhaps they found the ridge provided an excellent lookout for fauna traversing this stretch of the Teklanika River Valley. To date, this occupation event represents the first known in the valley. Toolstone procurement and selection centered on both nonlocal and local toolstone. Hunters seem to have retooled with some local fine-grained chert and FGI resources. Our data suggest component 1 was a visit by foragers relatively unfamiliar with the local lithic landscape and, therefore, represents landscape learners in this specific context (Kelly 2003; Meltzer 2003).

About 500 years later the site was revisited by hunters using different hunting technologies, based on presence of lanceolate bifacial points and perhaps microblade-osseous composite projectile technology as microblades and two microblade-core technical spalls were also discovered in the component 2 assemblage. Component 2 occupants may have stayed longer at the site because both hunting and processing tools were made, refurbished, and discarded there. Procurement and preference for mostly local toolstones indicate they knew the lithic landscape better than initial visitors half a millennium earlier.

Component 3 represents the third and final visit to Owl Ridge about 200 years following component 2 and by a group focused even more on processing activities. Though one bifacial point and one microblade were found, the rest of the tool assemblage consists of various processing tools. Very similar to component 2, toolstone procurement and selection were almost exclusively local raw materials. We are certain that the foragers visiting Owl Ridge during this final, early Holocene occupation episode knew the local lithic landscape well because they preferred, and relied on, the local raw materials. Perhaps they came to Owl Ridge to procure and use andesite from the alluvium as well as capture and process food resources other than medium-large game, given the composition of their toolkit. Though speculative, these data may indicate that women used the site at this time, given that northern hunter-gatherer groups are known to focus primarily on hunted resources with men contributing most directly to hunting, and women engaging in tasks more supportive in nature, such as procuring smaller game, preparing food and other hunted resources, and mending or fixing tools (Halperin 1980; Jarvenpa and Brumbach 2006; Waguespack 2005) and processing activities were the focus during this final visit.

Our results indicate the behaviors responsible for production and maintenance of tool technologies and procurement and selection of toolstone during both the component 2 and component 3 occupations were much more similar to each other than either was to those reflected in the component 1 assemblage. We find that Phippen's (1988; Hoffecker et al. 1996) separation of components 1 and 2 into two temporally and technologically distinct complexes, Nenana and Denali, was warranted chronologically, descriptively, and now behaviorally. Component 1 and components 2 and 3 represent two different toolstone procurement and selection strategies, a strategy employed prior to the local Younger Dryas event and one used immediately following it. This does not necessarily mean the site was visited by two different groups of people. In contrast, our data indicate Owl Ridge inhabitants became increasingly knowledgeable of their local (Teklanika valley) environment through time in a stepwise fashion. We contend these changes reflect gradual behavioral adaptation by hunter-gatherers to their surroundings, a settling in process, as they became part of a changing ecosystem responding to fluctuating terminal Pleistocene climatic conditions. We recognize the limitation of basing interpretations for a region on analyses from a single site; however, this study is unique and future work considering additional sites should either support or refute our hypothesis.

How Did Central Alaskans Respond to Changing Environments at the Pleistocene-Holocene Boundary?

Climatic data for central Alaska indicates that between about 14,000 and 10,000 cal BP, the region experienced several climatic shifts and associated environmental changes. In a nutshell, late glacial climate was first cold and dry, shifted to warmer and moister conditions during the Allerød, reversed to cool and arid conditions during the Younger Dryas, gradually warmed into the Holocene with the first Holocene millennium warm and arid, and increasingly warmer and wetter by the onset of the Holocene Thermal Maximum at ~10,000 cal BP. During this 4000-year period, the biome shifted from herb tundra to shrub tundra to open-forest parkland to closed boreal forest.

Though our data at the Owl Ridge site are not robust enough to provide detailed answers to the question of how central Alaskans responded to Pleistocene-Holocene boundary climatic and environmental change, it does support findings in the Nenana Valley of initial occupation of the Alaska Range foothills during the Allerød. In the Teklanika valley, they were beginning to learn the local lithic landscape when the Younger Dryas occurred. Given data from other sites in the region and Owl Ridge, these initial inhabitants were not manufacturing or maintaining lanceolate or microblade-composite spear technologies, but using thin triangular-shaped and teardrop-shaped bifacial points as weapon tips (Powers and Hoffecker 1989; Goebel et al. 1991; Pearson 1999; Graf and Goebel 2009). Their technologies were relatively expedient, and based on faunal data from the Broken Mammoth site in the Tanana Valley, foragers at this time were subsisting in a shrub-tundra biome, hunting a wide variety of small and large faunal resources (Yesner 2007).

Between about 13,000 and 12,500 cal BP, the Younger Dryas cold and dry period is evidenced at regional archaeological sites by the appearance of culturally sterile sand units (Bigelow et al. 1990; Goebel et al. 1996; Graf and Bigelow 2011; Graf et al. 2015). This period of colder and drier climate affected the distribution and composition of floral and faunal resources, perhaps limiting availability of subsistence resources and the presence of humans. After this brief dry period, however, we see people using the region again. In fact, at Owl Ridge component 2 artifacts are found in a paleosol, indicating development of a relatively stable land surface and slightly moister conditions than during the previous centuries. The hunting technology, lanceolate bifacial points and microblade-composite-tool technology, was strikingly different from that used by initial inhabitants and suggests a focus on larger-game hunting (Guthrie 2006; Graf and Bigelow 2011). Paleoecological data are still too coarse-grained to understand faunal resource composition and availability for this period, but perhaps relatively dry conditions from the Younger Dryas still prevailed, and bison, wapiti, and caribou were sought after in an open-forest parkland environment (Guthrie 2006; Graf and Bigelow 2011). Certainly, the changes in toolstone selection represent an increased familiarity of the Teklanika valley as hunter-gatherers settled into the region.

Stratigraphically between components 2 and 3, the Owl Ridge profile evidences a major colluvial depositional event when humans were not present. Deposition of 15–25 cm of colluvial sands in about 200 years indicates a brief period of torrential rains and likely relatively warm, wet conditions. Immediately following this, humans revisited the site one final time, but this time focused on other resources since they did not leave behind hunting tools as before. After 11,000 cal BP as climate became even warmer and more humid, boreal-forest vegetation and biome spread into the region, and humans never returned to Owl Ridge. Perhaps the spread of the boreal-forest vegetation limited views from the site so that it no longer provided an overlook of the river valley to humans.

Through the Pleistocene-Holocene transition, evidence suggests humans were present at sites like Owl Ridge until boreal forest spread into the region. We contend terminal Pleistocene hunter-gatherers in central Alaska were reasonably resilient, only leaving the immediate foothills during the coldest several centuries of the Younger Dryas interval. Given that occupation events immediately following the Younger Dryas evidence foragers with learned knowledge of the local lithic landscape, we assume these were the descendants of people who visited the Teklanika River before.

Conclusions

With this paper, we set out to compare the lithic assemblages of components 1, 2, and 3 from the Owl Ridge site to investigate how people were using lithic raw materials through time as they settled in the region and responded to climate change and local environmental shifts. The study of toolstone procurement and selection strategies helps us address how people responded to changing resource availability. Our results indicate that initial occupants were not as familiar with the local lithic landscape compared with later inhabitants. These later inhabitants had learned where to find local raw materials and obviously had become familiar with the landscape around them. Our findings confirm clear chronostratigraphic, technological, and land-use differences between Nenana complex and Denali complex assemblages in the greater Nenana Valley. We conclude that the differences in toolstone procurement and selection strategies and organization of technologies observed at Owl Ridge represent increased landscape familiarity as people settled in the region and responded to changing environmental conditions at the end of the Pleistocene.

Acknowledgments We would like to thank the editors of the volume, Erick Robinson and Fred Sellet, for inviting us to participate. We are grateful to the Owl Ridge field crew, including Heather Smith, John Blong, Josh Keene, Angela Younie, Takako Sakamoto, Tarah Marks, Anna Gilmer, Stuart White, Ted Goebel, Nancy Bigelow, and Mike Waters. Additionally, we would like to thank Peter Phippen for visiting our labs at CSFA during the fall of 2010 and helping us interpret field notes and catalog information from his excavations. We would like to acknowledge the University of Alaska Museum of the North for access to the 1970s and 1980s Owl Ridge collection housed there. We would also like to acknowledge and thank the Archaeology and Arctic Social Sciences programs at the National Science Foundation (Award Numbers 0917648 and 1029094) and the CSFA and Department of Anthropology at Texas A&M University for providing financial and logistical support of this project.

References

Abbott, M. B., Finney, B. P., Edwards, M. D., & Kelts, K. R. (2000). Paleohydrology of Birch Lake, central Alaska: A multiproxy approach to lake-level records. *Quaternary Research, 53*(1), 154–166.

Anderson, P. M., Edwards, M. E., & Brubaker, L. B. (2004). Results and paleoclimate implications of 35 years of paleoecological research in Alaska. In A. R. Gillespie & S. C. Porter (Eds.), *The Quaternary Period in the United States* (pp. 427–440). Amsterdam: Elsevier.

Andrefsky, W. R. (2009). The analysis of stone tool procurement, production, and maintenance. *Journal of Archaeological Research, 17*(1), 65–103.
Bamforth, D. B. (1991). Technological organization and hunter-gatherer land use: A California example. *American Antiquity, 56*(2), 216–234.
Barber, V. A., & Finney, B. P. (2000). Late quaternary paleoclimatic reconstructions for interior Alaska based on paleolake-level data and hydrologic models. *Journal of Paleolimnology, 24*(1), 29–41.
Bigelow, N. H., & Edwards, M. E. (2001). A 14,000 yr paleoenvironmental record from Windmill Lake, central Alaska: Lateglacial and Holocene vegetation in the Alaska range. *Quaternary Science Reviews, 20*(1), 203–215.
Bigelow, N. H., & Powers, W. R. (2001). Climate, vegetation, and archaeology 14,000–9,000 cal yr B.P. in central Alaska. *Arctic Anthropology, 38*(2), 171–195.
Bigelow, N. H., Powers, W. R., & Begét, J. (1990). Latest Pleistocene increase in wind intensity recorded in eolian sediments from central Alaska. *Quaternary Research, 34*(2), 160–168.
Binford, L. R. (1979). Organization and formation processes: Looking at curated technologies. *Journal of Anthropological Research, 35*(3), 255–273.
Binford, L. R. (1980). Willow smoke and dogs' tails: Hunter-gatherer settlement systems and archaeological site formation. *American Antiquity, 45*(1), 4–20.
Birks, H. H., Gelorini, V., Robinson, E., & Hoek, W. Z. (2015). Impacts of palaeoclimate change 60,000-8,000 years ago on humans and their environments in Europe: Integrating palaeoenvironmental and archaeological data. *Quaternary International, 378*, 4–13.
Bleed, P., & Bleed, A. (1987). Energetic efficiency and hand tool design: A performance comparison of push and pull stroke saws. *Journal of Anthropological Archaeology, 6*(2), 189–197.
Bowers, P. M. (1980). *The Carlo Creek Site: Geology and Archaeology of an Early Holocene Site in the Central Alaska Range.* Fairbanks, AK: Occasional Papers No. 27, University of Alaska, Fairbanks, Anthropology and Historic Preservation Cooperative Park Series Unit, Occasional Papers No. 27.
Bowers, P. M., & Reuther, J. (2008). AMS re-dating of the Carlo Creek site, Nenana valley, central Alaska. *Current Research in the Pleistocene, 25*, 58–61.
Brubaker, L. B., Anderson, P. M., Edwards, M. E., & Lozhkin, A. V. (2005). Beringia as a glacial refugium for boreal trees and shrubs: New perspectives from mapped pollen date. *Journal of Biogeography, 32*(5), 833–848.
Coffman, S. C. (2011). *Archaeology at Teklanika West (HEA-001): An upland archaeological site, central Alaska.* Unpublished M.A. thesis, Department of Anthropology, University of Alaska, Fairbanks.
Coffman, S. C., & Rasic, J. (2015). Rhyolite characterization and distribution in central Alaska. *Journal of Archaeological Science, 57*, 142–157.
Cook, J. P. (1996). Healy Lake. In F. H. West (Ed.), *American beginnings: The prehistory and paleoecology of Beringia* (pp. 323–327). Chicago: University of Chicago Press.
Cooper, J., & Sheets, P. (2012). *Surviving sudden environmental change*. Boulder: University Press of Colorado.
Crass, B. A., Kedrowski, B. L., Baus, J., & Behm, J. A. (2011). Residue analysis of bone-fueled Pleistocene hearths. In T. Goebel & I. Buvit (Eds.), *From the Yenisei to the Yukon: Interpreting lithic assemblage variability in Late Pleistocene/Early Holocene Beringia* (pp. 192–198). College Station: Texas A&M University Press.
Dikov, N. N. (1977). *Arkheologicheskie Pamiatniki Kamchatki, Chukotki i Verkhnei Kolymy.* Moscow: Nauka.
Dortch, J. M., Owen, L. A., Caffee, M. C., Li, D., & Lowell, T. V. (2010). Beryllium-10 surface exposure dating of glacial successions in the Central Alaska range. *Journal of Quaternary Science, 25*(8), 1259–1269.
Elston, R. G., & Brantingham, P. J. (2002). Microlithic technology in northern Asia: A risk-minimizing strategy of the late Paleolithic and early Holocene. *Archeological Papers of the American Anthropological Association, 12*(1), 103–116.
Erlandson, J., Walser, R., Maxwell, H., Bigelow, N., Cook, J., Lively, R., Adkins, C., Dodson, D., Higgs, A., & Wilber, J. (1991). Two early sites of eastern Beringia: Context and chronology in Alaskan interior archaeology. *Radiocarbon, 33*(1), 35–50.

Fitzhugh, B. (2012). Hazards, impacts, and resilience among hunter-gatherers of the Kuril Islands. In J. Cooper & P. Sheets (Eds.), *Surviving sudden environmental change* (pp. 19–42). Boulder: University Press of Colorado.

Gal, R. (2002). Providence and frugality: Tools for high latitude living. In *Paper presented at the 29th annual meeting of the Alaska Anthropological Association*. Alaska: Anchorage.

Goebel, T., Powers, W. R., & Bigelow, N. H. (1991). The Nenana complex of Alaska and Clovis origins. In R. Bonnichsen & K. L. Turnmire (Eds.), *Clovis origins and adaptations* (pp. 49–79). Corvallis: Center for the Study of the First Americans.

Goebel, T., Powers, W. R., Bigelow, N. H., & Higgs, A. S. (1996). Walker road. In F. H. West (Ed.), *American beginnings: The prehistory and paleoecology of Beringia* (pp. 356–363). Chicago: University of Chicago Press.

Goebel, T., Waters, M. R., & Dikova, M. (2003). The archaeology of Ushki Lake, Kamchatka, and the Pleistocene peopling of the Americas. *Science, 301*(5632), 501–505.

Goebel, T., Waters, M. R., & O'Rourke, D. H. (2008). The late Pleistocene dispersal of modern humans in the Americas. *Science, 319*(5869), 1497–1502.

Goebel, T., Slobodin, S. B., & Waters, M. R. (2010). New dates from Ushki-1, Kamchatka, confirm 13,000 cal BP age for earliest Paleolithic occupation. *Journal of Archaeological Science, 37*, 2640–2649.

Gómez Coutouly, Y. A. (2011). Identifying pressure flaking modes at Diuktai cave: A case study of the Siberian upper Paleolithic microblade tradition. In T. Goebel & I. Buvit (Eds.), *From the Yenisei to the Yukon: Interpreting lithic assemblage variability in Late Pleistocene/Early Holocene Beringia* (pp. 75–90). College Station: Texas A&M University Press.

Gómez Coutouly, Y. A. (2012). Pressure microblade industries in Pleistocene-Holocene interior Alaska: Current data and discussions. In P. M. Desrosiers (Ed.), *Emergence of pressure blade making: From origins to modern experimentation* (pp. 347–374). New York: Springer.

Graf, K. E. (2008). *Uncharted territory: Late Pleistocene hunter-gatherer dispersal in the Siberian mammoth-steppe*. PhD dissertation, Department of Anthropology, University of Nevada, Reno.

Graf, K. E. (2010). Hunter–gatherer dispersals in the mammoth-steppe: Technological provisioning and land-use in the Enisei River valley, south-central Siberia. *Journal of Archaeological Science, 37*(1), 210–223.

Graf, K. E., & Bigelow, N. H. (2011). Human response to climate during the younger Dryas chronozone in central Alaska. *Quaternary International, 242*(2), 434–451.

Graf, K. E., & Goebel, T. (2009). Upper Paleolithic toolstone procurement and selection across Beringia. In B. Blades & B. Adams (Eds.), *Lithic materials and Paleolithic societies* (pp. 55–77). London: Blackwell Publishers.

Graf, K. E., Blong, J., & Goebel, T. (2010). A concave-based projectile point form new excavations at the Owl Ridge site, central Alaska. *Current Research in the Pleistocene, 27*, 88–91.

Graf, K. E., DiPietro, L. M., Krasinski, K. E., Gore, A. K., Smith, H. L., Culleton, B. J., Kennett, D. J., & Rhode, D. (2015). Dry Creek revisited: New excavations, radiocarbon dates, and site formation inform on the peopling of eastern Beringia. *American Antiquity, 80*(4), 671–694.

Guthrie, R. D. (1983). Chapter 6: Paleoecology of the site and its implications for early hunters. In W. R. Powers, R. D. Guthrie, & J. F. Hoffecker (Eds.), *Dry Creek: Archaeology and Paleoecology of a Late Pleistocene Alaskan Hunting Camp* (pp. 209–283). Unpublished report.

Guthrie, R. D. (2006). New carbon dates link climatic change with human colonization and Pleistocene extinctions. *Nature, 441*(7090), 207–209.

Guthrie, R. D. (2017). Chapter 6: Paleoecology of the site and its implications for early hunters. In W. R. Powers, R. D. Guthrie, J. F. Hoffecker, & T. Goebel (Eds.), *Dry Creek: Archaeology and paleoecology of a late pleistocene Alaskan Hunting camp* (pp. 153–192). College Station: Texas A&M University Press.

Halperin, R. (1980). Ecology and mode of production: Seasonal variation and the division of labor by sex among hunter-gatherers. *Journal of Anthropological Research, 36*(3), 379–399.

Hamilton, T. D., & Goebel, T. (1999). Late Pleistocene peopling of Alaska. In R. Bonnichsen & K. L. Turnmire (Eds.), *Ice Age peoples of North America: Environments, origins and adaptations of the first Americans* (pp. 156–199). Corvallis: Oregon State University Press.

Hoffecker, J. F. (2001). Late Pleistocene and early Holocene sites in the Nenana river valley, central Alaska. *Arctic Anthropology, 38*(2), 139–153.
Hoffecker, J. F., & Elias, S. A. (2007). *The human ecology of Beringia*. New York: Columbia University Press.
Hoffecker, J. F., Powers, W. R., & Goebel, T. (1993). The colonization of Beringia and the peopling of the new world. *Science, 259*, 46–53.
Hoffecker, J. F., Powers, W. R., & Phippen, P. G. (1996). Owl ridge. In F. H. West (Ed.), *American beginnings: The prehistory and paleoecology of Beringia* (pp. 353–357). Chicago: University of Chicago Press.
Holmes, C. E. (2001). Tanana river valley archaeology circa 14,000 to 9000 BP. *Arctic Anthropology, 38*(2), 154–170.
Holmes, C. E. (2011). The Beringian and transitional periods in Alaska. In T. Goebel & I. Buvit (Eds.), *From the Yenisei to the Yukon: Interpreting lithic assemblage variability in ate Pleistocene/Early Holocene Beringia* (pp. 179–191). College Station: Texas A&M University Press.
Hu, F. S., Brubaker, L. B., & Anderson, P. M. (1993). A 12000 year record of vegetation change and soil development from Wien Lake, central Alaska. *Canadian Journal of Botany, 71*(9), 1133–1142.
Jarvenpa, R., & Brumbach, H. J. (2006). Revisiting the sexual division of labor: Thoughts on ethnoarchaeology and gender. *Archeological Papers of the American Anthropological Association, 16*(1), 97–107.
Kaufman, D., Ager, T., Anderson, N., Anderson, P., Andrews, J., Bartlein, P., & Duvall, M. (2004). Holocene thermal maximum in the western Arctic (0–180 W). *Quaternary Science Reviews, 23*(5), 529–560.
Kelly, R. L. (2003). Colonization of new land by hunter-gatherers: Expectations and implications based on ethnographic data. In M. Rockman & J. Steele (Eds.), *Colonization of unfamiliar landscapes: The archaeology of adaptation* (pp. 44–58). London: Routledge.
Kokorowski, H., Anderson, P., Mock, C., & Lozhkin, A. (2008). A re-evaluation and spatial analysis of evidence for a younger Dryas climatic reversal in Beringia. *Quaternary Science Reviews, 27*(17), 1710–1722.
Kuhn, S. L. (1995). *Mousterian lithic technology: An ecological approach*. Princeton: Princeton University Press.
Lloyd, A. H., Edwards, M. E., Finney, B. P., Lynch, J. A., Barber, V., & Bigelow, N. H. (2006). Holocene development of the Alaskan boreal forest. In F. S. Chapin III, M. W. Oswood, K. Van Cleve, L. A. Viereck, & D. L. Verhyla (Eds.), *Alaska's Changing Boreal Forest* (pp. 62–78). New York: Oxford University Press.
Mason, O. K., Bowers, P. M., & Hopkins, D. M. (2001). The early Holocene Milankovitch thermal maximum and humans: Adverse conditions for the Denali complex of eastern Beringia. *Quaternary Science Reviews, 20*(1), 525–548.
Meiri, M., Lister, A. M., Collins, M. J., Tuross, N., Goebel, T., Blockley, S., Zazula, G. D., van Doorn, N., Guthrie, R. D., Boeskorov, G. G., Baryshnikov, G. F., Sher, A., & Barnes, I. (2014). Faunal record identifies Bering isthmus conditions as constraint to end-Pleistocene migration to the New World. *Proceedings of the Royal Society B: Biological Sciences, 281*(1776), 20132167.
Melton, J. A. (2015). *Pieces to a puzzle: A lithic refit study evaluating stratigraphic and lithic components of the Owl Ridge site, central Alaska*. Unpublished B.A. thesis, Department of Anthropology, Texas A&M University.
Meltzer, D. J. (2003). Lessons in landscape learning. In M. Rockman & J. Steele (Eds.), *Colonization of unfamiliar landscapes: The archaeology of adaptation* (pp. 222–241). London: Routledge.
Meltzer, D. J. (2004). Modeling the initial colonization of the Americas: Issues of scale, demography, and landscape learning. In C. M. Barton, G. A. Clark, D. R. Yesner, & G. A. Pearson (Eds.), *The settlement of the American continents: A multidisciplinary approach to human biogeography* (pp. 123–137). Tucson: The University of Arizona Press.
Mochanov, Y. A. (1977). *Drevneishie Etapy Zaseleniia Chelovekom Severo-Vostochnoi Azii*. Novosibirsk: Nauka.

Morgan, C. (2009). Climate change, uncertainty and prehistoric hunter–gatherer mobility. *Journal of Anthropological Archaeology, 28*(4), 382–396.

Mulligan, C. J., & Kitchen, A. (2014). Three-stage colonization model for the peopling of the Americas. In K. E. Graf, C. V. Ketron, & M. R. Waters (Eds.), *Paleoamerican Odyssey* (pp. 171–181). College Station: Texas A&M University Press.

Nelson, M. C. (1991). The study of technological organization. *Archaeological Method and Theory, 3*, 57–100.

Pearson, G. A. (1999). Early occupations and cultural sequence at Moose Creek: A Late Pleistocene site in central Alaska. *Arctic, 52*(4), 332–345.

Phippen, P. G. (1988). *Archaeology at Owl Ridge: A Pleistocene-Holocene boundary age site in central Alaska*. Unpublished M.A. thesis, Department of Anthropology, University of Alaska, Fairbanks.

Pitulko, V. V. (2011). The Berelekh quest: A review of forty years of research in the mammoth graveyard in Northeast Siberia. *Geoarchaeology, 26*(1), 5–32.

Plaskett, D. C. (1976). *Preliminary report: A cultural resources survey in an area of the nenana and teklanika rivers of Central Alaska*. Unpublished manuscript.

Potter, B. A. (2005). *Site structure and organization in central Alaska: Archaeological investigations at Gerstle River*. Unpublished Ph.D. dissertation, Department of Anthropology, University of Alaska, Fairbanks.

Potter, B. A. (2007). Models of faunal processing and economy in Early Holocene interior Alaska. *Environmental Archaeology, 12*(1), 3–23.

Potter, B. A. (2008). Radiocarbon chronology of central Alaska: Technological continuity and economic change. *Radiocarbon, 50*(2), 181–204.

Potter, B. A., Irish, J. D., Reuther, J. D., Gelvin-Reymiller, C., & Holliday, V. T. (2011). A terminal Pleistocene child cremation and residential structure from eastern Beringia. *Science, 331*(6020), 1058–1062.

Potter, B. A., Holmes, C. E., & Yesner, D. R. (2014a). Technology and economy among the earliest prehistoric foragers in interior eastern Beringia. In K. E. Graf, C. V. Ketron, & M. R. Waters (Eds.), *Paleoamerican odyssey* (pp. 81–103). College Station: Texas A&M University Press.

Potter, B. A., Irish, J. D., Reuther, J. D., & McKinney, H. J. (2014b). New insights into Eastern Beringian mortuary behavior: A terminal Pleistocene double infant burial at Upward Sun River. *Proceedings of the National Academy of Sciences, 111*(48), 17060–17065.

Powers, W. R., & Hoffecker, J. F. (1989). Late Pleistocene settlement in the Nenana valley, central Alaska. *American Antiquity, 54*(2), 263–287.

Raghavan, M., Steinrücken, M., Harris, K., Schiffels, S., Rasmussen, S., DeGiorgio, M., Albrechtsen, A., Valdiosera, C., Ávila-Arcos, M. C., Malaspinas, A. S., Eriksson, A., Moltke, I., Metspalu, M., Homburger, J. R., Wall, J., Cornejo, O. E., Moreno-Mayar, J. V., Korneliussen, T. S., Pierre, T., Rasmussen, M., Campos, P. F., de Baros Damgaard, P., Allentoft, M. E., Lindo, J., Metspalu, E., Rodríguez-Varela, R., Mansilla, J., Henrickson, C., Seguin-Orlando, A., Malmström, H., Stafford, T., Jr., Shringarpure, S. S., Moreno-Estrada, A., Karmin, M., Tambets, K., Bergström, A., Xue, Y., Warmuth, V., Friend, A. D., Singarayer, J., Valdes, P., Balloux, F., Leboreiro, I., Vera, J. L., Rangel-Villalobos, H., Pettener, D., Luiselli, D., Davis, L. G., Heyer, E., Zollikofer, C. P., Ponce de León, M. S., Smith, C. I., Grimes, V., Pike, K. A., Deal, M., Fuller, B. T., Arriaza, B., Standen, V., Luz, M. F., Ricaut, F., Guidon, N., Osipova, L., Voevoda, M. I., Posukh, O. L., Balanovsky, O., Lavryashina, M., Bogunov, Y., Khusnutdinova, E., Gubina, M., Balanovska, E., Fedorova, S., Litvinov, S., Malyarchuk, B., Derenko, M., Mosher, M. J., Archer, D., Cybulski, J., Petzelt, B., Mitchell, J., Worl, R., Norman, P. J., Parham, P., Kemp, B. M., Kivisild, T., Tyler-Smith, C., Sandhu, M. S., Crawford, M., Villems, R., Smith, D. G., Waters, M. R., Goebel, T., Johnson, J. R., Malhi, R. S., Jakobsson, M., Meltzer, D. J., Manica, A., Durbin, R., Bustamante, C. D., Song, Y. S., Nielsen, R., & Willerslev, E. (2015). Genomic evidence for the Pleistocene and recent population history of native Americans. *Science, 349*(6250), aab3884.

Redman, C. L. (2005). Resilience in archaeology. *American Anthropologist, 107*(1), 70–77.

Reimer, P. J., Bard, E., Bayliss, A., Beck, J. W., Blackwell, P. G., Bronk Ramsey, C., Buck, C. E., Cheng, H., Edwards, R. L., Friedrich, M., Grootes, P. M., Guilderson, T. P., Haflidason, H., Hajdas, I., Hatté, C., Heaton, T. J., Hoffmann, D. L., Hogg, A. G., Hughen, K. A., Kaiser, K. F., Kromer, B., Manning, S. W., Niu, M., Reimer, R. W., Richards, D. A., Scott, E. M., Southon, J. R., Staff, R. A., Turney, C. S. M., & van der Plicht, J. (2013). Intcal13 and Marine13 radiocarbon age calibration curves 0-50,000 years cal BP. *Radiocarbon, 55*(4), 1869–1887.

Reuther, J. D., Slobodina, N. S., Rasic, J. T., Cook, J. P., & Speakman, R. J. (2011). Gaining momentum: Late Pleistocene and early Holocene archaeological obsidian source studies in interior and northeastern Beringia. In T. Goebel & I. Buvit (Eds.), *From the Yenisei to the Yukon: Interpreting lithic assemblage variability in Late Pleistocene/Early Holocene Beringia* (pp. 270–286). College Station: Texas A&M University Press.

Shott, M. J. (1986). *Settlement mobility and technological organization among Great Lakes Paleo-Indian foragers.* PhD Dissertation, Department of Anthropology, University of Michigan, Ann Arbor.

Tackney, J. C., Potter, B. A., Raff, J., Powers, M., Watkins, W. S., Warner, D., Reuther, J. D., Irish, J. D., & O'Rourke, D. H. (2015). Two contemporaneous mitogenomes from terminal Pleistocene burials in eastern Beringia. *Proceedings of the National Academy of Sciences, 112*(45), 13833–13838.

Tamm, E., Kivisild, T., Reidla, M., Metspalu, M., Smith, D. G., Mulligan, C. J., Bravi, C. M., Rickards, O., Martinez-Labarga, C., Khusnutdinova, E. K., Fedorova, S. A., Golubenko, M. V., Stepanov, V. A., Gubina, M. A., Zhadanov, S. I., Ossipova, L. P., Damba, L., Voevoda, M. I., Dipierri, J. E., Villems, R., & Malhi, R. S. (2007). Beringian standstill and spread of native American founders. *PloS One, 2*(9), e829.

Théry-Parisot, I., Costamagno, S., Brugal, J. P., Fosse, P., & Guilbert, R. (2002). The use of bone as fuel during the Palaeolithic, experimental study of bone combustible properties. In J. Mulville & A. Outram (Eds.), *The Zooarchaeology of fats, oils, milk and dairying* (pp. 50–59). Oxford: Oxbow Books.

Thorson, R. M. (1986). Late Cenozoic glaciation of the northern Nenana river valley. In K. M. Reed & R. M. Thorson (Eds.), *Glaciation in Alaska-the geologic record* (pp. 171–192). Anchorage: Alaska Geological Society.

Waguespack, N. M. (2005). The organization of male and female labor in foraging societies: Implications for early paleoindian archaeology. *American Anthropologist, 107*(4), 666–676.

Wahrhaftig, C. (1970). Geologic map of the Fairbanks A-4 quadrangle, Alaska U.S. Geological Survey map GQ-810.

Warhaftig, C. (1958). Quaternary geology of the Nenana River Valley and adjacent of the Alaska Range. U.S. *Geological Survey Paper, 293-A*, 1–68.

Warhaftig, C. (1970a). Geologic map of the Healy D-5 quadrangle, Alaska. U.S. Geological Survey map GQ-807.

Warhaftig, C. (1970b). Geologic map of the Healy D-4 quadrangle, Alaska. U.S. Geological Survey Map GQ-806.

Warhaftig, C., & Black, R. F. (1958). Engineering geology along part of the Alaska Railroad. U.S. *Geological Survey Paper, 293-B*, 69–118.

West, F. H. (1967). The Donnelly Ridge site and the definition of an early core and blade complex in central Alaska. *American Antiquity, 32*(3), 360–382.

Wygal, B. T. (2009). *Prehistoric colonization of southcentral Alaska: human adaptations in a post glacial world.* Unpublished Ph.D. dissertation, Department of Anthropology, University of Nevada, Reno

Wygal, B. T. (2011). The microblade/non-microblade dichotomy: Climatic implications, toolkit variability and the role of tiny tools in Eastern Beringia. In T. Goebel & I. Buvit (Eds.), *From the Yenisei to the Yukon: Interpreting lithic assemblage variability in Late Pleistocene/Early Holocene Beringia* (pp. 234–254). College Station: Texas A&M University Press.

Yesner, D. R. (1996). Human adaptation at the Pleistocene-Holocene boundary (circa 13,000 to 8,000 BP) in eastern Beringia. In L. G. Straus, B. V. Eriksen, J. M. Erlandson, & D. R. Yesner (Eds.), *Humans at the end of the Ice Age* (pp. 255–276). New York: Plenum Press.
Yesner, D. R. (2001). Human dispersal into interior Alaska: Antecedent conditions, mode of colonization, and adaptations. *Quaternary Science Reviews, 20*(1), 315–327.
Yesner, D. R. (2007). Faunal extinction, hunter-gatherer foraging strategies, and subsistence diversity among eastern Beringia. In R. B. Walker & B. M. Driskell (Eds.), *Foragers of the terminal Pleistocene in North America* (pp. 15–31). Lincoln: University of Nebraska Press.

Chapter 11
Technological Change from the Terminal Pleistocene Through Early Holocene in the Eastern Great Basin, USA: The Record from Bonneville Estates Rockshelter

Ted Goebel, Aria Holmes, Joshua L. Keene, and Marion M. Coe

Introduction

There may be no better place to investigate long-term change in technological organization among prehistoric hunter-gatherers than the Great Basin of western North America. Since the mid-twentieth century, anthropological research in the region has been founded on ecological and evolutionary theory (e.g., Steward 1938; Cressman 1942; Jennings 1957), and since the 1980s, some of the most successful archaeological applications of cultural-ecological and behavioral-ecological theories have been carried out there (e.g., Bettinger and Baumhoff 1982; O'Connell et al. 1982; Thomas 1983a, b; Madsen and Kirkman 1988; Grayson 1991; Janetski 1977; Madsen and Simms 1998; Kelly 1999; see also Zeanah and Simms 1999). Central to these developments has been "a close and extremely fruitful relationship" between archaeology and the Quaternary sciences (Rhode 1999: 29), a relationship that continues to flower through interdisciplinary cooperation of archaeologists, geomorphologists, sedimentologists, palynologists, plant-macrofossil specialists, and paleontologists (e.g., Madsen 2000; Oviatt et al. 2003; Adams et al. 2008; Jenkins et al. 2012).

In recent years, ecological/evolutionary research in the Great Basin has expanded to include studies of technology and material culture. Many of these studies have focused on raw material procurement and mobility, often applying geochemical

T. Goebel (✉) • J.L. Keene
Center for the Study of the First Americans, Department of Anthropology,
Texas A&M University, College Station, TX, USA
e-mail: goebel@tamu.edu

A. Holmes
Department of Anthropology, University of New Mexico, Albuquerque, NM, USA

M.M. Coe
Center for the Study of the First Americans, Texas A&M University,
College Station, TX, USA

E. Robinson, F. Sellet (eds.), *Lithic Technological Organization and Paleoenvironmental Change*, Studies in Human Ecology and Adaptation 9,
DOI 10.1007/978-3-319-64407-3_11

data from lithics (Graf 2002; Jones et al. 2003; Smith and Kielhofer 2011) but, in a few cases, ceramics (Eerkens et al. 2002; Eerkens 2003). A series of studies have sought to measure and explain Paleoindian or Archaic bifacial-point variability, some focusing on technology (Musil 1988; Beck 1995), some on style (Hildebrandt and King 2002), and still others on reduction and resharpening, the so-called Frison Effect (Flenniken and Wilke 1989; see also Zeanah and Elston 2001). Selectionist archaeological studies have contributed to our understanding of form and function of these and other classes of stone tools (Beck and Jones 1993; Beck 1995), and investigations of cultural transmission specifically have been used to explain variability in late Archaic arrow-point form (Bettinger and Eerkens 1999). Lithic technological organization has been considered from a behavioral-ecological perspective, too, especially in relation to the Paleoindian period. For example, Beck et al. (2002) applied a transport cost analysis to explain Paleoindian quarrying behavior and artifact transport in the central Great Basin, and Elston and Zeanah (2002) considered raw material selection in the context of diet breadth to predict differential scales of mobility among men and women during Paleoindian times. Studies of perishable technology (e.g., basketry, sandals, netting, and cordage) typically have focused on detecting ethnicity, population movements, and interregional interactions (e.g., Adovasio 1986; Adovasio and Pedler 1994; Fowler 1994; Connolly and Barker 2004; Hattori and Fowler 2009; Connolly 2013), but these materials have also been examined from technological (Benson et al. 2006; Adovasio et al. 2009) and subsistence perspectives (Janetski 1979; Barlow et al. 1993). Ceramic studies have focused on the late Archaic period (the earliest archaeological period with a record of pottery in the Great Basin), but they, too, have sometimes incorporated evolutionary models to explain change and innovation (Eerkens 2004; Eerkens and Lipo 2014).

Analyses of technological organization across large periods of time, in relation to changing environments, behaviors, or social relationships, also have been accomplished but infrequently. Beck (1995) and Hildebrandt and King (2002) investigated bifacial-point variability through the middle and late Archaic period, ~5000 to less than 1000 years ago, with Beck taking a functional approach and Hildebrandt and King a stylistic one. Goebel (2007) presented an introductory study of Paleoindian and early Archaic lithic technological activities represented at Bonneville Estates Rockshelter, considering change in the context of environmental, subsistence, and settlement change. A few such studies also have been presented for perishables (Geib and Jolie 2008; Jolie 2014).

In the western Bonneville basin of northwestern Utah and northeastern Nevada, the focus of our study, much archaeological work has been undertaken focusing on the prehistory of the Pleistocene-Holocene boundary. J. Jennings' (1957) classic study of Danger Cave was centered here, and more recent surveys and excavations at Smith Creek Cave (Bryan 1979, 1988), Camels Back Cave (Schmitt and Madsen 2005), and surface/near-surface localities along the Old River Bed (Oviatt et al. 2003; Rhode et al. 2005; Duke and Young 2007; Duke 2011, 2015) have demonstrated a rich and multidimensional record of the Paleoindian and early Archaic periods.

In this tradition of archaeology grounded in ecological and evolutionary theory, we consider here a case study of paleoenvironmental change and technological organization. We present the results of our ongoing study of lithic technology at

Bonneville Estates Rockshelter, located in the western Bonneville basin of easternmost Nevada, exploring change in technology from the Paleoindian to the early Archaic periods. We set the stage by first providing a backdrop of climatic and environmental change from around 15,000 to 8000 calendar years ago (cal BP) in the Bonneville basin, reviewing regional proxy records from geomorphology, palynology, plant-macrofossil analysis, and paleontology. We follow this with an examination of the lithic assemblages from Bonneville Estates Rockshelter, interpreting raw material provisioning and technological activities of the shelter's early inhabitants and exploring change in technological organization from the terminal Pleistocene to early mid-Holocene. We complete the analysis by discussing these patterns in the context of other proxy records of human behavior in the shelter, primarily drawing on the work of our colleagues Bryan Hockett, David Rhode, and David Schmitt. In a nutshell, we find that increased aridity and warming in the eastern Great Basin led to significant adaptive changes among the humans occupying the region. Across the Paleoindian-Archaic transition at Bonneville Estates, raw material procurement became locally oriented, and a greater variety of lithic-reduction activities were performed, likely reflecting reduced mobility and longer stays in the shelter.

Environments and Environmental Change in the Eastern Great Basin

The North American Great Basin is the arid region of the Intermountain West situated between the Rocky Mountains in the east and the Sierra Nevada and Cascade Mountains in the west and between the Snake River basin in the north and the Colorado River basin in the south. Hydrographically, the region is internally drained—in other words, its rivers do not flow to the sea but instead into interior "sumps" like Owens Lake (1084 m above sea level [asl]) in California, Humboldt Sink in western Nevada (1210 m asl), and Great Salt Lake (1277 m asl at its historic low) in North Central Utah. Today, because of the relatively warm and dry climate of the Holocene, many of these pluvial lake basins hold surface water only seasonally, while those that still contain perennial lakes have water levels that are much lower in elevation than they were in the past. This is definitely the case for the eastern Great Basin, which is dominated hydrographically by the Bonneville basin (Fig. 11.1). During the late Pleistocene, the Bonneville basin became filled with a huge lake, in surface area on par with North America's Lake Michigan, about 52,000 km^2 (Grayson 1993, 2011), but by the middle Holocene, it had contracted into a series of small sub-basin lakes (e.g., Great Salt Lake, Utah Lake, Bear Lake, and Sevier Lake) and dry playas. The change clearly signals significant warming and drying coincident with the onset of the Holocene. A series of paleoecological proxy records provides the details regarding aridification—when it transpired, how it affected lake levels, and how it affected plant and animal communities. These are reviewed briefly below, focusing on the western Bonneville basin, home to Bonneville Estates Rockshelter.

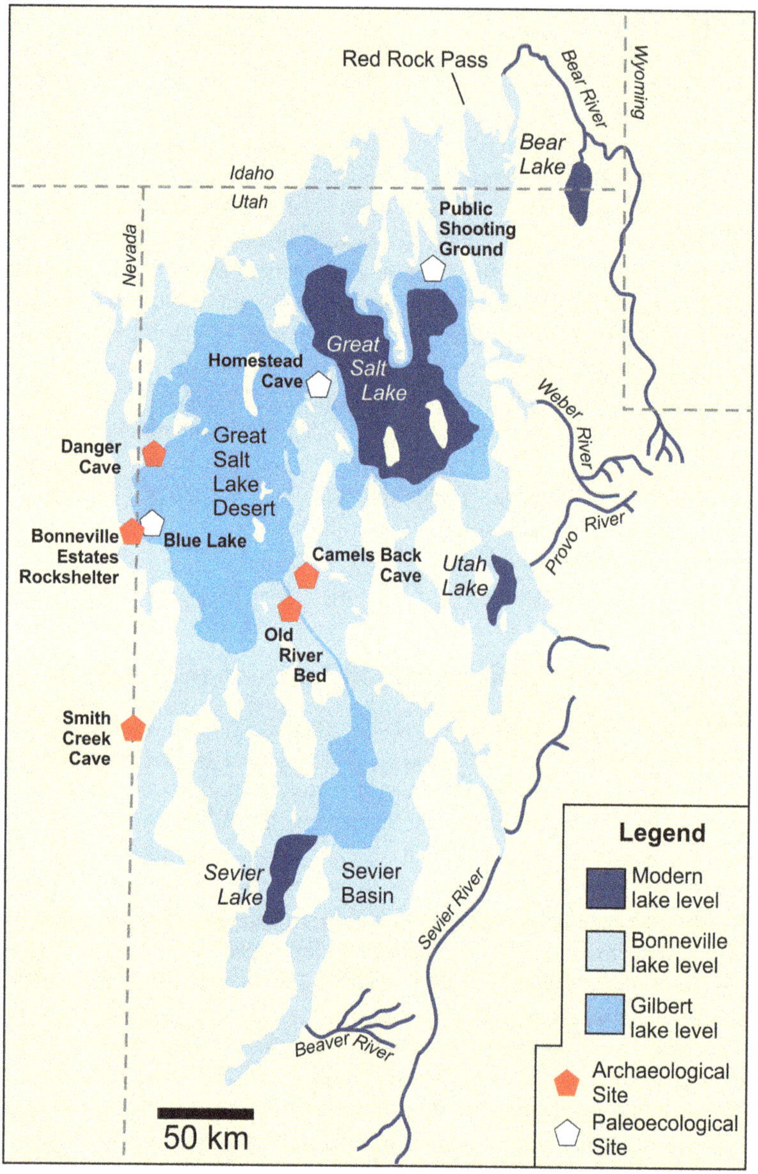

Fig. 11.1 Map showing geographic extent of Lake Bonneville during its late Pleistocene high stand (Lake Bonneville level) and the Younger Dryas chronozone (Gilbert Lake level), as well as paleoecological and archaeological localities mentioned in the text

The geomorphology of the Bonneville basin has been studied for more than a century. G. K. Gilbert (1890) was first to recognize that its hillslopes had been modified by a series of ancient lake-margin features (e.g., beach ridges or wave-cut terraces) and that these could be grouped into a series of topographically distinct shorelines, most notably the high Bonneville shoreline (at about 1585 m in eleva-

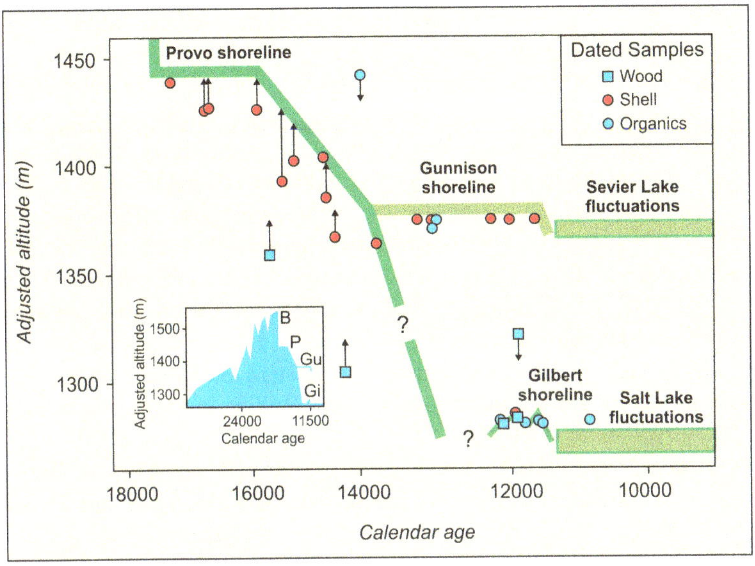

Fig. 11.2 The last transgressive/regressive cycle of Pleistocene Lake Bonneville, with a focus on the late-glacial record (After Benson et al. 2011) (in *inset B* refers to Lake Bonneville level, *P* Provo level, *Gu* Gunnison level, *Gi* Gilbert level)

tion) and a lower in elevation yet just as prominent Provo shoreline (~1465 m in elevation). Generations of geomorphologists—most notably D. Currey and J. Oviatt—have worked to refine the last major transgressive-regressive cycle of Lake Bonneville, applying careful mapping, radiocarbon-dating, and sedimentological analyses (Currey et al. 1984a, b; Oviatt 1988; Benson et al. 2011). The most recent iteration of this cycle is presented in Fig. 11.2, based on Benson et al. (2011) and Oviatt et al. (2003). Surface lake level began to rise significantly after 30,000 cal BP, and it steadily rose (albeit with some oscillations) over the course of the next ten millennia, reaching its peak about 18,500 cal BP. At this point, the lake found a sill near Zenda, Idaho, and began to flow into the Snake River basin to the north. Around 17,000 cal BP, it cut a gap into this threshold, unleashing a catastrophic flood resulting in an immediate 110-m drop in the lake's level, where it stabilized and created the Provo shoreline. The lake rested at the Provo level for nearly 2000 years (Benson et al. 2011), and around 15,200 cal BP, it began to decline rapidly again, this time because of lower precipitation rates or rising temperatures and evaporation rates or both. By 14,700 cal BP, Lake Bonneville had virtually ceased to exist, as smaller sub-basin lakes receded to historic levels and potentially even dried up (Oviatt et al. 1992; Benson et al. 2011). At the onset of the Younger Dryas chronozone, perhaps even several centuries before it (Benson et al. (2011) suggest 13,100 cal BP), water rose to the level of the Gilbert shoreline, about 1300 m above sea level, in Bonneville's northern sub-basins. The similarly positioned Gunnison shoreline formed in the Sevier basin to the south, although at a higher elevation due to this sub-basin's higher base level (Oviatt et al. 1992, 2003). Shortly after the Younger Dryas, Bonneville's sub-basin lakes appear to have receded from these shorelines.

Benson et al. (2011) conclude that the Gilbert Lake persisted until about 11,600 cal BP in the western Bonneville basin.

Alluvial histories offer a complementary history, especially for the Pleistocene-Holocene transition. During the Younger Dryas, when the Sevier sub-basin was filled to the level of the Gunnison shoreline, water flowed over its sill northward into the western Bonneville sub-basin, along a floodplain referred to as the "Old River Bed" (Oviatt et al. 2003). A series of radiocarbon dates gathered on organics from overbank, sand channel, and associated wetland deposits indicates that water flowed along the Old River Bed well into the Holocene, until about 9900–9700 cal BP (Oviatt et al. 2003). At that time, the shallow lake in the Sevier sub-basin regressed from its sill, and the wetlands along the Old River Bed's drainage dried up completely. Other small, isolated wetlands in the western Bonneville basin, however, continued to be locally sustained through the Holocene (e.g., Louderback and Rhode 2009; Benson et al. 2011).

Paleobotanical studies from pack rat middens and pollen cores in the Bonneville basin indicate a correlative record of warming and drying during the late Pleistocene and early Holocene. D. Rhode's (2000a, b; Rhode and Madsen 1995) model study of the macrobotanical remains recovered from 30-fossil pack rat middens—more than half from the hills surrounding the western Bonneville sub-basin—together provides signals of changing vegetation communities from today's sunbaked playa floor to the cooler subalpine mountain slopes of the Deep Creek, Goshute, Pequop, and Pilot Ranges. During full-glacial times (>17,000 cal BP), high-elevation areas were characterized by Engelmann spruce (*Picea engelmannii*) woodlands and sagebrush shrublands, indicating relatively cool and moist conditions (Rhode 2000b). Shortly after 17,000 cal BP, limber pine (*Pinus flexilis*) replaced spruce as the dominant conifer in the region's uplands, and limber-pine woodlands with a sagebrush (*Artemisia* sp.) understory began to expand across middle-elevation hillslopes, as far down as the Lake Bonneville shoreline and the site of Bonneville Estates Rockshelter (Rhode 2000b). The low-elevation limber pines persisted until 13,400–12,600 cal BP, when they were gradually replaced by a mosaic of dry-loving shrubs including shadscale (*Atriplex confertifolia*), saltbush (*A. canescens*), greasewood (*Sarcobatus vermiculatus*), horsebrush (*Tetradymia* sp.), snakeweed (*Gutierrezia sarothrae*), and rabbitbrush (*Chrysothamnus* sp.), not to mention sagebrush (Rhode 2000a, b). By about 9000–8700 cal BP, pinyon pine (*Pinus monophylla*) and Utah juniper (*Juniperus osteosperma*) dispersed across middle elevations, while sagebrush became underrepresented in low elevations, being replaced by chenopod shrub communities, an indication of continued warming and drying (Rhode 2000a).

The pollen record from Blue Lakes, situated along the Nevada-Utah border about 25 km south of the town of Wendover and 9 km east of Bonneville Estates Rockshelter, chronicles a regional vegetation history complementing Rhode's macrobotanical record. Late Pleistocene vegetation was dominated by conifers and sagebrush, but a clear shift from a sagebrush-dominated to shadscale-dominated shrubland occurred around 11,500 cal BP (Louderback and Rhode 2009). The shift to warm-, dry-loving shrubs in the Blue Lakes core, however, did not happen overnight but instead seems to have progressed gradually from about 11,800 to 9500 cal

BP. Noticeable, too, is a concomitant drop in conifers, a further indication of drying, warming climate (Louderback and Rhode 2009). Other palynological records in the greater Bonneville region portray similar warming trends, but none has such high resolution chronologically for the Pleistocene-Holocene boundary (e.g., Bright 1966; Spencer et al. 1984; Beiswenger 1991).

Although little is known about Pleistocene large-mammal extinctions in the western Bonneville sub-basin (but see Grayson 2011 and Hockett and Dillingham 2004 for recent regional reviews), changes in the records of fish and small mammals indicate at least two episodes of significant drying during the late Pleistocene and early Holocene. Central to the region's paleontological record is Homestead Cave, located near the Eardley threshold separating the western Bonneville and Great Salt Lake sub-basins (Madsen 2000). For its excellent paleontological record, the cave deserves special treatment here. Homestead Cave's lowest stratigraphic layer, stratum I, contained thousands of bones of fish, primarily from pellets of scavenging owls (Broughton 2000). In his analysis of these remains, J. Broughton identified 11 freshwater species, including cutthroat trout (*Oncorhynchus* cf. *clarki*) and whitefish (*Prosopium* sp.), as well as suckers (*Catostomus* sp.) and Utah chub (*Gila atraria*). Their strontium isotopic values suggested that a shallow but freshwater lake existed near the cave, no more than 6 km away given the daily range of owls (Broughton 2000). Behaviorally, owls are not known to hunt live fish, but they do scavenge dead fish; as Broughton reasons, therefore, the vast quantities of fish bones in stratum I must have resulted from a massive fish die-off, and the near disappearance of trout and whitefish after this date in the Homestead Cave record suggests that the die-off is related to a rapid decline in lake level, possibly even desiccation of the lake. Radiocarbon dates bracketing the die-off are 13,200 and 12,000 cal BP (Broughton 2000), so that, on the one hand, the continued presence of trout and whitefish to such a late date suggests that Lake Bonneville did not completely dry up prior to 13,200 cal BP and, on the other hand, desiccation occurred immediately before or after the Younger Dryas or both (Broughton 2000).

The small-mammal record from Homestead Cave indicates continued aridification in the early Holocene, with a "breaking point" being reached for some species ~9500 cal BP. Relative proportions of cottontails (both *Sylvilagus nuttallii* and *S. audubonii*), pygmy rabbits (*Brachylagus idahoensis*), voles (*Microtus* sp.), harvest mice (*Reithrodontomys* sp.), and yellow-bellied marmots (*Marmota flaviventris*) declined significantly by that time, at the expense of kangaroo rats (*Dipodomys* sp.) and desert woodrats (*Neotoma lepida*), which concomitantly rose in frequency (Grayson 1998, 2000). Moreover, the small-mammal assemblages from Homestead Cave prior to 9500 cal BP were taxonomically richer than those following this date (Grayson 1998, 2000). Together these paleontological changes in the Homestead Cave record suggest relatively cool and moist conditions persisted in the Bonneville basin until 9500 cal BP, but at that time significant warming and drying occurred (Grayson 1998, 2000). The small-mammal records from Camels Back Cave and Bonneville Estates Rockshelter present analogous faunal turnovers (Schmitt and Lupo 2012).

To sum up, humans inhabiting the Pleistocene-Holocene transition in the eastern Great Basin witnessed significant warming and aridification (Fig. 11.3). The pro-

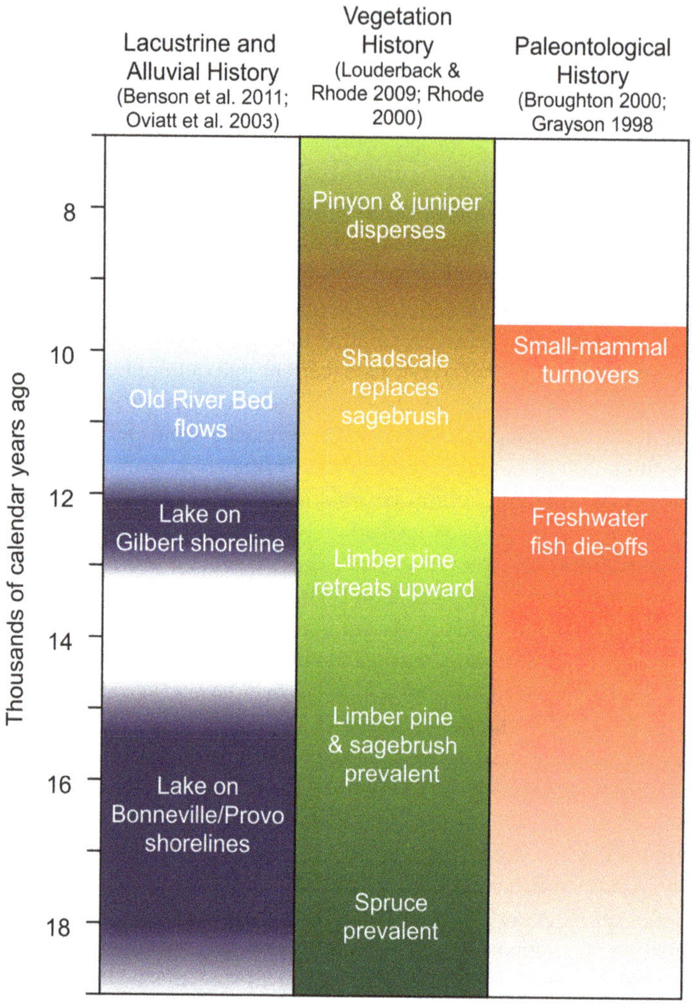

Fig. 11.3 Timeline of environmental change in the western Bonneville basin, 18,000–7000 cal BP

cess began around 15,000 cal BP, as Lake Bonneville quickly began to recede from its Provo shoreline. In the western Bonneville basin in particular, the lake may have dried completely by 14,700 cal BP, but it rebounded briefly during the Younger Dryas as a shallow yet extensive lake. Water flowed down the Old River Bed, and limber-pine woodlands persisted at fairly low elevations as well, indicating relatively cool but dry conditions. The century around 9500 cal BP appears to have been a threshold time, with all proxy records signaling significant drying, especially at low elevations—the shallow Younger Dryas lake receded, water ceased flowing down the Old River Bed, plant communities shifted from cool-loving sagebrush to hot-loving shadscale and chenopods, and cool-loving small mammals were replaced by hot-loving species.

Terminal Pleistocene/Early Holocene Archaeology at Bonneville Estates Rockshelter

Bonneville Estates Rockshelter (BER) is located about 30 km south of West Wendover, Nevada, in the Lead Mine Hills overlooking the western Bonneville basin (Fig. 11.1). First discovered by S. Dondero and T. Murphy of the US Department of Interior's Bureau of Land Management, our team (led by Goebel, K. Graf, B. Hockett, and D. Rhode) excavated at BER from 2000 through 2009, exposing a series of stratified cultural occupations spanning the last 13,000 years of prehistory, in an area reaching approximately 67 m^2 (Graf 2007) (Fig. 11.4).

Bedrock geology is a Pennsylvanian-aged dolomite. Wave action cut into the bedrock creating the rockshelter at some point in the past when Lake Bonneville sat at its 1580-m high stand. The basal stratigraphic layer within the shelter consists of well-rounded gravels of various clast sizes that may date to the lake's high stand, 18,500–17,000 cal BP. Overlying this is up to 1 m of loose angular rubble and silt, with little organic preservation. From just above this rubble layer we AMS radiocarbon dated an unidentified bone fragment to about 18,500 cal BP; if this bone was in a primary context, then it suggests that the basal gravels and mantling silt-and-rubble stratum deposited very quickly or that the basal gravels date to a much earlier time than the lake's last transgressive cycle. In the western area of our excavation, the silt-and-rubble stratum was mantled by a well-preserved woodrat midden with occasional artifacts (very small resharpening flakes, likely redeposited from younger layers) (Goebel 2007), and in the eastern area, it was mantled by more rubble.

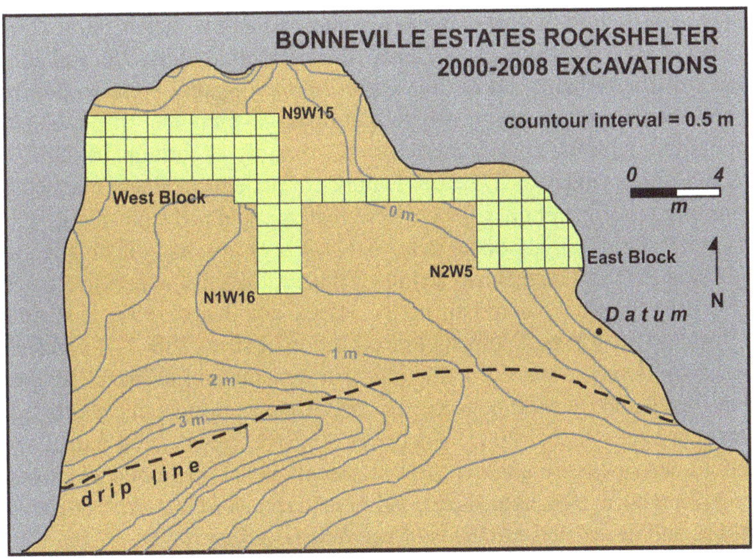

Fig. 11.4 Map of the interior of Bonneville Estates Rockshelter, showing extent of excavation during the 2000s

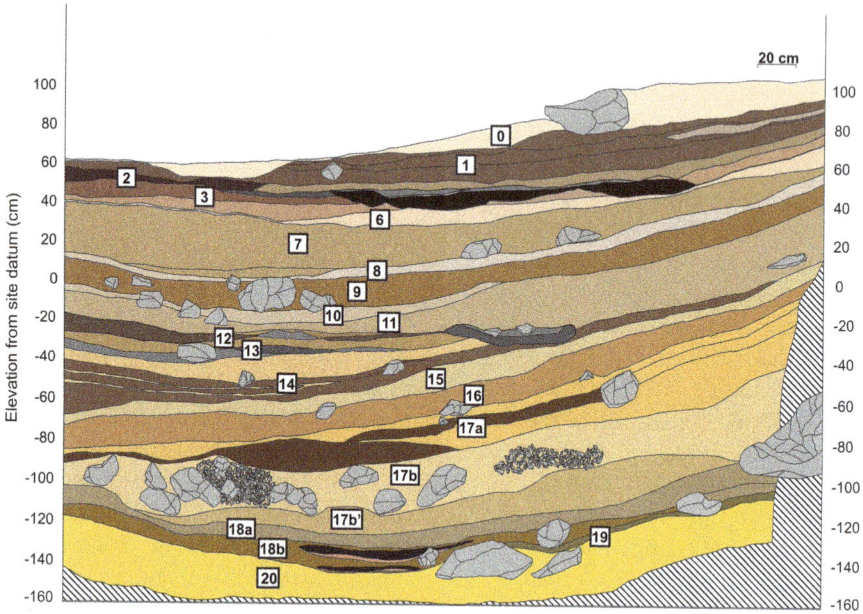

Fig. 11.5 Stratigraphic profile from near center of Bonneville Estates Rockshelter, N1-4 W14, showing Paleoindian (strata 18b–17b') and early Archaic (strata 17b–14) components (Profile by K. Graf)

The earliest clear evidence for humans in the shelter occurs in a packet of sediments referred to as substrata 18b, 18a, and 17b', which in the western area of the excavation were rich in macrobotanical remains and animal bones (Graf 2007; see also Goebel et al. 2007) (Fig. 11.5). During the excavation, strata 18a and 18b were not always distinguishable, and in the eastern area, organics were so poorly preserved and the deposit so full of rubble that distinguishing any of these substrata was impossible, with the zone being referred to simply as stratum 18 (Graf 2007). In these deposits, as many as 27 hearth features were delineated and mapped across the shelter floor. Radiocarbon dates on these features spanned from about 13,000 to 10,300 cal BP (Graf 2007; Goebel et al. 2011) (Fig. 11.6), and diagnostic artifacts include a small set of Great Basin stemmed points, the most expressive stone-tool forms for the region's Paleoindian period (Beck and Jones 1997; Hockett et al. 2008). Besides the stemmed points, our excavation of strata 18b–17b' yielded thousands of stone debitage pieces, a few biface fragments and unifacial scrapers, an eyed bone needle and several bone needle/awl fragments, and some short segments of twined cordage (Goebel 2007; Goebel et al. 2011). Associated faunal remains included numerous large-mammal bone fragments (some identified as artiodactyls including pronghorn, mountain sheep, and mule deer) as well as bones of hares and sage grouse and desiccated carcasses of katydids (Hockett 2007).

Across the shelter, the Paleoindian occupation was mantled by about 30 cm of rubble and silt, the result of an intensive period of roof fall and aeolian activity. Very few cultural remains occur in this deposit, labeled strata 17b and 17a, but four

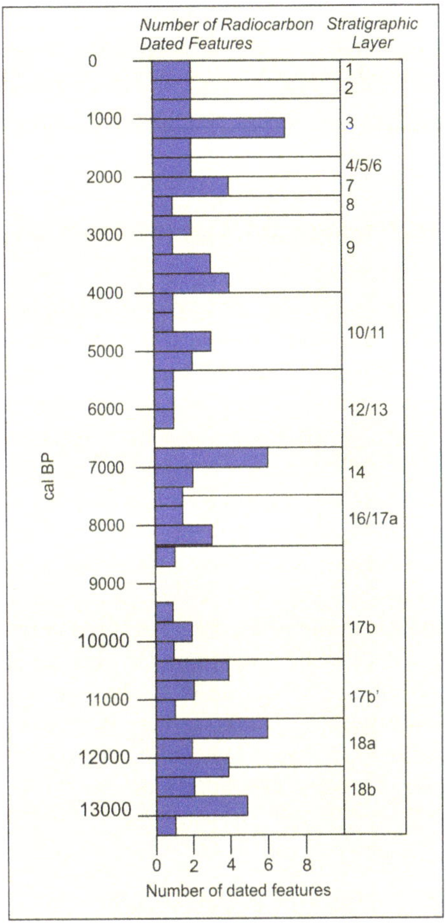

Fig. 11.6 Distribution of radiocarbon dates on cultural features from Bonneville Estates Rockshelter

hearths yielded radiocarbon ages ranging between 10,200 and 8520 cal BP, and a fifth, 8370–8160 cal BP. Associated cultural remains are chiefly stone flakes.

A strong pulse of human occupation began with the onset of deposition of stratum 16 and continued through deposition of stratum 14. These deposits are organic rich, with numerous plant macrofossils, some the result of human subsistence activities. Among these were 17 hearth features radiocarbon dated from 8200 to 6750 cal BP (Graf 2007). Besides thousands of stone debitage pieces, the associated lithic assemblage includes numerous large side-notched points, diagnostic of the region's early Archaic record, unhafted bifaces, unifacial scrapers, and ground-stone artifacts (i.e., manos and metates) (Goebel 2007). In addition, strata 16 and 14 yielded an interesting collection of cordage and baskets, including a fragment of one of the oldest coiled baskets yet known from Nevada (8160–7930 cal BP) (Graf 2007). Faunal remains are not decidedly different from the earlier Paleoindian layers, in that fragments of large-mammal bone dominate the assemblage, except that pronghorn represents the only identified large-mammal species (Hockett 2007). Besides this, a few hare and sage grouse bones also occur.

The early Holocene sediments in BER are capped by a series of organic-rich middle and late Holocene cultural layers interdigitated with bands of sterile silt and rubble, together reaching more than 1 m in thickness. These are not the focus of the analysis that follows so they are not presented further.

Technological Change at Bonneville Estates

As presented above, two cultural components occur within Bonneville Estates' terminal Pleistocene/early Holocene deposits. The lower component spans from 13,000 to 10,500 cal BP and is Paleoindian in character, while the upper component spans from 8200 to 6750 cal BP, perhaps as early as 8370 cal BP, and is early Archaic in character. Given that this record spans the period of terminal Pleistocene and early Holocene warming and aridification discussed in detail above, it presents an important case study of how humans adapted technology to significant environmental change.

Analysis of the lithic assemblage from BER is still in progress; however, the results presented below include almost all of the Paleoindian assemblage (>90%) and a majority of the early Archaic assemblage (>75%). Although multiple occupation events and stratigraphic layers are represented in the rockshelter's record, a previous study showed little variation in technological activities within each component (Goebel 2007), so here they are treated at this gross component level. The Paleoindian assemblage includes 1063 lithic artifacts and the early Archaic assemblage includes 3091 lithic artifacts.

Raw Material Procurement

Three major classes of lithic raw material are present in the BER assemblages—cryptocrystalline silicates (CCS) (i.e., cherts and chalcedonies), fine-grained volcanics (FGV) (i.e., basalt and andesite), and obsidian (Fig. 11.7a). In the Paleoindian component ($n = 1063$), no single raw material dominates, with CCS occurring about 45% of the time, FGV 30%, and obsidian 25%. Generally, the same trend holds for the early Archaic component ($n = 3091$), except that CCS occurs much more frequently, 65% of the time, with FGV and obsidian being much less frequent (about 20% and 15%, respectively). This difference is significant ($X^2 = 149.842$; df = 3; $p < 0.001$).

Of the three raw material classes, CCS occurs within 10 km of the rockshelter, in isolated outcrops in the foothills of the Goshute Mountains west of the site, and in small cobbles in Gilbert shoreline beach deposits and alluvial fan deposits southeast of the shelter around the "mouths" of Dead Cedar and Ferguson Washes. So far, we have not attempted to analyze these materials geochemically, but we hope to do so in the near future.

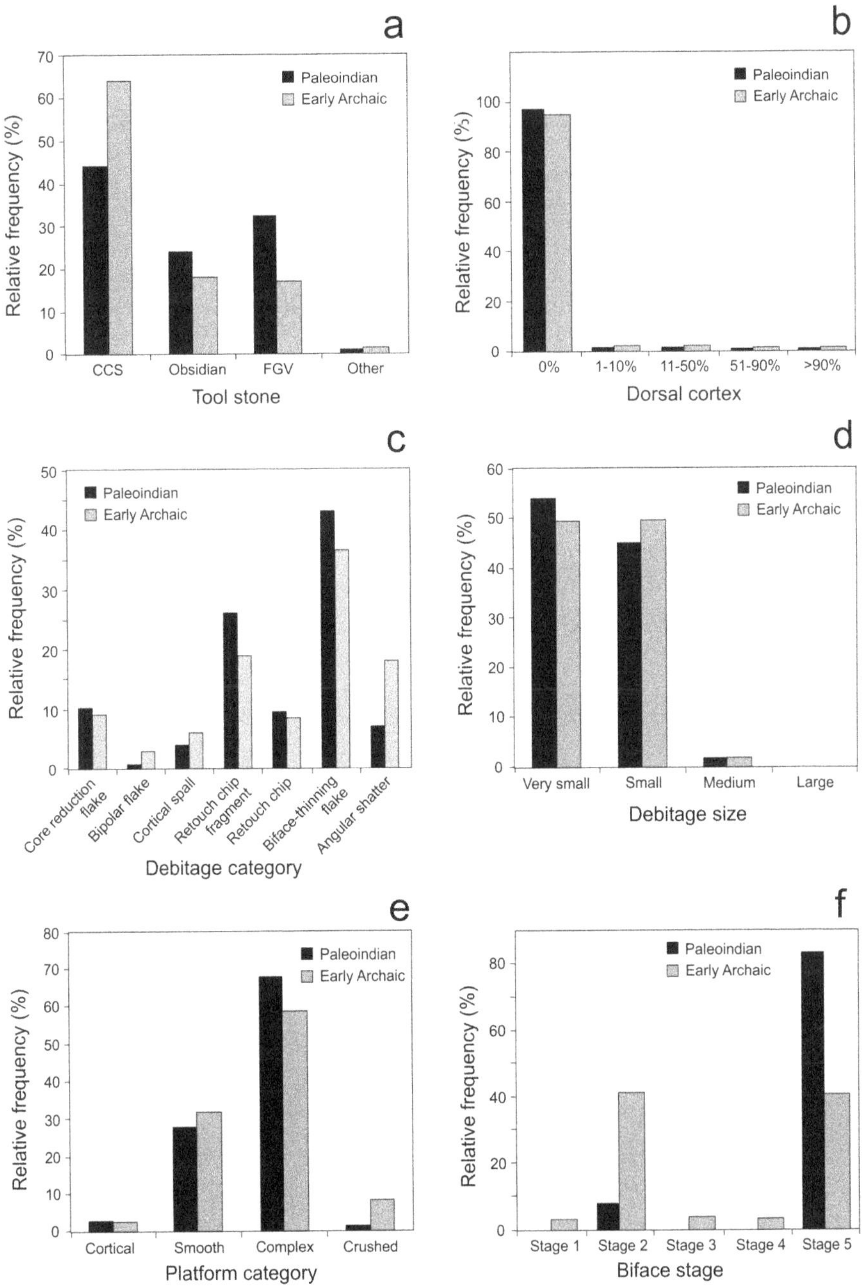

Fig. 11.7 Technological characteristics of the Paleoindian and early Archaic components at Bonneville Estates Rockshelter (**a**) lithic raw material; (**b**) amount of cortex on dorsal face; (**c**), debitage type; (**d**) debitage size; (**e**) platform surface; (**f**) biface stage

FGV occurs naturally in several isolated basalt flows around the southern rim of the western Bonneville basin and southwest of BER in neighboring Steptoe/Goshute Valley. Although as many as nine FGV sources have been geochemically distinguished (Page 2008), only recently did we begin to characterize the basalt and andesite artifacts from the shelter. First results presented here were analyzed by Craig Skinner at the Northwest Research Obsidian Studies Laboratory. So far, with a sample of 20 bifacial points, bifaces, and other tools (8 from the Paleoindian component and 12 from the early Archaic component), the Deep Creek A source dominates (62.5% and 75%, respectively), not surprising given that this is the nearest large-scale source of basalt to BER, occurring as cobbles in the alluvium of lower Deep Creek, as close as 25 km from the rockshelter (Fig. 11.8). A ninth Paleoindian artifact was made on Badlands A/Wildcat Mountain basalt, also likely from the lower Deep Creek basin, where it occurs in the same alluvial deposits as Deep Creek A, and a tenth was made on Flat Hills D andesite, from an alluvial source located about 100 km east of the rockshelter, across the Bonneville Salt Flats. Three unknown sources occur in the assemblage, too, three of them in the Paleoindian component and two in the early Archaic component, with no overlap between the components. This analysis is obviously just underway and more results will be presented at a later date.

Most of our geochemical work has focused on obsidian, much better known in the eastern Great Basin than FGV. Initially, Craig Skinner analyzed 112 Bonneville

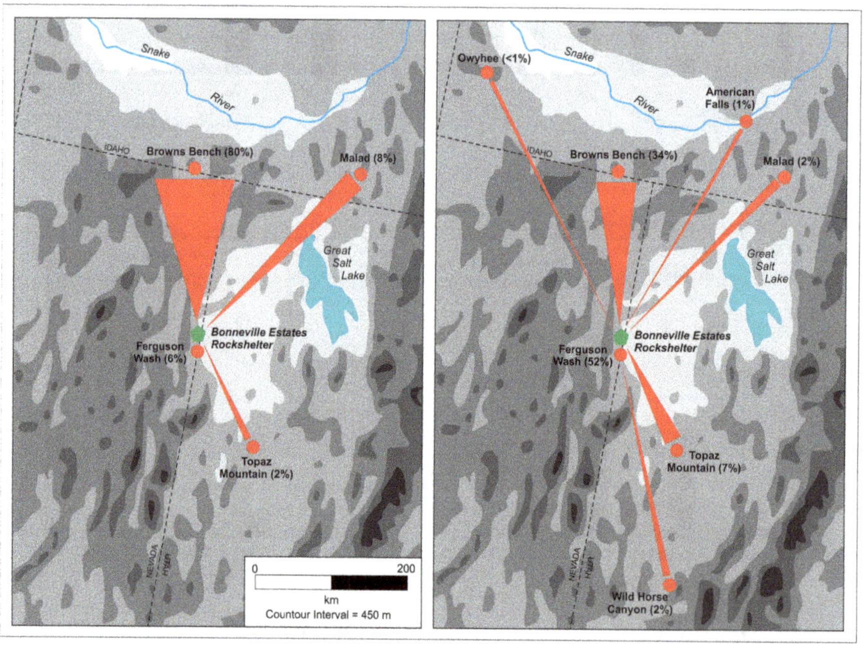

Fig. 11.8 Locations of obsidian sources identified in the Bonneville Estates Rockshelter assemblages (*left*, Paleoindian assemblages; *right*, early Archaic assemblages). Width of *arrows* generally represents proportions of sources in the assemblages, with the exception of Ferguson Wash, which is too close to the rockshelter to show an *arrow*

Estates artifacts, but more recently, we have analyzed 1205 artifacts using a Bruker TRACER III-V portable ED-XRF (pXRF) provided by Donny Hamilton of the Conservation Research Laboratory at Texas A&M University. The pXRF was equipped with a rhodium target and a 170 eV resolution Si-PIN detector with 13 μ Be detector window and a 45 kV x-ray generator. All samples were analyzed at 40 kV and 11.5 μA. The pXRF was equipped with 12 mil Al, 2 mil Ti, and 6 mil Cu filters used in conjunction with a software calibration using 40 international obsidian standards provided by Jeff Ferguson at the MURR Archaeometry Laboratory. This permitted calculation of quantitative ppm values for the following elements: Mn, Fe, Zn, Ga, Th, Rb, Sr, Y, Zr, Nb, and Rh (Ferguson 2012). Of these, bi-plots of Ga, Th, Rb, Sr, and Y were used to attribute source identifications, using GAUSS Runtime v. 8.8c software developed by the MURR Archaeometry Laboratory (University of Missouri 2014). Obsidian samples from regional obsidian sources were analyzed using our pXRF to create a reference collection. So far, these include eight to ten samples from each of the following obsidian sources: Brown's Bench (ID/NV/UT), Bear Gulch (ID), Big Southern Butte (ID), Cannonball I and II (ID), Cedar Butte (ID), Kelly Canyon (ID), Malad (ID), Obsidian Cliff (WY), Packsaddle (ID), Topaz Mountain (UT), Teton Pass (WY), American Falls (Walcott Tuff) (ID), Wild Horse Canyon (UT), Ferguson Wash (UT/NV), Owyhee (ID), and Timber Butte (ID). Reference samples for each of these obsidian types were either collected from their source or lent to us by Jeff Ferguson at the MURR Archaeometry Laboratory. Reference samples were scanned for 180 dead time-corrected seconds and were all greater than 1 cm in diameter and 4 mm thick.

All obsidian artifacts analyzed via XRF were chosen to meet basic criteria for stratigraphic integrity as well as for surface topography (Shackley 2010). Tools and debitage from Bonneville Estates were limited to specimens >5 mm in diameter and >2 mm thick. While sample thickness is a potential issue when considering infinite thickness (Shackley 2010; Ferguson 2012), calibrations and methods described by Ferguson (2012) made it possible to positively assign sources to smaller artifacts with ~95% accuracy. For the sake of the mass analysis of obsidian debitage from Bonneville Estates, it was necessary to scan each artifact for less time (30 dead time-corrected seconds). While precision of absolute ppm values was reduced, it was still possible to positively attribute obsidian sources to more than 90% of artifacts. The remaining artifacts with equivocal results, as well as artifacts smaller than 10 mm in diameter or 4 mm thick, were re-scanned at the standard 180 s.

Source locations used to determine distances from BER are based on maps provided by the Northwest Research Obsidian Studies Laboratory website (http://www.sourcecatalog.com/image_maps/index.html, accessed in 2014). In the case of diffuse sources such as Browns Bench and Owyhee, the minimum distance was used.

Of the 1317 obsidian artifacts so far analyzed, 50 come from the Paleoindian component and 387 come from the early Archaic component (results from later components will be presented at a later date). The disparity in sample size is a product of artifact metrics—much of the Paleoindian debitage assemblage is too small for reliable pXRF analysis. In the Paleoindian component, Browns Bench obsidian clearly dominates, making up 80% of the analyzed assemblage. Browns Bench is an extensive source, occurring in the eastern Jarbidge Mountains of northeast Nevada, northwest Utah, and southcentral Idaho. Browns Bench itself, a high ridge overlooking the town of Jackpot, Nevada, is about 180 km north-northwest of BER, as the raven flies. Other infrequently occurring obsidians come from Malad (8%), located in southeast Idaho near the northern edge of the Bonneville basin (about 260 km northeast of the shelter); Ferguson Wash (6%), located in the southern Lead Mine Hills (less than 10 km south-southeast of the shelter); and Topaz Mountain (2%), located in the southern Bonneville basin in western Utah (105 km southeast of the shelter). Besides these, 8% of the Paleoindian obsidians remain unknown to us. An important pattern in these data is that 88% of the tools and debitage analyzed originated from sources north of Bonneville Estates. Moreover, only 6% was from the local Ferguson Wash source.

In the early Archaic component, the local Ferguson Wash obsidian dominates, making up 52.2% of the analyzed obsidian assemblage. Browns Bench, however, is still common (34.4%), not surprising given that it is probably the highest-quality obsidian in the region, at least in terms of largest nodule size. Other infrequent occurrences include obsidian from Topaz Mountain (6.7%); Wild Horse Canyon (1.8%), located south of the Sevier Desert, western Utah (250 km south-southeast of BER); Malad (1.8%); American Falls (Walcott) (0.5%), located about 275 km north-northeast of BER in southern Idaho; and Owyhee (0.3%), located in southwestern Idaho about 330 km to the northwest. The early Archaic geographic trend is obviously oriented toward southern sources, with 60.7% of the obsidian assemblage originating from south of the shelter. Furthermore, the preponderance of the local source is an important feature of this assemblage.

We calculated weighted means for the minimum distances that obsidian was transported from sources to the shelter. For the Paleoindian component, the weighted mean distance was 174 km, while for the early Archaic component, it was 88 km. A Mann-Whitney test comparing mean ranks indicated this difference is significant ($U = 4360$; df = 1; $p < 0.001$), obviously due to the later focus on the local Ferguson Wash source.

Technological Activities

To explore differences in technological activities between the Paleoindian and early Archaic components, we analyzed the following variables: the presence of cortex, debitage category, debitage size, and platform preparation.

The presence of cortex is a useful measure of degree of local procurement represented in the lithic assemblages. For example, in an earlier study, we found a strong correlation between the presence of cortex and the local obsidian, Ferguson Wash, with 60% of that material type bearing cortex (Goebel 2007). Overall, for the Paleoindian component (n = 1061), 3.4% of artifacts had cortex preserved on their dorsal faces, whereas for the early Archaic component, 5.8% had cortex (n = 3088) (Fig. 11.7b). This difference was significant ($X^2 = 10.752$; df = 4; $p = 0.029$).

For analyzing debitage categories, we organized the assemblages into six types (not including fragments): core-reduction flake, bipolar flake, cortical spall, retouch chip (i.e., pressure flake), retouch chip fragment, biface-thinning flake, and angular shatter. Although both components have debitage assemblages with the same general character, the Paleoindian component (n = 1034) had more material representing secondary reduction activities (i.e., retouch chips and biface-thinning flakes), whereas the early Archaic component (n = 2974) had more material representative of primary-reduction activities (i.e., bipolar flakes, cortical spalls, and angular shatter) (Fig. 11.7c). Again these differences were significant ($X^2 = 92.566$; df = 6; $p < 0.001$).

Debitage size was analyzed using an ordinal scale of ranks (1–4, with 1 representing pieces less than 1 cm in length and width (called very small), 2 representing pieces 1–3 cm in length and width (small), 3 representing pieces 3–5 cm in length and width (medium), and 4 representing pieces greater than 5 cm in length and width (large)). Both assemblages (Paleoindian, n = 542; early Archaic, n = 1278) were dominated by the two smallest categories; however, a Mann-Whitney nonparametric test indicated a significant difference, with the Paleoindian component's debitage being significantly smaller (U = 1,314,043; $p = 0.019$) (Fig. 11.7d). This difference in size may be a function of differing raw material packages (i.e., with the early Archaic component representing reduction of bigger, more locally procured materials), differing technological activities (the early Archaic component representing more primary reduction), or both.

Platform preparation preserved on debitage pieces generally followed the same trend, with both components being dominated by smooth and complex platforms (Fig. 11.7e). However, the Paleoindian component (n = 520) had significantly more complex platforms, and the early Archaic component (n = 1622) had significantly more smooth and crushed platforms ($X^2 = 28.987$; df = 3; $p < 0.001$). The high frequency of complex platforms in the Paleoindian component likely represents a higher degree of bifacial reduction, whereas the high frequency of smooth and crushed platforms in the early Archaic component likely represents more simple core reduction and bipolar reduction.

Overall, the trends represented in the debitage assemblages indicate higher degrees of secondary working, specifically bifacial reduction, in the Paleoindian component, versus higher degrees of primary working, specifically simple core reduction and bipolar reduction, in the early Archaic component. Although these differences are a matter of degree, the repeated patterning and significant test statistics suggest they represent differential human behavior between the Paleoindian and early Archaic occupations.

Tool Production and Raw Material Selection

Tools recovered from the Paleoindian and early Archaic components were organized into four categories, with the goal of characterizing general trends in tool production: unifacial flake tools, bifaces, bifacial points, and ground-stone tools. The Paleoindian assemblage (n = 29) was dominated by unifacial flake tools (primarily flakes with marginal retouch) and finished (and broken) bifacial points, while the early Archaic assemblage (n = 117) was dominated by unhafted bifaces and had the only ground-stone artifacts. These proportional differences were significant ($X^2 = 14.315$; df = 3; $p = 0.003$).

We further organized the bifaces and bifacial points into five stages, with stage 1 representing obvious biface blanks, stage 2 representing edged bifaces, stage 3 representing thinned bifaces, stage 4 representing preforms, and stage 5 representing finished and hafted bifaces (following Andrefsky 1998). The Paleoindian component (n = 15) was characterized by mostly finished (stage 5) bifaces and a few edged (stage 2) bifaces, while the early Archaic component (n = 79) contained examples of every stage and similarly high proportions of finished and edged bifaces (Fig. 11.7f) (a Chi-square analysis was not conducted because of relatively low expected cell counts in three of the five stages). These data suggest obvious differences in bifacial-reduction activities at BER. While the Paleoindian occupants typically transported finished bifaces into the shelter, early Archaic occupants transported nodules or blanks to the shelter and manufactured bifaces there.

In the Paleoindian component, there was no obvious pattern of raw material selection in the production of bifaces, with relative proportions hovering between 27 and 40%. A closer look, however, suggests that unifaces were predominantly made on CCS (67%), likely a function of the expedient nature of their production on locally procured raw materials but possibly simply a result of human preference for CCS in flake-tool production. Moreover, all of the obsidian bifaces originated from distant sources (primarily Browns Bench), none on more nearby sources like Ferguson Wash. Early Archaic raw material selection differed from this, with an obvious preference toward the use of CCS in producing bifaces (71%), and although the same was true of unifaces (with 55% being made on CCS), there was an unexpectedly higher proportion of unifaces made on obsidian (32%), primarily local obsidian from Ferguson Wash. Given that sources of CCS, FGV, and obsidian are readily available near the rockshelter, these differences could represent real preferences in raw material use—Paleoindian occupants not preferring one tool stone over another for biface production but using locally available CCS for uniface production and early Archaic occupants preferentially selecting CCS (some locally procured and some perhaps not) for biface production and locally available CCS and obsidian for uniface production.

Discussion

From the ongoing analysis presented above, we have detected some differences in raw material procurement, tool-stone selection, and technological activities for the Paleoindian and early Archaic occupants of Bonneville Estates Rockshelter. For the Paleoindian component, technology centered on transporting nonlocal raw materials to the shelter, typically in the form of finished or nearly finished bifaces. Signs of primary reduction are uncommon, tool resharpening was a common practice, and obsidians came primarily from northern sources on the fringe of the western Bonneville basin and beyond. For the early Archaic component, technology centered on transporting unworked or minimally worked raw-material packages to the shelter, primarily from local sources. Signs of primary reduction, including bipolar reduction, are common, all stages of biface reduction are represented, and obsidians came primarily from more local, southern sources within the western Bonneville basin. These differences in tool-stone provisioning reflect distinctive strategies of settlement organization: Paleoindians repeatedly visited the shelter but stayed there for relatively short episodes, operating across a relatively large territory regularly extending outside the limits of the western Bonneville basin; early Archaic people had longer stays within the shelter, operating on a more local scale, typically within the western Bonneville basin. These interpretations, however, are still based on just a subset of the lithic record from Bonneville Estates, and they need to be checked with more complete analyses of the artifacts recovered from our excavations and from more extensive lithic-procurement studies of fine-grained volcanic and cryptocrystalline tool-stones.

Another interesting change in technology not analyzed here (but currently being investigated) is a change in bifacial-point function. The Paleoindian component is characterized by fragments of stemmed points inferred to have been used as hafted weapon tips as well as knives (e.g., Musil 1988; Beck and Jones 1997, 2009; Lafayette and Smith 2012). This multifunctionality seems to be evident on the BER points, too, in that they display hinge fragmentation resulting from impact, as well as post-breakage resharpening indicative of their recycled use as knives. The early Archaic component, however, is characterized by complete and fragmented side-notched points inferred to have been used solely as hafted weapon tips (Musil 1988). Impact damage on the BER side-notched points occurs in the form of distal breaks as well as damage to the points' corners. Macroscopically, they rarely bear signs of having been used for any other purpose than as weapon tips. Given that there is no overlap stratigraphically in these two projectile technologies and that no diagnostic points have been found in stratum 17a, we have to conclude that the technological transition in projectile weaponry occurred sometime between 10,300 and 8200 cal BP. Other sites suggest the emergence of side-notched points was closer to 8200 cal BP (e.g., Schmitt and Madsen 2005). Interestingly, the Old River Bed surface record has yielded yet another bifacial-point form, the corner- and basal-notched Pinto point, which does not appear in substantial numbers at Bonneville Estates. Perhaps this was the point form typically made and used by humans in the region during the 2000-year period intervening the disappearance of stemmed points and appearance of side-notched points at BER, the 2000 years represented by stratum 17a in the shelter.

Other technological differences occur in other Bonneville Estates record, too. Most importantly, in the early Archaic component ground-stone tools (i.e., manos and metates) appear for the first time, and there is a proliferation in textile forms—not just a variety of cordage and twine but also the earliest coiled baskets. With the exception of small snips of small-diameter twine, these technologies are absent in the Paleoindian component. Together this early Archaic proliferation of technologies represents an expansion of subsistence activities carried out from the shelter.

In a preliminary zooarchaeological and taphonomic analysis of faunal remains from the early layers at Bonneville Estates, Hockett (2007) recognized subtle differences in subsistence as well. During the Paleoindian occupation of the shelter, the food quest focused on artiodactyls, sage grouse, and hare. Carcasses of artiodactyls were heavily processed and burned, with long bones being fragmented for marrow extraction, so much so that only a few specimens could be identified to species—pronghorn (*Antilocapra americana*), bighorn sheep (*Ovis canadensis*), and deer (*Odocoileus hemionus*). Sage grouse (*Centrocercus urophasianus*) were systematically butchered in the shelter, evident from repeated anthropogenic breakage patterns and stone-tool-produced cut marks. Hare (*Lepus* sp.) long bones were cracked open to extract marrow, but few bore signs of burning. Besides these, Hockett (2007) also identified a single burnt bone of a black bear (*Ursus americanus*) and 18 well-preserved katydid carcasses. During the early Archaic, faunal diversity became reduced, with primary prey continuing to be artiodactyls—primarily pronghorn, mountain sheep, and deer but also infrequently bison (*Bison* sp.) (being represented by a single bone). Remains of hare and sage grouse were rarely consumed in the shelter, and no grasshoppers occur in the assemblage. This apparent reduction in species diversity may have been a function of increased aridity, not human preference (Hockett 2007). Schmitt and Lupo (2012) provide further evidence of the drying climate's effect on BER's fauna: five naturally occurring mesic-adapted rodent species (yellow-bellied marmot [*Marmota flaviventris*], bushy-tailed woodrat [*Neotoma cinerea*], sage vole [*Lemmiscus curtatus*], Great Basin pocket mouse [*Perognathus parvus*], and Western harvest mouse [*Reithrodontomys megalotis*]), common in the Paleoindian-aged deposits of the shelter, became rare in the early Archaic deposits.

Paleobotanical analyses of Bonneville Estates Rockshelter's hearths are still underway. Rhode and Louderback (2007) presented a preliminary analysis, reporting dietary data for the Paleoindian component but not the early Archaic component. Taphonomically, it was difficult to distinguish which plant macrofossils were brought into the shelter by humans versus other plant-collecting and plant-eating animals (e.g., woodrats), but the macrofossils' presence in hearths and charring was one way. In this respect, Paleoindians may have utilized a variety of seeds (e.g., ricegrass, goosefoot, and saltbush) as well as cactus pads, but it is important to remember that rodents like the bushy-tailed woodrat could have been the agent responsible for the accumulation of these remains. Bulrush and cattail seeds, however, would have been transported nearly 10 km from nearby marshes to the shelter, too far for BER's rodents, but not for humans, who could have brought them amidst fluff to be used as a fire starter (Rhode and Louderback 2007). Another way to tell that these seeds were used by humans is evidence of stone grinding; however, no such signs occur in the Paleoindian assemblage, and no ground-stone artifacts were found. In other words, if the Paleoindian inhabitants of the shelter were consuming seeds, they were not grinding them first, suggesting that they were not yet components of an intensive seed-focused subsistence system. Paleobotanical analyses of the early Archaic component are still underway, but during excavations we encountered numerous examples of small grass seeds, pine nuts, and charred cactus pads, often associated with grinding stones, suggesting a major transformation in human diets by the onset of the early Archaic occupation at Bonneville Estates. This pattern of a broad-spectrum diet including hard-to-process seeds is repeated at other early Archaic sites in the western Bonneville basin, for example, Danger Cave and Camels Back Cave (Schmitt and Madsen 2005; Rhode et al. 2006; Rhode and Louderback 2007).

The changes in technology and subsistence noted above occurred sometime between 10,300 and 8200 cal BP. Although a small assemblage of artifacts and related subsistence remains has been found with a few ephemeral hearth features in deposits dating between 10,300 and 9500 cal BP (Fig. 11.6), so far these cultural remains have proven inexpressive and the faunal and floral remains poorly preserved. In other words, even though the repeated activities that so characterized the Paleoindian occupation of Bonneville Estates had effectively ceased by 10,300 cal BP, humans continued to occasionally trickle in and out of the shelter for another 800 years. It was not until after an apparent hiatus in occupation between 9300 and 8700 cal BP that humans reentered Bonneville Estates with a new Archaic adaptation. Obviously this period of adjustment in technological, subsistence, and settlement organization occurred simultaneously to significant aridification of the western Bonneville basin: the drying up of the Old River Bed and its marshes 9900–9700 cal BP, the replacement of cool-/dry-adapted sagebrush by hot-/dry-adapted shadscale and other xerophytic shrubs in the Blue Lakes pollen record starting 11,800 and complete by 9500 cal BP, and the turnover in small-mammal fauna in Homestead Cave indicating significant aridification by 9500 cal BP. Obviously, in the case of Bonneville Estates Rockshelter, climate and environmental change was an important factor in human-adaptive reorganization of technology, subsistence, and settlement across the Paleoindian-early Archaic transition.

Conclusions

All things considered, the archaeological record of Bonneville Estates Rockshelter presents significant differences in the behavior of its Paleoindian and early Archaic inhabitants. Paleoindian technology in the shelter focused on resharpening and recycling bifaces which were transported great distances, whereas early Archaic technology in the shelter focused on producing a broader array of tools on locally procured tool stone. We infer that Paleoindian occupations were brief but frequent, while early Archaic occupations were much longer and reflected more technological and subsistent activities. Related subsistence analyses indicate, too, that early Archaic diet was much broader than Paleoindian diet, incorporating locally available high-cost plant foods that Paleoindians largely ignored. Although we cannot yet say with certainty precisely when these changes occurred, they seem to be related to an apparent ~600-year hiatus in the rockshelter's record, 9300–8700 cal BP. The cessation of the Paleoindian way of life at Bonneville Estates coincided with the "tap being shut off" in the western Bonneville basin: the primary source of freshwater into the basin—the Old River Bed—dried up, and vegetation and small-mammal communities in the basin's lowlands became dominated by xerophyllic species. We conclude, therefore, by inferring that the changes in raw material procurement and technological organization represented in the archaeological record of the Paleoindian-early Archaic transition at Bonneville Estates Rockshelter took place as a response to climate and environmental change. These technological changes occurred in tandem with subsistence and settlement shifts, which together seem to represent dramatic cultural reorganization in the face of significant aridification around 9500 cal BP.

Additional work needs to be done, however, to present a complete reconstruction of the Paleoindian-to-early-Archaic transition at Bonneville Estates Rockshelter. T. Goebel continues lithic artifact and tool-stone analyses, M. Coe continues perishable artifact analyses, B. Hockett continues zooarchaeological analyses, and D. Rhode continues archaeobotanical analyses. Hopefully these and other studies will provide a high-resolution perspective on environmental change and human adaptation 13,000–8000 cal BP and later. Furthermore, recent publication of D. Madsen's (2015) monograph on the archaeological record of the Old River Bed is providing us with an important complementary look at the regional events characterized in this paper. After all, Bonneville Estates represents just one stop among many for these mobile hunter-gatherers, and most assuredly the rockshelter's record will be best understood by considering it in the greater context of terminal Pleistocene and early Holocene land-use patterns across the entire western Bonneville basin and beyond. Madsen's team's results (2015; see also Duke and Young 2007; Rhode et al. 2005; Schmitt et al. 2007) suggest low-mobility levels, long occupations, and locally oriented lithic-procurement systems in the Old River Bed during the Paleoindian period, at odds with our record from Bonneville Estates Rockshelter. But even there, intense aridification after 10,000 cal BP led to significant reorganization in Paleoindian adaptation—human activity along the Old River Bed, for all intents and purposes, appears to have ceased when water stopped flowing.

References

Adams, K. A., Goebel, T., Graf, K. E., Smith, G. M., Camp, A. J., Briggs, R. W., & Rhode, D. (2008). Late Pleistocene and early Holocene lake-level fluctuations in the Lahontan Basin, Nevada: Implications for the distribution of archaeological sites. *Geoarchaeology, 23*(5), 608–643.

Adovasio, J. M. (1986). Prehistoric basketry. In W. L. d'Azevedo & W. C. Sturtevant (Eds.), *Handbook of north American Indians* (Vol. 11, pp. 194–205). *Great Basin*. Washington, DC: Smithsonian Institution.

Adovasio, J. M., & Pedler, D. R. (1994). A tisket, a tasket: Looking at the Numic speakers through the "lens" of a basket. In D. B. Madsen & D. Rhode (Eds.), *Across the west: Human population movement and the expansion of the Numa* (pp. 114–123). Salt Lake City: University of Utah Press.

Adovasio, J. M., Andrews, R. L., & Illingworth, J. S. (2009). Netting, net hunting, and human adaptation in the eastern Great Basin. In B. Hockett (Ed.), *Past, present and future issues in Great*

Basin archaeology: Papers in honor of Don D. Fowler (pp. 84–102). Reno: U.S. Department of Interior, Bureau of Land Management.

Andrefsky, W., Jr. (1998). *Lithics: Macroscopic approaches to analysis*. Cambridge: Cambridge University Press.

Barlow, K. R., Henriksen, P. R., & Metcalfe, D. (1993). Estimating load size in the Great Basin: Data from conical burden baskets. *Utah Archaeology, 6*, 27–38.

Beck, C. (1995). Functional attributes and the differential persistence of Great Basin dart forms. *Journal of California and Great Basin anthropology, 17*, 222–243.

Beck, C., & Jones, G. T. (1993). The multi-purpose function of Great Basin stemmed series points. *Current Research in the Pleistocene, 10*, 52–54.

Beck, C., & Jones, G. T. (1997). The terminal Pleistocene/early Holocene archaeology of the Great Basin. *Journal of World Prehistory, 11*, 161–236.

Beck, C., & Jones, G. T. (2009). *The archaeology of the eastern Nevada Paleoarchaic: The Sunshine locality*. Salt Lake City: University of Utah Press.

Beck, C., Taylor, A., Jones, G. T., Fadem, C. M., Cook, C. R., & Milward, S. A. (2002). Rocks are heavy: Transport costs and Paleoarchaic quarry behavior in the Great Basin. *Journal of Anthropological Archaeology, 21*, 481–507.

Beiswenger, J. M. (1991). Late Quaternary vegetational history of Grays Lake, Idaho. *Ecological Monographs, 61*(2), 165–182.

Benson, L. V., Hattori, E. M., Taylor, H. E., Poulson, S. R., & Jolie, E. A. (2006). Isotope sourcing of prehistoric willow and tule textiles recovered from western Great Basin rock shelters and caves – Proof of concept. *Journal of Archaeological Science, 33*, 1588–1599.

Benson, L. V., Lund, S. P., Smooth, J. P., Rhode, D. E., Spencer, R. J., Verosub, K. L., Louderback, L. A., Johnson, C. A., Rye, R. O., & Negrini, R. M. (2011). The rise and fall of Lake Bonneville between 45 and 10.5 ka. *Quaternary International, 235*, 57–69.

Bettinger, R. L., & Baumhoff, M. A. (1982). The Numic spread: Great Basin cultures in competition. *American Antiquity, 47*, 485–503.

Bettinger, R. L., & Eerkens, J. (1999). Point typologies, cultural transmission, and the spread of bow and arrow technology in the prehistoric Great Basin. *American Antiquity, 64*(2), 231–242.

Bright, R. C. (1966). Pollen and seed stratigraphy of Swan Lake, southeastern Idaho: Its relation to regional vegetation history and to Lake Bonneville history. *Tebiwa, 9*, 1–28.

Broughton, J. M. (2000). The Homestead Cave icthyofauna. In D. B. Madsen (Ed.), *Late Quaternary paleoecology in the Bonneville Basin, Bulletin 30* (pp. 103–121). Salt Lake City: Utah Geological Survey.

Bryan, A. L. (1979). Smith Creek Cave. In D. R. Tuohy (Ed.), *The archaeology of Smith Creek canyon, Nevada State Museum anthropological papers no. 17* (pp. 164–251). Carson City: Nevada State Museum.

Bryan, A. L. (1988). The relationship of the stemmed point and fluted point traditions in the Great Basin. In J. A. Willig, C. M. Aikens, & J. L. Fagan (Eds.), *Early human occupation in far western North America: The Clovis-Archaic Interface, Nevada State Museum anthropological papers no. 21* (pp. 53–74). Carson City: Nevada State Museum.

Connolly, T. J. (2013). Implications of new radiocarbon ages on coiled basketry from the northern Great Basin. *American Antiquity, 78*(2), 373–384.

Connolly, T. J., & Barker, P. (2004). Basketry chronology of the early Holocene in the northern Great Basin. In D. L. Jenkins, T. J. Connolly, & C. M. Aikens (Eds.), *Early and middle Holocene archaeology of the northern Great Basin, University of Oregon Anthropological Papers no. 62* (pp. 241–250). Eugene: Museum of Natural History and Dept. of Anthropology, University of Oregon.

Cressman, L. S. (1942). *Archaeological researches in the Northern Great Basin, Carnegie Institution of Washington publication no. 538*. Washington, DC: Carnegie Institution of Washington.

Currey, D., Atwood, R. G., & Mabey, D. R. (1984a). *Major Levels of Great Salt Lake and Lake Bonneville. Utah geological and mineral survey map 73*. Salt Lake City: Utah Geological Survey.

Currey, D. R., Oviatt, C. G., & Czarnomski, J. E. (1984b). Late Quaternary geology of Lake Bonneville and Lake Waring. *Utah Geological Association Publications, 13*, 227–238.

Duke, D. (2011). *If the desert blooms: A technological perspective on Paleoindian ecology in the Great Basin from the Old River Bed, Utah.* Ph.D. dissertation, University of Nevada, Reno.

Duke, D. (2015). Haskett spear weaponry and protein-residue evidence of proboscidean hunting in the Great Salt Lake Desert, Utah. *PaleoAmerica, 1*(1), 109–112.

Duke, D. G., & Young, D. C. (2007). Episodic permanence in Paleoarchaic basin selection and settlement. In K. E. Graf & D. N. Scmitt (Eds.), *Paleoindian or Paleoarchaic? Great Basin human ecology at the Pleistocene-Holocene transition* (pp. 123–138). Salt Lake City: University of Utah Press.

Eerkens, J. W. (2003). Residential mobility and pottery use in the western Great Basin. *Current Anthropology, 44*, 728–738.

Eerkens, J. W. (2004). Privatization, small-seed intensification, and the origins of pottery in the western Great Basin. *American Antiquity, 69*(4), 653–670.

Eerkens, J. W., & Lipo, C. P. (2014). A tale of two technologies: Prehistoric diffusion of pottery innovations among hunter-gatherers. *Journal of Anthropological Archaeology, 35*, 23–31.

Eerkens, J. W., Neff, H., & Glascock, M. (2002). Ceramic production among small-scale and mobile hunters and gatherers: A case study from the southwestern Great Basin. *Journal of Anthropological Archaeology, 21*, 200–229.

Elston, R. G., & Zeanah, D. W. (2002). Thinking outside the box: A new perspective on diet breadth and sexual division of labor in the Prearchaic Great Basin. *World Archaeology, 34*, 103–130.

Ferguson, J. R. (2012). X-ray fluorescence of obsidian: Approaches to calibration and the analysis of small samples. In A. N. Shugar & J. L. Mass (Eds.), *Studies in archaeological sciences: Handheld XRF for art and archaeology* (pp. 402–422). Leuven: Leuven University Press.

Flenniken, J. J., & Wilke, P. J. (1989). Typology, technology, and chronology of Great Basin dart points. *American Anthropologist, 91*, 149–158.

Fowler, C. S. (1994). Material culture and the proposed Numic expansion. In D. B. Madsen & D. Rhode (Eds.), *Across the west: Human population movement and the expansion of the Numa* (pp. 103–113). Salt Lake City: University of Utah Press.

Geib, P. R., & Jolie, E. A. (2008). The role of basketry in early Holocene small seed exploitation: Implications of a ca. 9,000 year-old basket from Cowboy Cave. *Utah. American Antiquity, 73*(1), 83–102.

Gilbert, G. K. (1890). *Lake Bonneville, U.S. Geological Survey monograph 1*. Washington: U.S. Government Printing Office.

Goebel, T. (2007). Pre-Archaic and early Archaic technological activities at Bonneville Estates Rockshelter: A first look at the lithic record. In K. E. Graf & D. N. Schmitt (Eds.), *Paleoindian or Paleoarchaic? Great Basin human ecology at the Pleistocene-Holocene transition* (pp. 156–184). Salt Lake City: University of Utah Press.

Goebel, T., Graf, K. E., Hockett, B., & Rhode, D. (2007). The Paleoindian occupations at Bonneville Estates Rockshelter, Danger Cave, and Smith Creek Cave (eastern Great Basin, U.S.A.): Interpreting their radiocarbon chronologies. In M. Kornfeld, S. Vasil'ev, & L. Miotti (Eds.), *On Shelter's ledge: Histories, theories and methods in Rockshelter research* (pp. 147–161). Oxford: BAR International Series.

Goebel, T., Hockett, B., Adams, K. D., Rhode, D., & Graf, K. E. (2011). Climate, environment, and humans in North America's Great Basin during the Younger Dryas, 12,900-11,600 calendar years ago. *Quaternary International, 242*, 479–501.

Graf, K. E. (2002). Paleoindian obsidian procurement and mobility in the western Great Basin. *Current Research in the Pleistocene, 19*, 87–89.

Graf, K. E. (2007). Stratigraphy and chronology of the Pleistocene and Holocene transition at Bonneville Estates Rockshelter, eastern Great Basin. In K. E. Graf & D. N. Schmitt (Eds.), *Paleoindian or Paleoarchaic? Great Basin human ecology at the Pleistocene-Holocene transition* (pp. 82–104). Salt Lake City: University of Utah Press.

Grayson, D. K. (1991). Alpine faunas from the White Mountains, California: Adaptive change in the late prehistoric Great Basin? *Journal of Archaeological Science, 18*, 483–506.
Grayson, D. K. (1993). *The Desert's past: A natural prehistory of the Great Basin*. Washington, DC: Smithsonian Institution Press.
Grayson, D. K. (1998). Moisture history and small mammal community richness during the latest Pleistocene and Holocene, northern Bonneville basin, Utah. *Quaternary Research, 49*, 330–334.
Grayson, D. K. (2000). The Homestead Cave mammals. In D. B. Madsen (Ed.), *Late Quaternary paleoecology in the Bonneville Basin, Utah geological survey, bulletin 130* (pp. 67–89). Salt Lake City: Utah Department of Natural Resources.
Grayson, D. K. (2011). *The Great Basin: A natural history*. Berkeley: University of California Press.
Hattori, E. M., & Fowler, C. S. (2009). Recent advances in Great Basin textile research. In B. Hockett (Ed.), *Past, present, and future issues in Great Basin archaeology: Papers in honor of Don D. Fowler* (pp. 117–138). Reno: U.S. Department of the Interior, Bureau of Land Management.
Hildebrandt, W. R., & King, J. H. (2002). Projectile point variability along the northern California-Great Basin interface: Results from the Tuscarora-Alturas projects. In K. R. McGuire (Ed.), *Boundary lands: Archaeological investigations along the California-Great Basin Interface, Nevada State Museum anthropological papers no. 24* (pp. 5–28). Carson City: Nevada State Museum.
Hockett, B. (2007). Nutritional ecology of late Pleistocene to middle Holocene subsistence in the Great Basin. In K. E. Graf & D. N. Schmitt (Eds.), *Paleoindian or Paleoarchaic? Great Basin human ecology at the Pleistocene-Holocene transition* (pp. 204–230). Salt Lake City: University of Utah Press, Salt Lake City.
Hockett, B., & Dillingham, E. (2004). *Paleontological investigations at Mineral Hill Cave, Contribution to the study of cultural resources technical report No. 18, U.S.* Reno: Department of the Interior, Bureau of Land Management.
Hockett, B., Goebel, T., & Graf, K. E. (2008). The early peopling of the Great Basin. In C. S. Fowler & D. D. Fowler (Eds.), *The Great Basin: People and place in ancient times* (pp. 35–44). Santa Fe: School for Advanced Research Press.
Janetski, J. C. (1977). Fremont hunting and resource intensification in the eastern Great Basin. *Journal of Archaeological Science, 24*, 1075–1089.
Janetski, J. C. (1979). Implications of snare bundles in the Great Basin and southwest. *Journal of California and Great Basin anthropology, 1*(2), 306–321.
Jenkins, D. L., Davis, L. G., Stafford, T. W., Jr., Campos, P. F., Hockett, B., Jones, G. T., Cummings, L. S., Yost, C., Connolly, T. J., Yohe, R. M., II, Gibbons, S. C., Raghavan, M., Rasmussen, M., Paijmans, J. L. A., Hofreiter, M., Kemp, B. M., Barta, J. L., Monroe, C., Gilbert, M. T. P., & Willerslev, E. (2012). Clovis age western stemmed projectile points and human coprolites at the paisley caves. *Science, 337*, 223–228.
Jennings, J. D. (1957). *Danger Cave, University of Utah Anthropological Papers no. 27*. Salt Lake City: University of Utah Press.
Jolie, E. A. (2014). Technology, learning, and innovation in textile arts: Integrating archaeological and ethnographic perspectives. *North American Archaeologist, 35*(4), 303–329.
Jones, G. T., Beck, C., Jones, E. E., & Hughes, R. E. (2003). Lithic source use and Paleoarchaic foraging territories in the Great Basin. *American Antiquity, 68*, 5–38.
Kelly, R. L. (1999). Theoretical and archaeological insights into foraging strategies among the prehistoric inhabitants of the Stillwater wetlands. In B. E. Hemphill & C. S. Larsen (Eds.), *Prehistoric Lifeways in the Great Basin: Bioarchaeological reconstruction and interpretation* (pp. 117–166). Salt Lake City: University of Utah Press.
Lafayette, L., & Smith, G. M. (2012). Use-wear traces on experimental (replicated) and prehistoric stemmed points in the Great Basin. *Journal of California and Great Basin anthropology, 32*, 141–160.

Louderback, L. A., & Rhode, D. (2009). 15,000 years of vegetation change in the Bonneville basin: The Blue Lake record. *Quaternary Science Reviews, 28*, 308–326.

Madsen, D. B. (2000). *Late Quaternary paleoecology in the Bonneville Basin, Utah Geological Survey, Bulletin 130*. Salt Lake City: Utah Department of Natural Resources.

Madsen, D. B. (2015). *The Paleoarchaic occupation of the Old River Bed Delta, University of Utah Anthropological Papers no. 128*. Salt Lake City: University of Utah Press.

Madsen, D. B., & Kirkman, J. (1988). Hunting hoppers. *American Antiquity, 53*, 593–604.

Madsen, D. B., & Simms, S. R. (1998). The Fremont complex: A behavioral perspective. *Journal of World Prehistory, 12*(3), 255–336.

Musil, R. R. (1988). Functional efficiency and technological change: A hafting tradition model for prehistoric North America. In J. A. Willig, C. M. Aikens, & J. L. Fagan (Eds.), *Early human occupation in far western North America: The Clovis-Archaic Interface, Nevada State Museum anthropological papers no. 21* (pp. 373–387). Carson City: Nevada State Museum.

O'Connell, J. F., Jones, K. T., & Simms, S. R. (1982). Some thoughts on prehistoric archaeology in the Great Basin. In D. B. Madsen & J. F. O'Connell (Eds.), *Man and environment in the Great Basin, SAA papers no. 2* (pp. 227–240). Washington, DC: Society for American Archaeology.

Oviatt, C. G. (1988). Late Pleistocene and Holocene lake fluctuations in the Sevier Lake basin, Utah, USA. *Journal of Paleolimnology, 1*, 9–21.

Oviatt, C. G., Currey, D. R., & Sack, D. (1992). Radiocarbon chronology of Lake Bonneville, eastern Great Basin, USA. *Palaeogeography, Palaeoclimatology, Palaeoecology, 99*, 225–241.

Oviatt, C. G., Madsen, D. B., & Schmitt, D. N. (2003). Late Pleistocene and early Holocene rivers and wetlands in the Bonneville basin of western North America. *Quaternary Research, 60*, 200–210.

Page, D. (2008). *Fine-grained volcanic toolstone sources and early use in the Bonneville Basin of Western Utah and Eastern Nevada*. M.A. thesis, Department of Anthropology, University of Nevada, Reno.

Rhode, D. (1999). The role of paleoecology in the development of Great Basin archaeology, and vice-versa. In C. Beck (Ed.), *Models for the millennium: Great Basin anthropology today* (pp. 29–49). Salt Lake City: University of Utah Press.

Rhode, D. (2000a). Holocene vegetation history in the Bonneville basin. In D. B. Madsen (Ed.), *Late Quaternary paleoecology in the Bonneville Basin, Utah geological survey, bulletin 130* (pp. 149–163). Salt Lake City: Utah Department of Natural Resources.

Rhode, D. (2000b). Middle and late Wisconsin vegetation in the Bonneville basin. In D. B. Madsen (Ed.), *Late Quaternary paleoecology in the Bonneville Basin, Utah geological survey, bulletin 130* (pp. 137–147). Salt Lake City: Utah Department of Natural Resources.

Rhode, D., & Louderback, L. A. (2007). Dietary plant use in the Bonneville basin during the terminal Pleistocene/early Holocene. In K. E. Graf & D. N. Schmitt (Eds.), *Paleoindian or Paleoarchaic? Great Basin human ecology at the Pleistocene-Holocene transition* (pp. 231–247). Salt Lake City: University of Utah Press.

Rhode, D., & Madsen, D. B. (1995). Late Wisconsin/early Holocene vegetation in the Bonneville basin. *Quaternary Research, 44*, 246–256.

Rhode, D., Goebel, T., Graf, K. E., Hockett, B., Jones, K. T., Madsen, D. B., Oviatt, C. G., & Schmitt, D. N. (2005). Latest Pleistocene-early Holocene human occupation and paleoenvironmental change in the Bonneville basin, Utah-Nevada. In J. Pederson & C. M. Dehler (Eds.), *Interior Western United States: Field Guide 6* (pp. 211–230). Boulder: Geological Society of America.

Rhode, D., Madsen, D. B., & Jones, K. T. (2006). Antiquity of early Holocene small-seed consumption and processing at Danger Cave. *Antiquity, 80*, 328–339.

Schmitt, D. N., & Lupo, K. D. (2012). The Bonneville Estates Rockshelter rodent fauna and changes in late Pleistocene-middle Holocene climates and biogeography in the northern Bonneville basin, USA. *Quaternary Research, 78*, 95–102.

Schmitt, D. N., & Madsen, D. B. (2005). *Camels Back Cave, Anthropological papers 125*. Salt Lake City: University of Utah Press.

Schmitt, D. N., Madsen, D. B., Oviatt, C. G., & Quist, R. (2007). Late Pleistocene/early Holocene geomorphology and human occupation of the Old River Bed Delta, western Utah. In K. E. Graf & D. N. Schmitt (Eds.), *Paleoindian or Paleoarchaic? Great Basin human ecology at the Pleistocene-Holocene transition* (pp. 105–119). Salt Lake City: University of Utah Press.

Shackley, M. S. (2010). Is there reliability and validity in portable X-ray fluorescence spectrometry (PXRF). *The SAA Archaeological Record, 10*(5), 17–20.

Smith, G. M., & Kielhofer, J. (2011). Through the High Rock and beyond: Placing the Last Supper Cave and Parman lithic assemblages into a regional context. *Journal of Archaeological Science, 38*, 3568–3576.

Spencer, R. J., Baedecker, M., Eugster, H. P., Forester, R. M., Goldhaber, M. H., Jones, B. F., Kelts, K., MacKenzie, J., Madsen, D. B., Rettig, S. L., Rubin, M., & Browser, C. J. (1984). Great Salt Lake and precursors, Utah: The last 30,000 years. *Contributions to Mineralogy and Petrology, 86*, 321–334.

Steward, J. H. (1938). *Basin-plateau aboriginal sociopolitical groups, Bureau of American Ethnology Bulletin no. 120*. Washington, DC: Bureau of American Ethnology.

Thomas, D. H. (1983a). *The archaeology of Monitor Valley 1: Epistemology, Anthropological papers Vol. 58, pt. 1*. New York: American Museum of Natural History.

Thomas, D. H. (1983b). *The archaeology of Monitor Valley 2: Epistemology, Anthropological papers Vol. 58, pt. 2*. New York: American Museum of Natural History.

Zeanah, D. W., & Elston, R. G. (2001). Testing a simple hypothesis concering the resilience of dart point styles to hafting element repair. *Journal of California and Great Basin anthropology, 23*, 93–124.

Zeanah, D. W., & Simms, S. R. (1999). Modeling the gastric: Great Basin subsistence studies since 1982 and the evolution of general theory. In C. Beck (Ed.), *Models for the millennium: Great Basin anthropology today* (pp. 118–140). Salt Lake City: University of Utah Press.

Chapter 12
My Flute Is Bigger Than Yours: Nature and Causes of Technological Changes on the American Great Plains at the End of the Pleistocene

Frédéric Sellet

Introduction

This essay proposes to explore fundamental similarities and differences in lithic technology between Clovis and Folsom groups on the American Great Plains as they relate to the climate changes and environmental dynamics of the Late Pleistocene-Holocene transition in general and the Younger Dryas (YD) in particular. Conventionally, Paleoindian archaeologists have often mentioned Folsom and Clovis stone tool systems in the same breath since they are both built around fluted points (Bradley 1991: 374; but see Bradley 2010: 481; Morrow 2015: 102). This emphasis on superficial resemblances (and presumed concomitant historical connections within a broader fluted point tradition) has drawn attention away from the specific organizational characteristics of these Paleoindian technologies, some of which may be traceable to climate dynamics (e.g., LaBelle 2012; Jennings 2015).

Outside of North America, the abrupt and dramatic paleoenvironmental transformations brought on by the YD have been linked to similarly spectacular adjustments by prehistoric people – such as the beginning of domestication – but in North America, the adaptive responses were arguably more subtle. Yet, because the switch from Clovis to Folsom in the American Great Plains region occurred at the onset of the YD (with seemingly only a slight temporal overlap if any between the two fluted point types), a causal relationship between the climate event and concomitant changes in the archaeological record is possible and merits exploration.

Of particular importance to the following discussion is the question of how the shifting abundance and spatial distribution of targeted food resources in general, and large game in particular (Hill 2008; Bement and Carter 2015), may have shaped the mobility patterns and, ultimately, the lithic systems of these early Paleoindian

F. Sellet (✉)
Department of Anthropology, University of Kansas, Lawrence, KS, USA
e-mail: fsellet@ku.edu

E. Robinson, F. Sellet (eds.), *Lithic Technological Organization and Paleoenvironmental Change*, Studies in Human Ecology and Adaptation 9,
DOI 10.1007/978-3-319-64407-3_12

groups. We argue that such technological reorganizations were significant and likely not a reliable measure of cultural relatedness or historical connections. In fact, a switch in subsistence patterns that included a greater focus on bison may have occurred within Clovis, in tandem with the onset of the YD. Since this change took place late in the Clovis sequence, it may have produced an archaeological signature that was too weak and limited in scope (and possibly in space) to visibly and significantly impact traditional interpretations and reconstructions of Clovis adaptations and technological organization. In other words, our ability to recognize these shifts in subsistence practices within Clovis has been limited by the small number of sites.

By contrast, the Folsom tradition blossomed once the YD was fully felt and vanished when the transformations associated with that climatic episode weakened, an overlap that points to a potential nexus. An appraisal of technological organizations along the criteria of resource distribution and related mobility strategies will bolster our growing understanding of the influence the YD had globally on human adaptations on the American Great Plains.

Statement of the Problem

In North America specifically, the nature and force of the impact the YD exerted on Paleoindians has been a source of scholarly disagreement. Some have argued in favor of a catastrophic event (Firestone et al. 2006, 2007; Anderson et al. 2011), while others have championed an uneventful conclusion. Meltzer and Holliday (2010: 31), for example, state that:

> All things considered, it is likely that across most of North America south of the retreating ice sheets Paleoindians were not constantly scrambling to keep up with Younger Dryas age climate changes.

Such disagreement regarding the effects of the YD is not surprising in itself, given that assessing the consequences of past climate changes on humans in general is a difficult task, in part because of the coarse resolution of paleoclimatic data (e.g., Carlson 2010; Miller and Gingerich 2013). This hindrance is further compounded by a series of analytical hurdles related to poor chronological contexts at most archaeological sites and by the fact that climate, faunal and floral changes, and human subsistence activities all operate at different temporal and spatial scales.

While Paleoindian groups may not have been scrambling to counter the effects of temperature and precipitation changes on the environment, they certainly had to make routine adjustments in their quest for food. Even if such fine-tunings were unconscious or unplanned by the individuals who implemented them, their archaeological signature would have grown incrementally over many generations and therefore should represent a legitimate target of technological analyses. Therein lies the value of contrasting Folsom and Clovis technologies: these past technological organizations should highlight broad evolutionary shifts in adaptation that could reveal significant data reflecting differential resource distribution within a landscape at a climatic pivot point. (Alley et al. 1993)

In doing so, the following discussion takes the obligatory stance that stone tools represent pragmatic solutions to the procurement of food resources. Prehistoric lithic technological systems were fundamentally (but not exclusively) organized to respond to the temporal and spatial discrepancies between food and people, on one hand, and tool manufacture activities and the required lithic raw materials on the other. Given that both Plains Clovis and Folsom groups had access to the same raw materials (which is not to say they used them in a similar fashion), if the types of food resources available to Clovis groups (or their abundance) varied from the ones targeted by Folsom groups, then one can expect the organizational differences in mobility patterns and subsistence planning to have filtered down to their lithic systems. If, on the contrary, the YD was a nonevent, then reorganization of the technologies should be minimal or superficial, considering that it would provide little evolutionary benefit.

The first part of the following discussion will address the changes in resource availability and distribution that occurred in Clovis and Folsom times. The second part will focus more specifically on organizational characteristics and argue that the contrasts in organization reflect the scope of Late Pleistocene-early Holocene paleoenvironmental changes.

Younger Dryas Dynamics

The YD, which started at around 12.9k cal BP, was a widespread climatic event that lowered temperatures significantly and thereby represented a reversed trend to the Late Glacial. A marked rapid cooling, in sharp contrast with the general warming that preceded it, has been recorded in the Greenland ice cores and elsewhere (Alley et al. 1993). Although there is abundant evidence for rapid environmental responses (Severinghaus et al. 1998; Shuman et al. 2005), the types and magnitude of these changes have long been a topic of contention (Broecker et al. 2010).

That debate has been further muddied by the fact that significant ecological changes were already happening during the Last Glacial (Webb et al. 2003), chief among them: the extinction of multiple genera of large mammals. In North America, these Late Pleistocene climatic and environmental dynamics alone were substantial enough that they would have entailed significant adjustments by humans. Hence, comparing adaptations between two given sites, one Clovis and the other Folsom, may not tell the entire story. It is likely that during both Clovis and Folsom times, groups were living in environments that were undergoing significant transformations. Portraying each Paleoindian tradition as a uniform adaptation is bound to be a fallacy. There was not one archetypal Clovis adaptation (Meltzer 1993), and neither was there a single Folsom one (Kornfeld 1989, 2002; Andrews et al. 2008; LaBelle 2012). In fact there may have been organizational similarities between the two, given Clovis and Folsom groups who targeted similar resources.

Clovis sites range from roughly 11,500 to 10,800 ^{14}C yr BP in age, thus preceding the YD. Although the oldest Clovis ages, such as the ones from the Aubrey site in Texas (Ferring 2001), have been a topic of contention (Waters and Stafford 2007; Prasciunas and Surovell 2015), it is widely recognized that most Clovis ages fall within the range of a 250-year-long calibration plateau (Fiedel 2015; Prasciunas and Surovell 2015:13), which creates imprecision in assessing the age span of the Clovis tradition. This being said, Clovis as it is usually defined, overlaps the late Allerød interglacial, with the youngest Clovis context from the aforementioned Jake Bluff site in Oklahoma dating to the onset of the YD at about 10,800 ^{14}C yr BP. This clearly places the tail end of Clovis within the range of the earliest Folsom sites (see Haynes 1993; Meltzer and Holliday 2010 for a list of Folsom ages). It should be noted though that the current sample of radiocarbon dates does not equivocally demonstrate that the two traditions coexisted (Fiedel 1999). At localities where both components are found in stratigraphic succession, Folsom has been found repeatedly above Clovis, suggesting that it may be slightly younger.

The youngest Clovis site, the Jake Bluff site, stands out in that it depicts a subsistence practice not seen elsewhere among Clovis sites. It is strictly a bison kill site where hunters used an arroyo to trap and kill a small herd (Bement 2009; Bement and Carter 2010). Its setting and function are similar to the nearby Cooper site (Bement 1999; Johnson and Bement 2009) which contains multiple Folsom kills. The Jake Bluff radiocarbon age of 10,800 ^{14}C yr BP (versus 10,500 to 10,600 ^{14}C yr BP for Cooper) places it at the beginning of the YD (Bement et al. 2014).

At the site, Clovis hunters opportunistically embraced what is generally portrayed as a typical Folsom bison hunting strategy (Bement and Carter 2015), fulfilling a scenario in which subsistence strategies alter toward the end of the Clovis period. This is not to say that bison was not a prey targeted by Clovis groups elsewhere: multiple animals were processed at the Murray Springs (Hemmings 2007) and Blackwater Draw sites (Hester 1972), for instance, but it begs the question of its importance to Clovis adaptations. Ultimately, a model of prey (or plant resource) availability and abundance is needed to evaluate shifts in landscape utilization by humans.

McDonald (1981) tracked changes in relative megafaunal abundance at the end of the Pleistocene via a number of available radiocarbon ages for each taxon. While the method has a number of limitations, the results are nonetheless interesting. They show few available radiocarbon ages for megafauna prior to 16k ^{14}C yr BP. After that time, there is a steady increase in the representation of mastodon, *Equus*, and bison. The trends reverse abruptly around 12k ^{14}C yr BP, at which time bison population quadruples while other taxa show little change. After that, while most megafauna become extinct, bison population remains constant. This suggests that bison population was expanding in Clovis times (Scott 2010) concomitant with the extinction of other large faunal resources (mammoth, mastodon, horse, etc.).

This observation is consistent with Guthrie's (1984:262) assumption that many complex plant associations and biomes were fractioned into homogeneous zones at the end of the Pleistocene and that large mammal biomass declined as a result of decreased vegetational mosaic and increased vegetational zonation. This too is in line with Lundelius' portrayal of the Late Pleistocene (Lundelius 1967, 1974) according to which the major biotic communities of the Pleistocene were more like modern ecotones in their habitat diversity. Altogether, Late Pleistocene communities were characterized by exceptional biotic diversity, with complex species association. By contrast, the Holocene represents a simplification, or zonation, across the spectrum of plant and animal associations.

Herbivore populations would have felt significant changes in vegetation type and plant distribution at the Pleistocene/Holocene boundary. Not only were nutrients no longer sufficient for the groups that became extinct, these nutrients were increasingly locked up in plant taxa to which these extinct lines were not adapted (Guthrie 1984:283). Nonetheless, large ruminant grazers like bison can survive in monotonous summer range of just a few plant species. As they were not highly selective feeders, their population was not affected in the same way as mammoth, for instance.

Based on these observations, it is clear that Clovis groups could have targeted bison regularly, since it was increasingly plentiful. Nevertheless, it was not their sole option: a variety of large mammals were available *within the same ecological niche* (albeit in dwindling numbers). Folsom hunters by contrast had fewer choices: horse, camel, mammoth, mastodon, and several other large mammals were now extinct. Accordingly, the landscapes traveled by Folsom groups were more redundant in terms of potential prey.

It may well be that late Clovis groups were already facing a similar situation, and Jake Bluff may be an illustration of such a scenario (Bement and Carter 2015). Little information on bison procurement in Clovis time is available, but the Gault site in Texas shows a similar pattern: mammoth, horse and bison were hunted early in the Clovis interval, but only bison were hunted near the end of that time (Collins 2002).

A precise measure of the overall abundance and size of bison herds that Paleoindian hunters would have encountered, or even the herds' dispersion, is difficult to obtain. It seems reasonable, though, to assume that bison location would have varied across seasons, more so during the YD and early Holocene (when the climate became less equable) than during the Late Pleistocene. Bison has a tendency to select C3 grass over C4 (Chisholm et al. 1986); therefore migration and herd distribution across the plains might have been highly dependent on the seasonal availability and distribution of C4 plants relative to their C3 counterparts (Carlson and Bement 2013). A seasonal movement of modern bison has been observed for the Northern Plains, for instance (albeit under significantly unrelated climatic and ecological conditions), where Morgan suggests that bison would form large aggregates in winter around areas of larger productivity while dispersing in the summer to spread over the less productive grasslands (Morgan 1980). Similarly, bison in Alberta undertook seasonal migrations between the mixed prairies and surrounding parklands and foothill regions (Chisholm et al. 1986).

A projection of seasonal movements of herds for the period considered is therefore dependent on reconstructions of past vegetation patterns (Chisholm et al. 1986). Unfortunately, extant pollen, phytolith and stable isotope records are of limited use since they do not offer a fine-grained resolution of vegetation changes (and in some cases even indicate contrasting or varied local signatures). In spite of these limitations, a rough estimate of landscape changes for the Plains points toward a stronger signature of C4 plants in the YD (Balakrishnan et al. 2005; Bement et al. 2007; Cordova et al. 2011; Nordt et al. 2008; Williams et al. 2004).

The aforementioned Jake Bluff Clovis site, for example, produced a robust signal of C4 phytoliths, which was attributed by Bement (2009) to the disproportionate signature of grasses present in the animal's gut piles. Although the data contrast with the record obtained at Beaver Creek in Oklahoma, the case can be made that YD vegetation patterns were already in place at the time of the kill, which would also help explain the focus on bison by Clovis groups in this area. Jake Bluff does not typify a Clovis adaptation as much as it illustrates an adjustment to YD environmental dynamics (Bement and Carter 2015).

Interestingly, Meltzer and Holliday's recent survey of the pollen record at multiple Great Plains localities indicates that the most significant vegetation changes occurred during the middle of the YD period rather than at its onset (Meltzer and Holliday 2010:20). This suggests a potential temporal lag between variations in temperature and precipitation on one hand and their impact on ecosystems on the other (or at least their visible signature in the sedimentological record).

The pace of YD paleoenvironmental changes is only one of the many factors controlling human adaptive responses; how abruptly the YD dynamics were felt by plants and animal populations, a given ecosystem's resilience (Gunderson 2000) and, whether these changes were affecting species synchronously in a given ecosystem are other important elements to consider. Climate changes were complex and their effects difficult to assess, as are vegetational responses to climate changes (Webb 1986). The intricacies of these causal relationships are underscored by the fact that the latter are not unidirectional but function in a feedback loop wherein vegetation patterns affect temperatures as well (Notaro et al. 2006). Additionally, in North America, the YD is not marked by distinct lithostratigraphic or biostratigraphic signatures and may have resulted in varied local conditions and outcomes (Holliday et al. 2011:530). Unfortunately, current archaeological and paleoenvironmental reconstructions are still too coarse to sketch anything but an incomplete picture, forcing us to rely on generalized models to assess the impact YD dynamics had on people.

Modeling Clovis and Folsom Technological Organizations

In bridging ecological data and human behavior, optimal foraging theory (OFT) (Krebs 1978) provides a useful canvas for assessing past human mobility patterns. In particular, it offers working models of travel between and within patches of resources. As stated above, type, distribution, and abundance of prey within said patches are important parameters to consider when modeling Folsom and Clovis subsistence strategies. Nonetheless, the most striking contrast between the two subsistence patterns (notwithstanding the caveat already mentioned) is the degree to which bison becomes a focal resource during the YD (Collins 2007:81). This fundamental shift may, in turn, have shaped discrete lithic technological organizations (see also Collins 1999a).

Charnov's marginal value theorem (Charnov 1976) provides useful insights into how environmental conditions may have affected Clovis and Folsom group mobility strategies. The theorem considers energetic return of patches, travel time, and time spent in patches. Given these variables, and because of the patchy distribution of resources in the Late Pleistocene relative to the YD, the theorem predicts that Clovis groups would have to travel further than Folsom to reach a productive area (note that productive in the context of this discussion refers to abundance of large game) and, as travel time increases, spend more time in patch until the optimal energetic return point was reached (after which there were diminishing incentives to remain in the patch, everything else being equal). The models are illustrated in Fig. 12.1. Table 12.1 summarizes the main organizational differences between Clovis and Folsom groups.

It can be assumed that at a broad temporal scale, the archaeological signature of Clovis and Folsom settlement systems would have reflected these organizational patterns (Surovell 2003). Haynes summarizes the potential impact of this organization on the archaeological record:

> If the patches were increasingly isolated refuges, distant from each other and limited in number, Clovis foragers (assuming groups sizes were fairly consistent) would have spent relatively more time in them, creating either larger sites, more sites, or sites containing more artifacts and more kinds of artifacts (Haynes 2002:222)

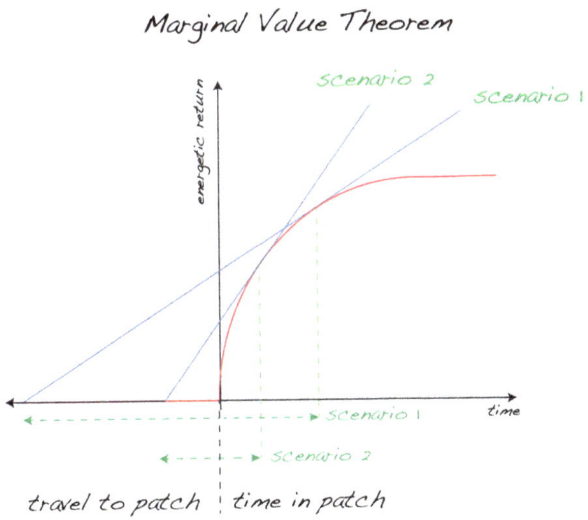

Fig. 12.1 Marginal value theorem. As travel time to patch decreases (*scenario 2*), time spent in patch before optimum energetic return is reached decreases as well. The model predicts that since resource patches were more widely spaced, Clovis hunter (*scenario 1*) had to travel longer to reach patches and spent more time in their patch relative to Folsom (*scenario 2*)

Table 12.1 Differences in organization between Clovis and Folsom groups

Clovis	Folsom
Target a variety of large mammals (mammoth, horse, bison, camel, etc.)	Target bison
Movements are more predictable (target rich and diverse patches)	Movements are unpredictable (follow bison)
Move longer distances between patches	Move shorter distances between patches
Stay in patch longer	Stay in patch for shorter periods of time

Aside from these broad factors determining organizational variability, the temporal and spatial distributions of prey within patches (particularly bison) are also critical to consider when assessing strategies of production, transport, and use of lithic tool kits. Ancient bison were less gregarious than modern species (McDonald 1981), and as argued earlier, environmental and climatic factors may have kept herds more dispersed. This partially explains why Paleoindian communal hunts were smaller and may not have occurred with the same frequency as in late prehistoric times (Bamforth 1985:244). Given these conditions, Bamforth comments on how bison distribution affected the organization of the technology. He notes that, given the unpredictable location of bison resources, hunters would have organized accordingly:

> This uncertainty would have made it difficult to foresee the distance hunters would have had to travel, the duration of the trip and the equipment which would be required. The simplest way to exploit this pattern of availability was to carry only the tools needed and to make use of local resources when needed (Bamforth 1985:256)

Hence, a diet that primarily targeted bison likely required a tool kit that was easily transportable and that could, at the same time, sustain long episodes of use (long tool use life) to avoid repeated trips to the quarry. The transported tools also had to be flexible in their function (multipurpose tools) in order to minimize transport costs. It should be noted, however, that the reliance on local material suggested by Bamforth represents an option of dubitable sustainability and would have been implemented only in the context of limited mobility and of lengthy exploitation of a given ecological zone with suitable lithic resources.

Realized Technological Organizations

With the exception of the reliance on local materials (at least for the Plains), most of the organizational characteristics of bison oriented systems mentioned by Bamforth are present in Folsom technology. This has led some to describe it as a liberating technology, according to which groups depend on long episodes of tool use without concern for immediate availability of lithic raw materials (Hofman 2003) – an adaptation that can be tied to high mobility. Folsom hunters also relied on the staging of projectile point manufacture, made possible by the transport of suitable blanks in the tool kit. Consequently, Folsom points were replaced as they wore out, and in a few cases, the entire tool kit is replenished in excess of immediate needs via gearing up (Sellet 2013).

Although Clovis technological systems are also based on a highly flexible tool kit (Smallwood 2010), there are fundamental differences that distinguish them from the above-described Folsom ones. These contrasts are discussed below.

Projectile Point Designs

While both projectile point types are fluted, some design elements differentiate Clovis points from their Folsom counterparts. Clovis points are larger and, thus, could be maintained more easily since they will sustain many more episodes of resharpening or refurbishing. They also vary considerably in size at any given locality (e.g., Smallwood 2010), suggesting they may be adapted to a variety of functions and possibly to a greater diversity in prey sizes. Folsom points' design elements have been described as reliable. Ahler and Geib (2000), for instance, have proposed that they could be pushed up in the foreshaft as they wear out, similar to modern retractable blade knives. Hence, the Folsom point has all the trademarks of a highly specialized weapon.

Finally, it should be mentioned that even though both types are fluted, archaeological analyses have shown a higher failure rate for Folsom point manufacture: two to four times higher than Clovis (Ellis and Payne 1995; Sellet 2004). Those numbers are noteworthy since a higher failure rate would have influenced raw material procurement and point production strategies for Folsom more so than for Clovis and required even greater planning depth.

Tool-Caching Behavior

Differences in planning depth and mobility patterns would have also impacted tool-caching strategies. A small number of Clovis caches have been identified (Huckell and Kilby 2014). They typically include oversize projectile points and bifaces, blades, and, sometimes, beveled bone, ivory, or antler rods. If these caches were utilitarian rather than symbolic deposits, they might represent the provisioning of a landscape for future consumption. Such deposits typically would have been placed in areas poor in raw materials, in anticipation of future visits. The presence or absence of caches in a landscape might thus be linked to abundance of suitable lithic raw materials and to anticipated mobility (the ability to predict specific territory utilization).

Caches have not yet been reported for Folsom. If we assume similar access to lithic raw materials for Clovis and Folsom groups, their absence may denote fundamental differences in mobility patterns and technological organizations. As argued above, the dispersion of prey and the unpredictability of prey location may have been critical in explaining organizational differences. Folsom groups may not have been able to forecast with any consistency the location of bison herds (or at least the location of future kills) and therefore the future location of tool use. Additionally, the flexibility of the Folsom tool kit would have alleviated to some degree the need to provision a landscape with tools for future use.

Specialized Tool Forms

Although there is a great deal of variability in tool assemblages within each technological tradition, some basic differences in tool forms are informative of the overall adaptations. Clovis tool kits contain forms that are absent from Folsom, such as blades and blade cores (Collins 1999b), large bifaces in caches (Jennings 2013), and heavy bifacial tools such as adzes (Collins and Hemmings 2005; Smallwood 2010:2419). These various tools do not appear uniformly at all Clovis localities and thus may map specialized activities across a landscape. Adzes, for example, may denote woodworking (Shoberg 2010). The greater number of tool forms for Clovis conforms to Haynes' aforementioned prediction regarding site size and assemblage diversity. On the same note, it should be mentioned that some particularly large Clovis sites such as Gault, Carson Conn Short, or Topper indicate heavy exploitation and reliance on local lithic resources (see also Koldehoff and Loebel 2009; Speer 2014), a pattern not visible for Folsom (at least not on the Plains).

True blades, in the technological sense, occur in Clovis assemblages but are absent from Folsom and other Paleoindian tool kits. Clovis blades have been linked to the exploitation of plant resources (Collins 2007) but are not found ubiquitously nor with the same frequency. Considerable evidence of blade production can be found at massive workshops such as the Gault site or Carson Conn Short site, which are located in lithic-rich areas. Elsewhere, blades occur in smaller numbers in the discarded tool kit. They may have been curated tools, which were transported and cached in small numbers in strategic locations (Collins 1999b; Kilby 2015).

Folsom tool kits contain idiosyncratic forms as well, some of which are absent from Clovis'. Ultrathins, which squarely express Folsom's focal adaptation to bison, have been interpreted as fileting tools, used in the preparation of meat prior to drying (Jodry 1999). These curated tools, unlike Clovis bifaces, were not cached, which indicates that the location of their utilization probably could not or would not have been predicted in advance.

Concluding Thoughts

In broad terms, the observed organizational differences between Clovis and Folsom lithic technologies are in line with the projections of the OFT model. These fundamental variations can be reconciled and explained by disparities in subsistence activities (the type and distribution of targeted prey) and ancillary requirements in human mobility (time traveled between patches and time spent in patches). All in all the evidence cited above suggests that the Clovis technological system was tethered to a place (one or possibly several specific ecological niches), while Folsom was organized around the exploitation of a resource (bison).

Both Clovis and Folsom organizations reflect deep planning and flexibility as a response to the anticipated temporal and spatial availability of resources (food, materials, water, etc.), but the differences in lithic technologies suggest that Clovis adaptations rely on a more predictable pattern of mobility. What's more, Clovis groups likely spent additional time in any given patch, targeting more varied resources. This picture of subsistence practices is in line with Collins' (2007:85) depiction of Clovis adaptations in Central Texas as being analogous to Archaic adaptations. If we take into consideration a wider geographic range to overcome the limitations of a small number of sites in the Plains/Rockies, then an even greater variety of faunal remains can be seen. The Clovis record indicates the consumption of fish, turtles, various reptiles, rabbits, birds, horse, camel, proboscideans, bison, deer, caribou, etc. (see Collins 2002; Haynes 2002:177, Table 5.1 for a more complete list of faunal association with Clovis artifacts). Other regions also provide a blueprint for what Clovis adaptation may have looked like on the Great Plains. In the Southwest, for example, Holliday's (2015:198) analysis of Clovis subsistence practices in the San Pedro Valley reveals a reliance on relatively abundant and maybe even circumscribed large game populations, tethered to water sources or specific plant biomes. That amount of regional contextualization is still missing for the Great Plains, but it is likely that the San Pedro Valley strategies would have been implemented there as well.

While there is variability in Folsom diets (Kornfeld 1989, 2002), a similar amount of breadth is not seen. Folsom adaptations are more focused, centering around bison, to the point where a case can be made, as Hofman has suggested that Folsom was the first true Plains-oriented adaptation (Hofman 2004:37).

In sum, Clovis and Folsom technologies should be seen as the result of distinct subsistence practices rather than unique and/or divergent cultural traditions. This is a view that contrasts with lithic studies that focus on specific aspects of the technology such as fluting to emphasize either a symbolic dimension or an evolutionary cultural link between the two traditions. MacDonald, for example, explains the fluting of projectile points as an adaptive mechanism. A similar claim about the value of fluting for socializing a changing landscape has been made for Clovis by Kornfeld et al. (2001).

MacDonald describes Folsom fluting as follows:

> The symbolic importance of fluting clearly emerged full force within Folsom culture, likely associated with the emerging importance of bison as one of the last remaining big game species available for procurement. Good hunters, thus, fluted the entire length of the point, perhaps as a means to show-off or signal their affiliation to the newly encultured landscape (MacDonald 2010:46)

This essay argues that environmental changes should not be measured in the length of a flute but rather in technological reorganizations. There is also abundant literature that phylogenetically links Clovis and Folsom via the flute. The idea is ancient and deeply rooted in Paleoindian archaeology; it stems from the fact that all fluted points were once referred to as Folsom. Prior to the Second World War, all Clovis points were labeled Folsomoid, Folsom-like, or generalized Folsom (Sellet 2011; Wormington 1939). It was not until the 1941 Santa Fe meetings that true Folsom was recognized as a distinct type. At the same meetings, it was also decided that both Folsom and Clovis points would remain lumped under the general term "fluted points" (Wormington 1944). As the stratigraphic and temporal relationships between Clovis and Folsom became established, typological attributes were used to justify those sequences. The flute was the most readily identifiable characteristic of all Paleoindian types and consequently played a critical role in assessing chronological and cultural relationships.

The original sequence of Folsom-Clovis established purely on typological grounds (in that order) was explained as a regression: it was thought that the flute shortened over time before eventually disappearing. Once the correct stratigraphic order was discovered, it became clear that Clovis was in fact older. Fluting was subsequently said to demonstrate an increasing mastery of technological skills, as the flute's extension increased from Clovis to Folsom. Fluting was used as a post hoc accommodative argument. Historical inertia and the preeminence of chrono-cultural questions in Paleoindian archaeology in the years following the war explain why fluting has been given such consideration. As an easily identifiable typological marker that was associated with culturally shared and transmitted knowledge, it fulfilled the need to find some continuity between Paleoindian complexes, underscoring a probable phylogenetic link between two lithic traditions. By the same token, it also placed an undue emphasis on a minute aspect of Paleoindian lithic technology, to the detriment of more significant organizational shifts. To paraphrase Collins, "in spite of these modest similarities, Clovis and Folsom technologies differ in most respects" (Collins 1999b:31).

Shott (2013:158) has recently pointed out the theoretical vacuum surrounding studies that attempt to untangle historical relatedness between lithic types. While the ecological approach advocated here, and elsewhere in this edited volume, will not elucidate all sources of archaeological variability, the connections between climate, resource distribution, and the organization of lithic technology addressed herein warrant further exploration.

It is noteworthy that the spatiotemporal distribution of Folsom sites accommodates the above ecological models. Collard et al. (2010) investigated the Clovis-Folsom transition and concluded that Folsom originated in the Northern Plains and spread north and south. They also called attention to a temporal gap between Clovis and Folsom on the Southern Plains and proposed that the transition in that region at least was the result of population expansion rather than cultural diffusion (Collard, et al. 2010:2519). Albeit bearing the limitations of small sample size and the usual problems with radiocarbon calibration, these observations are in line with the models presented here, in which Folsom represents a specific adaptation to changing environments (Irwin-Williams and Haynes 1970; Surovell 2003).

Yet many questions remain unresolved and will require a satisfactory answer before we can grasp the true nature and timing of the transition. Chief among the issues that need to be probed are the degree of organizational variability within Clovis and Folsom systems, the scope of environmental changes, their regional or temporal variation, and finally the speed of the human response to resource reshuffling. The Clovis and Folsom records provide a useful canvas for investigating human responses to a significant climatic event from a pluri-disciplinary point of view; it is up to us to keep building adequate methodologies and analytical approaches to disentangle all possible causal relationships.

In sum, this discussion offers a framework to contextualize technological organizations; because it is painted in broad strokes, it is not meant to accurately depict their internal complexity. From the above, it seems likely that the fundamental organizational differences between Clovis and Folsom reflect major paleoenvironmental changes and a reshuffling of resources, in particular bison, but whether these alterations were primarily triggered by the YD or were the consequences of broader shifts that began in the Late Pleistocene and were subsequently amplified by the YD remains to be determined. Resolving this question will require us to tease apart the connections between the archaeological and paleoenvironmental records at the regional or possibly even at the local level. The Great Plains have never been a completely homogeneous landscape, and the biotic response to climate change therein may have varied considerably from one region to another, making the true picture of Paleoindian adaptation likely to be significantly more intricate than the one sketched here.

References

Ahler, S. A., & Geib, P. R. (2000). Why flute? Folsom point design and adaptation. *Journal of Archaeological Science, 27*(9), 799–820.

Alley, R. B., Meese, D. A., Shuman, C. A., Gow, A. J., Taylor, K. C., Grootes, P. M., White, J. W. C., Ram, M., Waddington, E. D., Mayewski, P. A., & Zielinksi, G. A. (1993). Abrupt increase in Greenland snow accumulation at the end of Younger Dryas event. *Nature, 362*, 527–529.

Anderson, D. G., Goodyear, A. C., Kennett, J., & West, A. (2011). Multiple lines of evidence for possible human population decline/settlement reorganization during the early Younger Dryas. *Quaternary International, 242*(2), 570–583.

Andrews, B. N., LaBelle, J. M., & Seebach, J. D. (2008). Spatial variability in the Folsom archaeological record: A multi-scalar approach. *American Antiquity, 73*(3), 464–490.

Balakrishnan, M., Yapp, C. J., Meltzer, D. J., & Theler, J. L. (2005). Paleoenvironment of the Folsom archaeological site, New Mexico, USA, approximately 10,500 ^{14}C yr B.P. as inferred from the stable isotope composition of fossil land snail shells. *Quaternary Research, 63*(1), 31–44.

Bamforth, D. B. (1985). The technological organization of Paleo-Indian small-group bison hunting on the Llano Estacado. *Plains Anthropologist, 30*(109), 243–258.

Bement, L. C. (1999). *Bison hunting at Cooper Site*. Norman: University of Oklahoma Press.

Bement, L. C. (2009). Clovis sites, gut piles, and environmental reconstructions in northwest Oklahoma. *Plains Anthropologist, 54*(212), 325–331.

Bement, L. C., & Carter, B. J. (2010). Jake Bluff: Clovis bison hunting on the Southern Plains of North America. *American Antiquity, 75*(4), 907–933.

Bement, L. C., & Carter, B. J. (2015). From mammoth to bison, changing Clovis prey availability at the end of the Pleistocene. In A. M. Smallwood & T. A. Jennings (Eds.), *Clovis: On the edge of a new understanding* (pp. 263–275). College Station: Texas A&M University Press.

Bement, L. C., Carter, B. J., Varney, R. A., Cummings, L. S., & Sudbury, J. B. (2007). Paleo-environmental reconstruction and bio-stratigraphy, Oklahoma Panhandle, USA. *Quaternary International, 169*, 39–50.

Bement, L. C., Carter, B. J., Jelley, P., Carlson, K., & Fine, S. (2014). Badger Hole: Towards defining a Folsom bison hunting complex along the Beaver River, Oklahoma. *Plains Anthropologist, 57*(221), 53–62.

Bradley, B. A. (1991). Flaked stone technology in the northern High Plains. In G. C. Frison (Ed.), *Prehistoric hunters of the High Plains* (pp. 369–395). New York: Academic Press.

Bradley, B. A. (2010). Paleoindian flaked stone technology on the Plains and in the Rockies. In M. Kornfeld, G. C. Frison, & M. L. Larson (Eds.), *Prehistoric hunter-gatherers of the High Plains and Rockies* (pp. 463–497). Walnut Creek: Left Coast Press.

Broecker, W. S., Denton, G. H., Edwards, R. L., Cheng, H., Alley, R. B., & Putnam, A. E. (2010). Putting the Younger Dryas cold event into context. *Quaternary Science Reviews, 29*(9), 1078–1081.

Carlson, A. E. (2010). What caused the Younger Dryas cold event? *Geology, 38*(4), 383–384.

Carlson, K., & Bement, L. (2013). Organization of bison hunting at the Pleistocene/Holocene transition on the Plains of North America. *Quaternary International, 297*, 93–99.

Charnov, E. L. (1976). Optimal foraging, the marginal value theorem. *Theoretical Population Biology, 9*(2), 129–136.

Chisholm, B., Driver, J., Dube, S., & Schwarcz, H. P. (1986). Assessment of prehistoric bison foraging and movement patterns via stable-carbon isotopic analysis. *Plains Anthropologist, 31*(113), 193–205.

Collard, M., Buchanan, B., Hamilton, M. L., & O'Brien, M. J. (2010). Spatiotemporal dynamics of the Clovis–Folsom transition. *Journal of Archaeological Science, 37*(10), 2513–2519.

Collins, M. B. (1999a). Clovis and Folsom lithic technology. In D. S. Amick (Ed.), *Folsom lithic technology* (pp. 12–38). Ann Arbor: International Monographs in Prehistory.

Collins, M. B. (1999b). *Clovis blade technology: A comparative study of the Keven Davis Cache*. Austin: University of Texas Press.

Collins, M. B. (2002). The Gault Site, Texas, and Clovis research. *Athena Review, 3*(2), 31–42.

Collins, M. B. (2007). Discerning Clovis subsistence from stone artifacts and site distributions on the Southern Plains periphery. In R. Walker & B. N. Driskell (Eds.), *Foragers of the terminal Pleistocene in North America* (pp. 59–87). Lincoln: University of Nebraska Press.

Collins, M. B., & Hemmings, C. A. (2005). Lesser known Clovis diagnostic artifacts I: The bifaces. *La Tierra, 32*(2), 9–20.

Cordova, C. E., Johnson, W. C., Mandel, R. D., & Palmer, M. W. (2011). Late Quaternary environmental change inferred from phytoliths and other soil-related proxies: Case studies from the central and southern Great Plains, USA. *Catena, 85*(2), 87–108.

Ellis, C., & Payne, J. H. (1995). Estimating failure rates in fluting based on archaeological data: Examples from N.E. North America. *Journal of Field Archaeology, 22*, 459–474.

Ferring, C. R. (2001). *The archaeology and paleoecology of the Aubrey Clovis site (41DN479) Denton County, Texas*. DTIC Document.

Fiedel, S. J. (1999). Older than we thought: Implications of corrected dates for Paleoindians. *American Antiquity, 64*(1), 95–115.

Fiedel, S. J. (2015). The Clovis-era radiocarbon plateau. In A. M. Smallwood & T. A. Jennings (Eds.), *Clovis: On the edge of a new understanding* (pp. 11–19). College Station: Texas A&M University Press.

Firestone, R., West, A., & Warwick-Smith, S. (2006). *The cycle of cosmic catastrophes: How a Stone-Age comet changed the course of world culture*. Rochester: Bear and Company.

Firestone, R. B., West, A., Kennett, J. P., Becker, L., Bunch, T. E., Revay, Z. S., Schultz, P. H., Belgya, T., Kennett, D. J., Erlandson, J. M., & Dickenson, O. J. (2007). Evidence for an extraterrestrial impact 12,900 years ago that contributed to the megafaunal extinctions and the Younger Dryas cooling. *Proceedings of the National Academy of Sciences, 104*(41), 16016–16021.

Gunderson, L. H. (2000). Ecological resilience – In theory and application. *Annual Review of Ecology and Systematics, 31*, 425–439.

Guthrie, R. D. (1984). Mosaics, allelochemics and nutrients. An ecological theory of Late Pleistocene megafaunal extinctions. In P. S. Martin & R. G. Klein (Eds.), *Quaternary extinctions* (pp. 259–298). Tucson: The University of Arizon Press.

Haynes, C. V., Jr. (1993). Clovis-Folsom geochronology and climatic change. In O. Soffer & N. D. Praslov (Eds.), *From Kostenki to Clovis: Upper Paleolithic-Paleo-Indian adaptations* (pp. 219–236). New York: Springer.

Haynes, G. (2002). *The early settlement of North America: The Clovis Era*. Cambridge: Cambridge University Press.

Hemmings, E. T. (2007). Buried animal kills and processing localities, areas 1–5. In C. V. Haynes Jr. & B. B. Huckell (Eds.), *Murray Springs: A Clovis Site with multiple activity areas in the San Pedro Valley, Arizona* (pp. 83–137). Tucson: The University of Arizona Press.

Hester, J. J. (1972). *Blackwater Locality No. 1: A stratified early man site in Eastern New Mexico*. Rancho de Taos: Fort Burgwin Research Center Publication 8.

Hill, M. E. (2008). Variation in Paleoindian fauna use on the Great Plains and Rocky Mountains of North America. *Quaternary International, 191*, 34–52.

Hofman, J. L. (2003). Tethered to stone or freedom to move: Folsom biface technology in a regional perspective. In M. Soressi & H. L. Dibble (Eds.), *Multiple approaches to the study of bifacial technologies* (pp. 229–249). Philadelphia: University of Pennsylvania Museum of Archaeology and Anthropology.

Hofman, J. L. (2004). Evolution of human occupation on the High Plains: Grass, bison, and people. In *AMQUA 2004, American Quaternary Association, 18th Biennal Meetings* (pp. 35–40), Lawrence, Kansas.

Holliday, V. T. (2015). Clovis landscapes in the greater Southwest of North America. In A. M. Smallwood & T. A. Jennings (Eds.), *Clovis: On the edge of a new understanding* (pp. 205–241). College Station: Texas A&M University Press.

Holliday, V. T., Meltzer, D. J., & Mandel, R. (2011). Stratigraphy of the Younger Dryas chronozone and paleoenvironmental implications: Central and southern Great Plains. *Quaternary International, 242*(2), 520–533.

Huckell, B. B., & Kilby, J. D. (2014). Clovis caches: Discoveries, identification, lithic technology, and land use. In B. B. Huckell & J. D. Kilby (Eds.), *Clovis caches. Recent discoveries and new research* (pp. 1–9). Albuquerque: University of New Mexico Press.

Irwin-Williams, C., & Haynes, C. V., Jr. (1970). Climatic change and early population dynamics in the southwestern United States. *Quaternary Research, 1*(1), 59–71.

Jennings, T. A. (2013). The Hogeye Clovis cache, Texas: Quantifying lithic reduction signatures. *Journal of Archaeological Science, 40*(1), 649–658.

Jennings, T. A. (2015). Clovis adaptations in the Great Plains. In A. M. Smallwood & T. A. Jennings (Eds.), *Clovis: On the edge of a new understanding* (pp. 277–296). College Station: Texas A&M University Press.

Jodry, M. A. B. (1999). *Folsom technological and socioeconomic strategies: Views from Stewart's Cattle Guard and the Upper Rio Grande Basin, Colorado*. Ph.D. Thesis, Department of Anthropology, American University.

Johnson, E., & Bement, L. C. (2009). Bison butchery at Cooper, a Folsom site on the Southern Plains. *Journal of Archaeological Science, 36*(7), 1430–1446.

Kilby, D. (2015). A regional perspective on Clovis blades and caching behavior. In A. M. Smallwood & T. A. Jennings (Eds.), *Clovis: On the edge of a new understanding* (pp. 145–159). College Station: Texas A&M University Press.

Koldehoff, B., & Loebel, T. J. (2009). Clovis and Dalton: Unbounded and bounded systems in the midcontinent of North America. In B. Adams & B. Blades (Eds.), *Lithic materials and Paleolithic societies* (pp. 270–287). Chichester: Wiley-Blackwell.

Kornfeld, M. (1989). The Rocky Folsom site: A small Folsom assemblage from the northwestern plains. *North American Archaeologist, 9*(3), 197–222.

Kornfeld, M. (2002). Folsom technological organization in the Middle Park of Colorado: A case for broad spectrum foraging. In J. E. Clark & M. B. Collins (Eds.), *Folsom technology and lifeways* (pp. 47–67). Tulsa: University of Tulsa.

Kornfeld, M., Harkin, M., & Durr, J. (2001). Landscapes and peopling of the Americas. In *On being first: Cultural innovation and environmental consequences of first peopling: Proceedings of the 31st annual Chacmool conference* (pp. 149–162). Calgary: The Archaeological Association of the University of Calgary.

Krebs, J. R. (1978). Optimal foraging: Decision rules for predators. In J. R. Krebs & N. B. Davies (Eds.), *Behavioural ecology: An evolutionary approach* (pp. 23–63). Chichester: John Wiley and Sons.

LaBelle, J. M. (2012). Hunter-gatherer adaptations of the central Plains and Rocky Mountains of western North America. In M. I. Eren (Ed.), *Hunter-gatherer behavior: Human response during the Younger Dryas* (pp. 139–164). Walnut Creek: Left Coast Press.

Lundelius, E. L., Jr. (1967). Late-Pleistocene and Holocene faunal history of central Texas. In P. S. Martin & H. E. Wright (Eds.), *Pleistocene extinctions: The search for a cause* (pp. 288–319). New Haven: Yale University Press.

Lundelius, E. L., Jr. (1974). The last fifteen thousand years of faunal change in North America. In C. C. Black (Ed.), *History and prehistory of the Lubbock Lake site* (pp. 144–160). Lubbock: Texas Tech University Press.

MacDonald, D. H. (2010). The evolution of Folsom fluting. *Plains Anthropologist, 55*(213), 39–54.

McDonald, J. N. (1981). *North American Bison: Their classification and evolution*. Berkeley: University of California Press.

Meltzer, D. J. (1993). Is there a Clovis adaptation? In O. Soffer & N. D. Praslov (Eds.), *From Kostenki to Clovis: Upper Paleolithic-Paleo-Indian adaptation* (pp. 293–310). New York: Springer.

Meltzer, D. J., & Holliday, V. T. (2010). Would North American Paleoindians have noticed Younger Dryas age climate changes? *Journal of World Prehistory, 23*(1), 1–41.

Miller, D. S., & Gingerich, J. A. M. (2013). Regional variation in the terminal Pleistocene and early Holocene radiocarbon record of eastern North America. *Quaternary Research, 79*(2), 175–188.

Morgan, R. G. (1980). Bison movement patterns on the Canadian plains: An ecological analysis. *Plains Anthropologist, 25*, 143–160.

Morrow, J. E. (2015). Clovis-era point production in the Midcontinent. In A. M. Smallwood & T. A. Jennings (Eds.), *Clovis: On the edge of a new understanding* (pp. 83–107). College Station: Texas A&M University Press.

Nordt, L. C., Von Fischer, J., Tieszen, L., & Tubbs, J. (2008). Coherent changes in relative C4 plant productivity and climate during the Late Quaternary in the North American Great Plains. *Quaternary Science Reviews, 27*(15), 1600–1611.

Notaro, M., Liu, Z., & Williams, J. W. (2006). Observed vegetation–climate feedbacks in the United States. *Journal of Climate, 19*(5), 763–786.

Prasciunas, M. M., & Surovell, T. A. (2015). Reevaluating the duration of Clovis: The problem of non-representative radiocarbon. In A. M. Smallwood & T. A. Jennings (Eds.), *Clovis: On the edge of a new understanding* (pp. 21–35). College Station: Texas A&M University Press.

Scott, E. (2010). Extinctions, scenarios, and assumptions: Changes in latest Pleistocene large herbivore abundance and distribution in western North America. *Quaternary International, 217*(1–2), 225–239.

Sellet, F. (2004). Beyond the point: Projectile manufacture and behavioral inference. *Journal of Archaeological Science, 31*(11), 1553–1566.

Sellet, F. (2011). Fallen giants: The history of Paleoindian point types on the North American Plains. In D. Vialou (Ed.), *Peuplements et Préhistoire en Amériques* (pp. 97–105). Paris: CTHS.
Sellet, F. (2013). Anticipated mobility and its archaeological signature: A case study of Folsom retooling strategies. *Journal of Anthropological Archaeology, 32*(4), 383–396.
Severinghaus, J. P., Sowers, T., Brook, E. J., Alley, R. B., & Bender, M. L. (1998). Timing of abrupt climate change at the end of the Younger Dryas interval from thermally fractionated gases in polar ice. *Nature, 391*(6663), 141–146.
Shoberg, M. (2010). Functional analysis of Clovis tools. In B. A. Bradley, M. B. Collins, & A. Hemmings (Eds.), *Clovis technology* (pp. 138–156). Ann Arbor: International Monographs in Prehistory.
Shott, M. J. (2013). Human colonization and Late Pleistocene lithic industries of the Americas. *Quaternary International, 285*, 150–160.
Shuman, B., Bartlein, P. J., & Webb, T. (2005). The magnitudes of millennial-and orbital-scale climatic change in eastern North America during the Late Quaternary. *Quaternary Science Reviews, 24*(20), 2194–2206.
Smallwood, A. M. (2010). Clovis biface technology at the Topper site, South Carolina: Evidence for variation and technological flexibility. *Journal of Archaeological Science, 37*(10), 2413–2425.
Speer, C. A. (2014). LA-ICP-MS analysis of Clovis period projectile points from the Gault Site. *Journal of Archaeological Science, 52*, 1–11.
Surovell, T. (2003). *The behavioral ecology of Folsom lithic technology*. Ph.D. Thesis, Department of Anthropology, University of Arizona.
Waters, M. R., & Stafford, T. W. (2007). Redefining the age of Clovis: Implications for the peopling of the Americas. *Science, 315*(5815), 1122–1126.
Webb, T. (1986). Is vegetation in equilibrium with climate? How to interpret Late-Quaternary pollen data. *Vegetatio, 67*(2), 75–91.
Webb, T., Shuman, B., & Williams, J. W. (2003). Climatically forced vegetation dynamics in eastern North America during the Late Quaternary period. *Developments in Quaternary Sciences, 1*, 459–478.
Williams, J. W., Shuman, B. N., Webb, T., Bartlein, P. J., & Leduc, P. L. (2004). Late-Quaternary vegetation dynamics in North America: Scaling from taxa to biomes. *Ecological Monographs, 74*(2), 309–334.
Wormington, H. M. (1939). *Ancient man in North America*. Denver: Denver Museum of Natural History.
Wormington, H. M. (1944). *Ancient man in North America*. Denver: Denver Museum of Natural History.

Chapter 13
The Peopling of Southeastern South America: Cultural Diversity, Paleoenvironmental Conditions, and Lithic Technological Organization During the Late Pleistocene and Early Holocene

Rafael Suárez

Introduction

The peopling of southeast of South America can only be understood within the general framework of the peopling of South America. This region includes parts of La Plata River Basin, the present coastline of Uruguay, and the Atlantic Forest of Southern Brazil, with 14 stratified Paleo-American sites and more than 60 early radiocarbon ages (Miller 1987; Suárez 2011a; Bueno et al. 2013).

The Uruguay and Paraná Rivers were probably two major natural routes of entry for the first human migrations into the continent from the Atlantic Coast (Miotti 2006; Suárez 2011a). The middle Uruguay River valley is an excellent setting to develop archaeological research on Paleo American sites and to study lithic technological organization during the peopling of this region of the continent. There is evidence of bifacial technology oriented toward the production of bifacial cores, points, preforms, and bifacial knives. Additionally, there is also significant variation in the style and design of Paleo-American points during the Pleistocene-Holocene Transition (Suárez 2011a), which yields clues to different aspects of projectile point rejuvenation, lithic resource exploitation, composition of lithic stone tool kits, and mobility strategies of human groups during the initial colonization of the southeast of South America.

One of the earliest sites in the region is Pay Paso 1 (Fig. 13.1, 2) (Suárez 2011a: 65–166). This site presents a solid chronology that includes 34 ^{14}C ages determined by four different laboratories, including 32 by AMS. The excavation of Pay Paso 1 provided evidence for a cultural sequence that helps researchers understand the human adaptations and Paleo-American technology that occurred during the Pleistocene-Holocene

R. Suárez (✉)
Depto. Arqueología – FHCE, Universidad de la República and Sistema Nacional de Investigadores – ANII, Avenue Uruguay 1695, Montevideo, Uruguay
e-mail: rsuarez@fhuce.edu.uy

E. Robinson, F. Sellet (eds.), *Lithic Technological Organization and Paleoenvironmental Change*, Studies in Human Ecology and Adaptation 9,
DOI 10.1007/978-3-319-64407-3_13

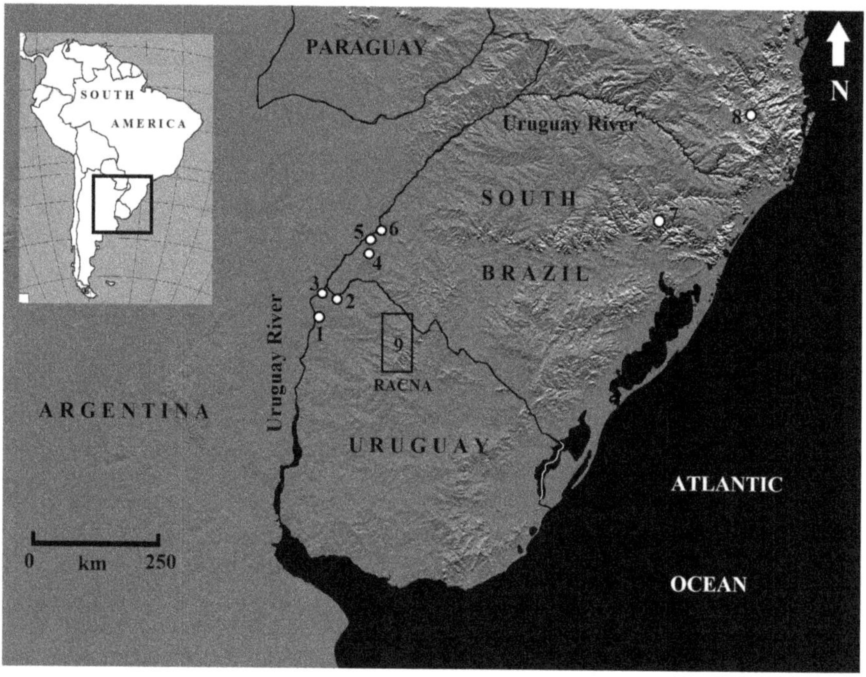

Fig. 13.1 Map of the main early sites with Tigre and Pay Paso points located in the Uruguay Middle River and Southern Brazil: *1* Tigre site (K87); *2* Pay Paso 1 site; *3* Laguna Canosa site; *4* RS-I-66; *5* RS-I-70; *6* RS-I-69; *7* das Flechas site; *8* Avencal Baixo site

transition. The site exposed three successive cultural components, including two with Paleo-American points dated during the Pleistocene-Holocene Transition and early Holocene. Lithic technological analysis of buried contexts from Pay Paso 1 indicates an industry oriented toward bifacial production (preforms, bifacial knives, and points), unifaces with high standardization of their shapes, as well as a blade production technology (Suárez 2011a, 2015a, b).

The excavation of Pay Paso 1 allowed recovery of the first record of Pleistocene fauna at an archaeological site in the Uruguay River area. At present, the 186 bones recovered provide the only faunal collection from an early archaeological site in the southeast of South America. The contextual, stratigraphic, and chronological associations between artifacts and fauna include remains of two species of extinct mammals of the Pleistocene, *Glyptodon* sp. and *Equus* sp., and three other species of present-day fauna, including a fish, *Leporinus* sp. (boga); a large flightless bird, *Rhea americana* (ñandú); and a medium-sized mammal, *Myocastor* (otter). The archaeological context, with remains of *Glyptodon* sp. as well as *Equus* sp., was dated between 9585 and 9120 years ^{14}C BP, an age which suggests survival of these herbivores until the early Holocene, similar to what occurred approximately 800 km farther south, in the Pampa of Argentina (Gutiérrez and Martínez 2008).

The present study has two main objectives: first, to analyze and document different aspects of the organization of lithic technology related to the peopling and colonization of this region and, second, to discuss how the changing climate parameters during the Pleistocene-Holocene Transition generated innovations and technological changes that are reflected in the emergence of different point styles.

Landscape and Paleoenvironmental Conditions Since the Last Glacial Maximum

A feature of southeast of South America is the extensive grassland landscape occupying part of the east of Argentina (Pampa and Entre Ríos provinces), Southern Brazil (Rio Grande do Sul state), and the entire territory of Uruguay. This region has a gently rolling landscape with smooth slopes where mainly fluvial action during the Tertiary has shaped landforms. Uruguay's maximum elevation reaches 513 meters, and the northeastern Rio Grande do Sul (Brazil) plateau is about 1410 meters. The landscape features rolling hills with elongated shapes; there is a wide branching fluvial network with lots of rivers, streams, and creeks. All watersheds empty into the Atlantic Ocean. The earliest known archaeological sites in this region are open-air sites, caves, and rock shelters, which are always located next to watercourses like creeks, arroyos, lakes, or rivers.

Uruguay, Southern Brazil, and the Pampa (Argentina) remained free of ice sheets during Pleistocene glaciations, in a periglacial zone. The main effect of glaciation is recorded with the accumulation of loess and sand dunes generated by wind deposition (Iriondo 1999; Rabassa et al. 2005). The end of the Last Glacial Maximum (LGM) caused a series of climatic events, with periods of glacial retreat and advance registered in southern Patagonia and Tierra del Fuego (Rabassa and Clapperton 1990; Heusser 1998; Rabassa et al. 2000; Coronato et al. 2004), which influenced the climate of the continent. The first glacial retreat began around 19,000 cal BP (Rabassa et al. 2005). Later, between 18,000 and 16,000 cal BP, there was a glacial advance; this pulse was repeated between ca. 13,250 and 11,900 cal BP with a reverse Antarctic cold event (Hajdas et al. 2003). This event appears to correlate to the Northern Hemisphere Younger Dryas (YD) (Heusser and Rabassa 1987), but in the middle latitudes of the Southern Hemisphere, it would have started earlier. From the end of the LGM, the sea level began to rise, and generalized climate conditions were becoming less rigorous: from dry cold to slightly cold-arid and subhumid/humid in Uruguay, the Pampa, and Southern Brazil (Iriondo 1999; Prieto 2000; Behling et al. 2005).

In Southern Brazil (Paraná, Santa Catarina, and Rio Grande do Sul states), high-resolution palynological studies indicate the dominance of grassland in both lowland and highland before and during the LGM, the Late Pleistocene, and early-middle Holocene (Behling et al. 2005; Behling and Pillar 2008). The lowland treeless vegetation of this period indicates cold and dry climatic conditions, with repeated frosts and winter minimum temperatures below -10 °C during the end of the LGM. The driest periods during the LGM are recorded when abundant pollen of *Eryngium* is present, a species which indicates very dry conditions. Seasonal climatic conditions developed during the LGM, with long annual seasonal drought periods that extended to the end of the Pleistocene (Behling and Pillar 2008). The grassland vegetation with pasture prevailed until the mid-Holocene in Southern Brazil.

In the northwest of Uruguay, the palynological record of three sites, at Pay Paso 1, Pay Paso 0 (peat), and Pay Paso 2, indicates that soon after the date of 13,000 cal BP, the dry and cool climate of the Late Pleistocene was gradually replaced by more humid and temperate conditions around 12,500 cal BP, marking the beginning of the Holocene (Suárez 2011a). Before 13,000 cal BP, grassland environment prevailed with abundant pastures; by 12,900 cal BP, records of grass—Chenopodiaceae-*Amaranthus*—began to appear. The biggest observed change in the paleovegetation was when *Amaranthus* was replaced on the riverbanks by several subtropical tree species that presently do not grow in this area, such as *Jacaranda*, *Jacaratia*, *Astronium*, *Protium*, bushes of *Ambrosia* and *Rhus*, plants adapted to humid conditions such as *Cyperaceae* and *Typha* ("totora"), and plants adapted to conditions of high humidity such as ferns of *Selaginella* and Araceae. These data would suggest the beginning of the expansion of the gallery forest along the riverbanks of the middle Uruguay River and the Cuareim River at 30°S 58°W, associated with an increase in temperature, humidity, and precipitation from ca. 12,300 cal BP (Suárez 2011a). The pollen study in Pay Paso 1 additionally allows the observation that two edible fruits, *Celtis* (local name "Tala") and *Psidium* (local name "Arazá"), were available for human consumption in the riparian forest that began expanding along the shores of the Cuareim River by 12,000 cal BP; these fruits could potentially be collected. Tree records (*Jacarantia*, *Ficus*), aquatic plants (Cyperaceae), and tropical herbs (*Aristolochia*) seem to indicate that the warm and wet conditions from the early Holocene continue during the middle Holocene.

The sediments at the Pay Paso 1 site postdate the LGM, exhibiting evidence of important climatic and environmental fluctuations during the last 13,000 years. The beginning of sedimentary accretion at the site indicates that environmental stability began soon after 13,000 cal BP, with the transition of sedimentary deposits from gravelly conglomerates to deposits of well-sorted fine to medium sand at the mouth of the Cuareim River. Subsequently, a laminated sedimentary unit (U2), composed of a succession of a sandy and clay-lime strata, began to form around 12,900 cal BP, coincident with the beginning of human occupation at the site.

Cultural Diversity and Technological Organization

This section will explore topics such as the availability of lithic resources, maintenance, rejuvenation of weaponry, and recycling from other functions, all of which can help us better understand the technological organization behind the peopling of southeast South America.

During the earliest human settlement in this region of the continent, an interesting cultural diversity emerges, represented by different cultural complexes. To date, we have recognized the Fishtail, Tigre, and Pay Paso techno-complexes, defined by their particular projectile points (Suárez 2015a). However, recent evidence indicate that prior to the emergence of Fishtail groups (ca. 12,900 cal BP), the Southern Plains and northern Uruguay were populated by humans ca. 14,000 cal BP, 1000 years before the appearance of Fishtail technology (Suárez 2014, 2015b; Meneghin 2015).

Chronological and archaeological evidence of these early human dispersal pulses across the plains of Uruguay come from two sites, Urupez 2 and Tigre (K87) located in the south and north of Uruguay, respectively. Two ages obtained for Urupez 2 site indicate human presence on the plains of the Atlantic Coast by 13,998–13,627 cal BP (12,000 ± 40 yr ^{14}C yr BP Beta 394639) and 13,708–13,292 cal BP (11,690 ± 80 yr ^{14}C yr BP Beta 211938) (Meneghin 2015). On the other hand, recent data indicate that the Uruguay River was colonized by 13,236–13,057 cal BP (11,320 ± 30 yr ^{14}C yr BP UCIAMS 145429). Current evidence suggests the presence of human occupation on the plains of Uruguay with pre-Fishtail archaeological sites. The two sites above described in the north and south of Uruguay have three radiocarbon ages of 14,000, 13,700 and 13,240 cal BP, which are similar to those obtained on the site Arroyo Seco 2 (Pampa Argentina) (Politis and Steele 2014).

There is very little that we can advance on patterns of mobility, lithic technology, and economics of these initial pulses dispersed across the plains of Uruguay by 14,000–13,200 cal BP. Further excavations are required to provide more archaeological and chronological evidence to corroborate more robustly the retrieved data for this initial period of South American prehistory.

Raw material sources have attracted interest because they reveal differences in the technological organization and mobility of hunter-gatherers (Binford 1980; Shott 1986; Nelson 1991; Amick 1996; Borrero and Franco 1997; Flegenheimer et al. 2003; Sellet 2004, 2013; Suárez 2011b; Miotti and Terranova 2015; Borrero 2015). In northern Uruguay, a 100 km long and 40 km wide corridor was recently defined in which workshops of extensive silicified lithic resources in the form of outcrops of translucent agate, silicified sandstone, jasper, and opal were found (Suárez 2010). We named this area *Región Arqueológica Catalanes Nacientes Arapey* (RACNA) (Fig. 13.1, 9). These new data allow us to model access to raw materials and transport of tool stone and artifacts.

The translucent agate helps us to better understand some aspects linked to mobility and the use of space during the Paleo-American Period in this region. There are three early residential sites on the middle Uruguay River that offer good information on the use of translucent agate during the Pleistocene-Holocene Transition. The Tigre site, located on the Uruguay River at the mouth of the Arroyo del Tigre (Fig. 13.1, 1), was dated at 10,420 ± 90 yr ^{14}C BP (MEC 1989:60) and contains stone tool kits of translucent agate in the earliest levels (MEC 1989: 50–54; Suárez 2011b). The Pay Paso 1 site, located on the left bank of the Cuareim River (Fig. 13.1, 2) and 53 km from the Tigre site, has very good archaeological, chronological, and stratigraphic resolution (Suárez 2011a). At the Pay Paso 1 site, during the beginning of the occupation (ca. 12,800 to 12,500 cal BP), a technology oriented toward blade production from pyramidal cores and pebble cores has been recorded (see Suárez 2011a: 141–151; Suárez 2011b: 365–372, Fig. 5–7; Suárez 2015a). Three early components were recognized and dated with more than 30 radiocarbon dates between ca. 12,800 cal BP and 10,100 cal BP (Suárez 2011a: 93, Table 5.2). Translucent agate made up 14.41% of the total artifacts and 10.71% of the debitage at the site. It is interesting to point out that in the earliest occupation component at Pay Paso I (12,800–12,500 cal BP), translucent agate represents 25.47% of all artifacts, whereas the proportion of this material drops to 11.33% (ca. 12,100 to 11,800 cal BP) and

Fig. 13.2 Early bifacial knives from the middle Uruguay River and Cuareim River: (**a**) on jasper with dark strong patina from Salto Grande region; (**b**) on translucent agate from Pay Paso 4 site; (**c**) on translucent silicified limestone (chert) from the Salto Grande region

11.19% (from 11,100 to 10,100 cal BP) in the other two early components (Suárez 2011b: 366). In the Laguna Canosa site (Fig. 13.1, 3) located 47 km north of the Tigre site and 20 km west of the Pay Paso 1 site, in an archaeological level dated to 11,200 cal BP, translucent agate artifacts with a rejuvenated Tigre point were recovered. The early levels of the Tigre, Pay Paso 1, and Laguna de Canosa sites thus share the presence of bifacial artifacts manufactured in translucent agate (Suárez 2011b).

The presence of large formal bifacial artifacts in translucent agate (ca. 110 mm long) in the middle Uruguay River and lower Cuareim residential sites (Fig. 13.2b) suggests the use of preforms and blanks that could be at least 180–150 mm long. The source of these rocks is located 170 km east of the residential sites. In the RACNA, translucent agate geodes were available that would permit the reduction and/or thinning of a complete series of bifaces culminating in the production of artifacts exceeding 110 mm in length, such as the ones recovered from the residential sites on the middle Uruguay River (Fig. 13.2b). The direction of transport from the sources to the sites indicates an east–west route and direct procurement of the tool stone.

Paleo-American Points of the Southeast of the South American Plains

In southeast South America, there is a morphological diversity of points that are chronologically delimited to the Pleistocene-Holocene Transition. One design is traditionally known as the Fishtail point; however, new designs emerged during the settlement of this region of the continent. We recently identified two new designs of Paleo-American points, called Tigre and Pay Paso, dated to the Pleistocene-Holocene Transition and early Holocene (Suárez 2003, 2011a), which are described below.

Fishtail Points

Fishtail points are the classic diagnostic artifacts of the earliest humans who populated the Southern Cone of South America (Fig. 13.3) by 12,800 to 12,200 cal BP (Waters et al. 2015). However, other authors suggest a more extensive chronology of 12,900 to 10,600 cal BP (11,000 to 9500 yr ^{14}C BP) (Flegenheimer et al. 2013: 365) and 12,900 to 11,250 cal BP (11,000 to 10,000 yr ^{14}C BP) (Nami 2007: 164). They are frequently recorded in both surface and stratigraphic contexts in the Pampa and in Patagonia (Argentina and Chile) (Flegenheimer et al. 2013; Miotti and Terranova 2015), Southern Brazil (Miller 1987; Loponte et al. 2015), and Uruguay (Suárez 2003, 2011a, 2015a; Nami 2007). These points show morphological and stylistic variability both between and within different regions (e.g., Pampa-Patagonia-Uruguay). There are two morphological variations of Fishtail points: one with rounded shoulders and another with very pronounced shoulders (Suárez 2003). As to the treatment of the base of a Fishtail point, the Uruguay sample indicates that 38.6% ($n = 34$) have fluting and 61.4% ($n = 54$) have no fluting but show basal thinning by retouch; only 10.5% ($n = 12$ of 88) have fluting on both sides of the stem.

An interesting measurement that has relevance to the maintenance, rejuvenation, discard, and subsequent recycling of projectile points is the penetration angle (PA). For projectile points to be effective, they must have a high degree of penetration, allowing the point to pass through the prey's fur and hide and then cause a substantial wound in the flesh. To date, the PA has not been taken into consideration in

Fig. 13.3 Fishtail points in different stages of life use (ca. 12,900–12,200 cal BP): (**a**) classic Fishtail point; (**b**) and (**c**) end of life-use-like weapon; (**d**) Fishtail recycled to bifacial knife or cutting tool, note the lack of penetration power of this artifact

previous analyses of Fishtail points. The PA is a valid tool to discuss issues such as maintenance and recycling in these artifacts, as well as for comparisons between samples from different regions of South America, and with other early points with a similar chronology from North America.

The PA is formed where the two sides of the blade edge converge, forming an angle that can be measured in degrees (Fig. 13.4): the smaller the angle, the greater the penetration power of the point. For example, points such as Folsom have PA ranges varying from 75° for the points that have not been rejuvenated and between 85° and 110° for the rejuvenated points (Ahlers and Geib 2000:809, Fig. 6).

Considering these observations, I measured the PA in Paleo-American points of Uruguay (Fishtail, Tigre, and Pay Paso; see below). The results for Fishtail points are presented in Fig. 13.4, in which there are three main groups. The points with higher penetrating power have angles between 60° and 75° ($n = 14$), with an average 69.2 ° (group A in Fig. 13.4). Another group of points has angles between 80° and 120° ($n = 32$) and an average of 96.9° PA (group B in Fig. 13.4). Finally, another

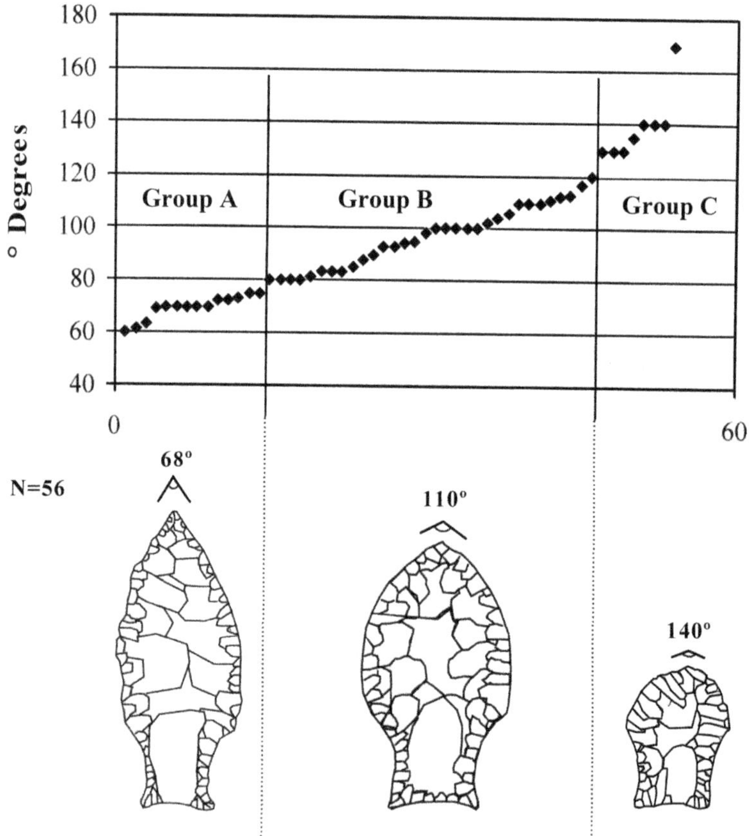

Fig. 13.4 Penetration angle (PA) for Fishtail points of Uruguay

group of points has the largest PA, between 130° and 170° ($n = 8$) and averaging 139° (group C in Fig. 13.4). The PA was measured in a sample of 56 complete points, resulting in a total average of 96°. But, if we eliminate group C, because possibly they were no longer weapons and have another function—i.e., bifacial knives or cutting tools (see Fig. 13.3d)—the sample is reduced with a PA average of 88.5°.

This pattern enables us to make the following observations regarding the PA. First, points with greater penetration angles may be more effective in hunting activities; these have an average angle of 69.2°, indicating that perhaps an angle close to 70° is the most effective with regard to higher penetration power of Fishtail points. Second, the great majority of Fishtail points have an average PA of 96.9°; the points with angles near to 120° would be at the end of their useful life as weapons, due to low penetration power. Third, the group C of points with a PA average of 139° and less penetrating power should have been recycled to other functional forms such as knives or cutting artifacts, losing their function as weapons. Fishtail technological organization indicates a clear intention to recycle points after they ceased to be effective as weapons. The practice of recycling stone artifacts such as the case of Fishtail points was recorded in other Paleo-American contexts of North America (Amick 2015; Shott and Seeman 2015).

The cause for the maintenance of these artifacts was not a shortage in raw materials, since rejuvenated and recycled points have been recovered where high-quality raw materials are present, such as at the middle Negro River (Fig. 13.3a, b, c). Fishtail points are highly curated, not only as part of the weaponry when they are constantly resharpened, but once they are no longer effective as weapons, they are recycled to achieve another function. Therefore, although they were no longer weapons, these artifacts were still preserved; some were not discarded immediately, and they are reconditioned for other functions such as hafted cutting artifacts (Fig. 13.3d).

Recently Discovered Paleo-American Points

Research on the Uruguayan side of the Uruguay River basin (PayPaso 1, Tigre, Laguna Canosa sites) (Suárez 2011a), compared with regional information mainly obtained in Southern Brazil (sites RS-I-66, RS-I-69, and RS-I-70 and das Flechas site) (Miller 1987; Corteletti 2008), indicates that during the time of the peopling of this area, there existed an interesting change in point forms, and the same point designs were being used throughout the middle Uruguay River area and the southeast of South America during the Pleistocene-Holocene Transition.

Research in Uruguay from stratigraphic, chronological, and technological evidence suggests the existence of two Paleo-American point designs during the Pleistocene-Holocene Transition. The name of each design (Tigre and Pay Paso) derives from the site in which the points were dated and recovered in buried context for the first time. The distribution of these points includes the present territory of Uruguay and Southern Brazil. The Argentinean margin of the middle Uruguay River has not yet been systematically investigated, but it is expected that future research in this area will identify the same types of points.

Tigre Points

This point type is geographically distributed throughout north–central Uruguay and Southern Brazil. Tigre points were first recognized at the Tigre (K87) site (MEC 1989; Hilbert 1991) and better defined via archaeological, stratigraphic, and chronological evidence from the Pay Paso 1 and Laguna de Canosa sites (Suárez 2011a:185–188). The chronological data indicate that this type dates between ~12,000 and 11,213 cal BP (10,200 to 9730 ^{14}C yr BP) (Table 13.1). In Southern Brazil these points have been recovered in Rio Grande do Sul state at the RS-I-69 and RS-I-70 sites (see Miller 1987:54 Fig. 13 points b, c, and d) and at the das Flechas site (Corteletti 2008: Figs. 1, 4–7). To date we have recorded a total of 63 Tigre points in Uruguay.

The main characteristics of Tigre points are a wide stem with straight or slightly convex sides, pronounced notches in the edges of ca. 70–90°, a short or long triangular blade, a convex base thinned by retouch, and complete bifacial thinning (Fig. 13.5). The Tigre points recovered in stratigraphic contexts have had a long-life use and were rejuvenated (Fig. 13.5a, c). They probably represent a late-stage result of a curated technology. The main tool stones that were used to manufacture these points are silicified sandstone (44%, $n = 28$), followed by silicified wood (17.46%, $n = 11$), silicified limestone (15.87% $n = 10$), opal (12.69% $n = 8$), and agate and jasper (9.5% $n = 6$). The bases have been thinned by retouch (96.8%, $n = 61$) or by fluting (3.17%, $n = 2$).

When comparing the mean values of length and width of the stems of Fishtail and Tigre points, the data indicate 19.99 mm (Fishtail, $n = 88$) and 18.76 mm (Tigre, $n = 63$) for the average stem length in both sets and 17.21 mm (Fishtail, $n = 82$) and 15.14 mm (Tigre, $n = 56$) for the average stem widths. Thus, there is a higher average stem length for the Fishtail points of 1.7 mm and a higher average width of 2.07 mm. Despite these small variations, the size and dimensions of the stem are roughly similar between the two assemblages of points. The PA of the Tigre points indicates values between 39° and 100°, with an average of 63.4 ° ($n = 50$).

Recently López Mazz (2013:99) renamed Tigre points and calls them "stemmed triangular with wings" (STW), omitting without a justified reason the name we originally gave these points, defined in the PhD dissertation thesis of R. Suárez (2010)

Table 13.1 Ages of Tigre points recovered in buried, stratigraphic contexts

Site	Age ^{14}C (year BP)	2σ Age cal. (year BP[a])	Laboratory number	Reference
Pay Paso 1	10,205 ± 35	12,008–11,629	UCIAMS 21632	Suárez (2011a)
Pay Paso 1	10,180 ± 20	11,974–11,623	UCIAMS 21634	Suárez (2011a)
Pay Paso 1	10,115 ± 25	11,795–11,399	UCIAMS 21633	Suárez (2011a)
Pay Paso 1	9,890 ± 90	11,700–10,887	n/i	Austral (1995)
Laguna Canosa	9,730 ± 30	11,213–10,826	UCIAMS 27739	Suárez (2011a)

Note: *n/i* no information
[a]Calibration with Calib Rev 7.0.0. SHcal13.14c, Hogg et al. (2013)

and published 1 year later (see Suárez 2011a:185–188). Previously, in other papers on the early occupation of Uruguay, López Mazz et al. (2009a, b) do not mention the existence of the STW point type but write that they took the name from Taddei (1969, 1987). The true roots of this name are found during the initial archaeological publications of a pioneer and avocational archaeologist José H. Figueira (1892: 198–199) who made the first classification of projectile points of Uruguay and defined several point types, among which one type he called "stemmed with wings" (*puntas pedunculadas con aletas* in Spanish). At the end of the nineteenth century, this was an important advance. After 121 years of the original publication of José H. Figueira, López Mazz (2013:99) improves that definition including the word "triangular." However, today the STW name is very imprecise for naming a type of Paleo-American projectile point. First, it is based on very general attributes such as "wings" or notches and general forms of blade such as "triangular" and "stem." The notches and the shape of the blade edge are attributes that are modified during the life of these artifacts due to the constant resharpening (Fig. 13.5a–d). Second, points described as STW have been recovered in Uruguay from different chronological periods: from the Pleistocene-Holocene Transition (Suárez 2011a), mid-Holocene archaeological contexts dated between 4190 and 3460 yr ^{14}C BP (Iriarte 2003: Fig. 4.17), and recent Holocene contexts (Hilbert 1991). Third, it is an ambiguous term. Most types of projectile points recovered from surface and stratigraphic contexts in Uruguay are STW points; and they have significant variability in the morphology of the stem, shape of the notch, form of the blade edge, and chronology. Therefore, the name "stemmed triangular with wings" is general, imprecise, and confusing. In addition, it is possible to include within this same type earlier points

Fig. 13.5 Tigre points (ca. 12,000–11,100 cal BP): (**a**) recovered in Pay Paso 1 site excavation area on silicified wood; (**b**) from Pay Paso 1 site surface on silicified sandstone; (**c**) from Laguna Canosa site; (**d**) possible prototype from surface Pay Paso 1 site

and later points from recent periods. If such general criteria had been followed to name the different types of Paleo-American points in North America, for example, the name "stemless unnotched lanceolate" points would include Clovis, Folsom, Agate Basin, Mesa, and Goshen points (Stanford 1999).

Pay Paso Points

The Pay Paso point was defined by Suárez (2003) and is represented by 10 radiocarbon ages between 11,081 and 10,119 cal BP (9585–8570 ^{14}C yr BP) (Table 13.2). To date we have surveyed a total of 40 Pay Paso points in Uruguay and 5 in the south of Brazil.

The design and main techno-morphological characteristics of Pay Paso points (Fig. 13.6) include a short stem, profoundly concave stem base, divergent concave stems expanded toward the base, convex or straight blade sides, regular laminar retouch of the blade, and very careful basal thinning of the stem by removal of short triangular flakes.

We recorded Pay Paso points (n = 40) at various early sites in the middle Uruguay, Cuareim River, and the Negro and Tacuarembó River areas. Furthermore, the same type has been identified in Southern Brazil (Figs. 1, 7, and 8) in the "das Flechas site" (29° 8′25 0.29 ″S, 51°7′49.46″ W) in Rio Grande do Sul state and the Avencal Site Baixo 1 (Urubici) in Santa Catarina state (28° 1′27 0.73 ″S 49°37′4.34″ W)(Corteletti 2008:41, Corteletti 2013). The Southern Brazil sites with Pay Paso points are located 802 and 609 km, respectively, from the Pay Paso type site. This distribution shows the vast territory occupied by hunter-gatherers who manufactured Pay Paso points. Like the Fishtail and Tigre points, many Pay Paso points suffered an intensive process of maintenance and rejuvenation (Figs. 13.2, 13.5, 13.6).

Table 13.2 Ages of Pay Paso points recovered in buried, stratigraphic contexts

Site	Age ^{14}C (year BP)	2σ Age cal. (year BP[a])	Laboratory number	Location
Pay Paso 1	9585 ± 25	11,081–10,711	UCIAMS 21641	Río Cuareim
Pay Paso 1	9555 ± 25	11,070–10,685	UCIAMS 21642	Río Cuareim
Pay Paso 1	9550 ± 20	11,069–10,679	UCIAMS 21647	Río Cuareim
Pay Paso 1	9545 ± 20	11,068–10,666	UCIAMS 21635	Río Cuareim
Pay Paso 1	9545 ± 20	11,068–10,666	UCIAMS 21646	Río Cuareim
Pay Paso 1	9525 ± 20	11,064–10,595	UCIAMS 21640	Río Cuareim
Pay Paso 1	9525 ± 20	11,064–10,595	UCIAMS 21638	Río Cuareim
Pay Paso 1	9280 ± 200	11,124–9,901	Uru-248	Río Cuareim
Pay Paso 1	9120 ± 40	10,373–10,176	Beta-156973	Río Cuareim
Pay Paso 1	8570 ± 150	10,119–9,093	Uru 246	Río Cuareim

Note: *n/i* no information

[a]Calibration with Calib Rev 7.0.0. SHcal13.14c, Hogg et al. (2013).

Fig. 13.6 Pay Paso points (ca. 11,000–10,100 cal BP): (**a**) in early stage of use life from surface Río Tacuarembó Grande; (**b**) intermediate stage of use life from Pay Paso 1 excavation area (dated ca. 11,000 cal BP); (**c**) very resharpened at the end of use life, from the surface of Pay Paso 1 site

Two main raw materials were utilized for manufacturing Pay Paso points: silicified sandstone (45%, n = 18) and silicified limestone (27.5%, n = 11). These raw material types were followed by jasper (7.5%, n = 3) and silicified wood and opal (5% n = 2), while silicified wood, riolite, and agate (7.5% n = 3) were used in lower proportions. The treatment of the base of the Pay Paso points was very careful and presents a distinctive feature, a "false" or "pseudo flute," obtained from extracting a short but deep triangular flake which thinned the stem base.

Pay Paso points have a small–medium size. The width is constant with values between ca. 20 and 30 mm, averaging 21.28 mm (n = 39). The length of the completed points range in value from 26.6 mm for the smallest to 68.2 mm for the largest, averaging 42.5 mm (n = 39). The average thickness is 7.09 mm (n = 40). Another distinctive feature of the Pay Paso points is the width of the stem, which is greater than the length, and causes the overall design to look like a point with a short and narrow stem. The angle of penetration of the Pay Paso points has values between 41° and 93°, with an average of 54.9° (n = 32).

Lithic Resources and Raw Material Used in Paleo-American Point Assemblages

Fishtail points from Uruguay were manufactured primarily from silicified limestone (53.40%, n = 47) from the Queguay Formation (Martínez et al. 2015). Other raw materials, respectively, represent less than 10% of the specimens: jasper 10%

(n = 9), quartzite 7.95% (n = 7), chert 6.8% (n = 6), opal 7.95% (n = 7), quartz 6% (n = 5), silicified sandstone 3.4% (n = 3), and agate and chalcedony 2.27% (n = 2). For Tigre points, the use of silicified sandstone reaches 44% (n = 28) followed by silicified wood 17.46% (n = 11), silicified limestone 15.87 (n = 10), opal 12.69% (n = 8), and agate and jasper 9.5% (n = 6). The Pay Paso points were preferentially manufactured from silicified sandstone (45%, n = 18), followed by silicified limestone (27.5%, n = 11), jasper (7.5%, n = 3), and other raw materials that include opal, silicified wood, and quartz (7.5%, n = 3). There are no Fishtail points manufactured from silicified wood, which was a raw material used to produce Tigre and Pay Paso points. This absence does not reflect a lack of knowledge of these materials by Fishtail groups (see below). The differences in raw material preference show a clear trend toward the use of silicified limestone for Fishtail technology; however, post-Fishtail groups producing Tigre and Pay Paso points preferred to manufacture their points from silicified sandstone.

The absence of Fishtail points manufactured from silicified wood certainly is not related to the lack of knowledge of specific locations where to obtain this material. There are two main sources for silicified wood in Uruguay: one is located along the Uruguay River, and the other is located in the Yaguarí Formation in the Negro River basin, where the highest density of Fishtail points in Uruguay has been recovered, so it would seem the users of Fishtail technology ignored silicified wood for making these artifacts. Another main difference in the use of resources is the presence of crystal quartz in Fishtail points (Nami 2009), a raw material apparently not used to produce Tigre points.

Discussion

The colonization and earliest settlement of the southeast of South America begins from the Atlantic Ocean into the interior by 14,000 yr cal BP. Current evidence indicates that at least three small populations with an unsophisticated flaked stone technology and a foraging economy were exploiting different ecosystems in the Southern Cone by 14,600–14,000 cal BP. The areas related to the Pacific Coastal Plain represented by the Monte Verde site (Dillehay et al. 2008), the open plains in the Pampa of Argentina (Politis and Steele 2014), and the grassland plains of Uruguay with the Tigre (K87) and Urupez 2 sites (Suárez 2014; Meneghin 2015) are the three highly productive regions which should have attracted the people who explored the Southern Cone during the Late Pleistocene. Regarding the stone material utilized during this earliest period of colonization, no technological homogeneity can be recognized in the morphology of the artifacts. Except for several bifaces at Monte Verde II, the technology of these four sites apparently has no standardized morphology in the artifacts, which were made mostly of local rock of medium to good flaking quality. The only diagnostic projectile points were recovered at Monte Verde II, and these can serve to guide the search for similar artifacts, such as those recently discovered in Uruguay (Suárez 2014) that may be examples of pre-Fishtail occupations in the Atlantic Coast zone of South America. Later, we recognize a bifacial Paleo-American tradition around 12,900–10,100 cal BP associated with Fishtail, Tigre, and Pay Paso techno-cultural complexes (Suárez 2015a).

There is a significant change in technology with the emergence of the Fishtail complex, when a new design of projectile point begins to expand across the Southern Cone by 12,900 cal BP. These human groups shared technological knowledge such as bifacial reduction, fluting of the base of the stem, and the same design in the style of the blade of Fishtail points. Furthermore, people began to use colorful and better-quality tool stone such as silicified limestone, translucent agate, jasper, opal, chalcedony, chert, and silicified sandstone, among others, with different strategies of lithic resource transfer from the sources to residential camps (Flegenheimer et al. 2003; Suárez 2011b). The Fishtail technology features high-maintenance recycling, retooling, resharpening, and hafting (Politis 1991; Suárez 2003, 2011a).

The presence of the Pay Paso and Tigre points has important ramifications for understanding adaptations during the Pleistocene-Holocene Transition. These data suggest an interesting technological diversity for the Paleo-American Period of southeast of South America that is beginning to be recognized in the different point types that circulated along the watershed of the Uruguay River. Human groups using this technology were spread over a vast region of grassland and pastures associated with fluvial environments in the southeast of South America. Around 12,200–12,000 cal BP, in the middle Uruguay River area, a technological reorganization occurred with several innovations such as the Tigre points and the technology associated with them. Technological restructuring includes the decreasing use of the flute as a technique for thinning the base and morphological changes in the shape of blade edges, stem, and base of the projectile point. There was a drastic reduction in the fluting of the base, with the percentage of occurrences descending abruptly from 38.6% for Fishtail points ($n = 34$ of 88) to 3.1% for Tigre points ($n = 2$ of 63). This change could indicate a transition between the two technologies, with a tendency to refrain from fluting the bases of the points. Other changes in the morphology of the stem and base are observed. However, the size of the stem—length and width—remains constant without major changes in Fishtail and Tigre points. This could be related to the maintenance of the same hafting technology for weaponry.

Another important technological difference is the change in the shape of the blade edge and the resulting increase in the penetration angle (PA). When comparing the PA average of 88.55° for the Fishtail points with the 63.4° from Tigre points, we observe that the Tigre points have a PA that is 25.15° smaller. This feature is also related to the form of the blade edges, which change from convex in Fishtail points to straight in Tigre points. The decrease in the PA and the change in the form of the blade edges are probably technological responses to the extinction of Pleistocene fauna and the subsequent reorganization toward the hunting of new prey that must have occurred at the beginning of the Holocene. This hypothesis should be confirmed in the future, when new and more data on the economic preferences of both groups are obtained. This technological innovation does not necessarily indicate a cultural replacement and could represent the reconditioning of the weaponry as a response to changes in paleoenvironmental parameters, climate, and fauna associated with the expansion of gallery forests around the middle Uruguay River and a relatively warm-humid climate phase that begins over 12,200–12,000 cal BP, just when Tigre points appear.

The lithic provisioning strategies indicate that during the initial settlement of this region of the continent, human groups moved in search of tool stone for distances of hundreds of kilometers. There are two well-documented case studies. Translucent agate in northern Uruguay indicates movement over distances of 170 km between the concentrations of these resources in the *Región Arqueológica Catalanes Nacientes Arapey* to residential sites in the middle Uruguay River region (Suárez 2011b). The long distances traveled by the hunter-gatherers to obtain translucent agate suggest logistic mobility (Binford 1980; Shott 1986) and provisioning circuits planned along access to specific tool stone sources (Amick 1996). Perhaps the translucent quality gave to this material a meaning or symbolic social value that made it particularly attractive for the manufacture of stone tool kits.

On the other hand, Flegenheimer et al. (2003) report the transport of silicified limestone over distances of 400–500 km that would have occurred through exchange between groups in south-central Uruguay with groups in the Pampas in Argentina during the Pleistocene-Holocene Transition. They suggest the possibility that small groups from the two regions maintained regular social relationships and shared different goods, including lithic source material (Flegenheimer et al. 2003:60–61). Thus, there were two different strategies related to long-distance forays by Paleo-Americans to acquire materials for the production of reddish silicified limestone and translucent agate artifacts during the peopling of the southeast of South America.

Conclusions

The peopling of southeast of South America started in small waves that gradually explored and colonized the region by 14,000 cal BP, one millennium before the Fishtail technology emerges in the South Plains. The population in the southeast of South America during the period between 14,000 and 13,200 cal BP must have been very small, with a small number of sites and with limited contact between groups and little social interaction. By 12,900 cal BP, the Fishtail complex emerged, population density increased, and technological innovations (stemmed points) were introduced. This development is reflected in the increase in the number of sites for this period, with Fishtail points frequent in such different regions of South America as Ecuador, northern Peru, Southern Brazil, northern and southern Chile, and the Pampa and Patagonia in Argentina and Uruguay.

The valleys of the major rivers like the Uruguay River, where significant concentrations of resources (tool stone, game animals, food, wood, water, etc.) could be found, were the main routes of entry, dispersal, and communication for the early explorers who ventured into the interior of the continent from the Atlantic Coast (Miotti 2006; Suárez 2011a). The access via numerous river systems may have minimized the risks of exploring a region previously unknown and uninhabited. The populations seem to have been very small and very sparsely distributed across large territories, with low-level visibility in the archaeological record. Fishtail point technology represents one of the initial and more successful human exploration-adaptation processes in the Southern Cone. Traces of human groups with this technology have been recognized across a broad and varied range of Late Pleistocene environments in South America. The Tigre and Pay Paso Paleo-American technology have a more restricted geographical distribution than Fishtail technology. However, these two types of points still circulated across large areas, such as the headwaters and middle Uruguay River in what is now Uruguay and Southern Brazil. Pay Paso points, for example, have been recorded more than 800 km away from the type site. It seems that a priority for colonization was a favorable environment with a wide variety of resources to exploit, as is usually found along major rivers (Dillehay 2012; Miotti 2006). The hunter-gatherers living in this area from 12,900 to 10,100 cal BP had a high residential mobility (characterized by frequent moves between sites) and logistical mobility to access specific resources (indicated by the long-distance transport of lithic source material).

The chronological spread and cultural variability expressed in the different Paleo-American point technologies can be explained as a response to internal social reorganization of hunter-gatherer groups, forced by the need to deal with climate change and the faunal and ecologic shifts that occurred during the Late Pleistocene, the Pleistocene-Holocene Transition, and early Holocene. The technological innovations seen in projectile point style changes may not necessarily signify cultural replacement but may be connected to a technological transition caused by paleoenvironmental changes, associated with the advance of the gallery forests along riverbanks during the early Holocene. The lifeways of human groups probably remained the same in regard to their social and political strategies.

The southern boundary of the Tigre and Pay Paso technologies and adaptations correlated with the mouth of the Uruguay River (34°10′S, 58°09′W). There is currently no archaeological evidence that these human groups occupied the present territory of the Pampa in Argentina. This distribution supports the idea that Tigre and Pay Paso technologies represented human adaptations to fluvial environments related to grasslands and to the gallery forests that began to spread along the Uruguay River during the Pleistocene-Holocene Transition.

Acknowledgments Thanks to Frédéric Sellet and Erick Robinson for inviting me to participate with this chapter of this book. Research on the early prehistory in northern Uruguay was performed by the National Geographic Society (USA), The Wenner-Gren Foundation for Anthropological Research (USA), CONICYT (Uruguay), Ministry of Culture (Spain), National Agency for Research and Innovation (ANII) (Uruguay), and CSIC-Udelar. Rafael Corteletti generously provided original photographs for the Tigre and Pay Paso points from Southern Brazil. Jill Patton, Ruth Gruhn, and Fréderic Sellet helped with the English translation.

References

Ahler, S. A., & Geib, P. R. (2000). Why flute? Folsom point design and adaptation. *Journal of Archaeological Science, 27*, 799–820.

Amick, D. (1996). Regional patterns of Folsom mobility and land use in the American Southwest. *World Archaeology, 27*, 411–426.

Amick, D. (2015). The recycling of material culture today and during the Paleolithic. *Quaternary International, 361*, 4–20.

Austral, A. (1995). Los cazadores del sitio estratificado Pay Paso hace 10,000 años. In M. Consens, J. M. L. Mazz, & C. del Carmen Curbelo (Eds.), *Arqueología en el Uruguay* (pp. 212–218). Montevideo: Banco Comercial.

Behling, H., & Pillar, V. D. (2008). Vegetation and fire dynamics in Southern Brazil during the Late Pleistocene and their impoliations for conservation and management of modern grassland ecosystem. In H. G. Schröder (Ed.), *Grasslands: Ecology, management and restoration* (pp. 181–194). New York: Nova Science Publishers.

Behling, H., Pillar, V., & Bauermann, S. G. (2005). Late Quaternary grassland (Campos), gallery forest, fire and climate dynamics, studied by pollen, charcoal and multivariate analysis of the São Francisco de Assis core in western Rio Grande do Sul (southern Brazil). *Review of Palaeobotany and Palynology, 133*, 235–248.

Binford, L. (1980). Willow smoke and dog's tails: Hunter-gatherer settlement system an archaeological site formation. *American Antiquity, 45*, 4–20.

Borrero, L. A. (2015). The process of human colonization of Southern South America: Migration, peopling and "the archaeology of place". *Journal of Anthropological Archaeology, 38*, 46–51.

Borrero, L., & Franco, N. (1997). Early Patagonian hunter-gatherers: Subsistence and technology. *Journal of Anthropological Research, 53*, 219–239.

Bueno, L., Schmidt, A., & Steele, J. (2013). The Late Pleistocene/Early Holocene archaeological record in Brazil: A geo-referenced database. *Quaternary International, 301*, 74–93.

Coronato, A., Meglioli, A., & Rabassa, J. (2004). Glaciations in the Magellan Straits and Tierra del Fuego, southernmost South America. In J. Ehelers & P. L. Gibbard (Eds.), *Quaternary glaciations-extent and chronology* (Vol. 1, pp. 45–48). Amsterdam: Elsevier.

Corteletti, R. (2008). *PatrimonioArqueológico de Caxias do Sul*. Porto Alegre: Nova Prova.

Corteletti, R. (2013). *Projeto Arqueológico Alto Canoas – PARACA; Um Estudo da Presença Jê no Planalto Catarinense*. Unpublished PhD dissertation, University of São Paulo. http://www.teses.usp.br/teses/disponiveis/71/71131/tde-19042013-093054/pt.br.php

Dillehay, T., Ramírez, C., Pino, M., Rossen, J., & Pino-Navarro, D. (2008). Monte Verde: Seaweed, food, medicine and the peopling of South America. *Science, 320*, 784–786.

Dillehay, T. (2012). Climate, Technology and Society during the Terminal Pleistocene Period in South America. In M. I. Eren (Ed.), *Hunter-gatherer behavior, human response during the Younger Dryas* (pp. 25–55). California: Left Coast Press.

Flegenheimer, N., Bayón, C., Valente, M., Baeza, J., & Femenías, J. (2003). Long distance tool stone transport in the argentine pampas. *Quaternary International, 109*, 49–64.

Flegenheimer, N., Miotti, L., & Mazzia, N. (2013). Rethinking early objects and landscapes in the Southern Cone: Fishtail point concentrations in the Pampas and Northern Patagonia. In K. Graf, C. Ketron, & M. Waters (Eds.), *Paleoamerican Odyssey* (pp. 359–376). College Station: Texas A&M University Press.

Gutiérrez, M. A., & Martínez, G. A. (2008). Trends in faunal human exploitation during the Late Pleistocene and early Holocene in the Pampeanregión (Argentina). *Quaternary International, 191*, 53–68.

Hajdas, I., Bonani, G., Moreno, P. I., & Ariztegui, D. (2003). Precise radiocarbon dating of late-glacial cooling in mid-latitude South America. *Quaternary Research, 59*, 70–78.

Heusser, C. J. (1998). Deglacial paleoclimates of the Americas sector of the Southern Ocean: Late Glacial-Holocene records from the latitude of Canal the Beagle (55°S), Argentina, Tierra del Fuego. *Palaeogeography, Palaeoclimatology, Palaeoecology, 141*, 277–301.

Heusser, C. J., & Rabassa, J. (1987). Cold climatic episode of Younger Dryas age in Tierra del Fuego. *Nature, 328*, 609–611.
Hilbert, K. (1991). *Aspectos de la Arqueología en el Uruguay*. Mainz: Verlag Philipp Von Zabern.
Hogg, A. G., Hua, Q., Blackwell, P. G., Niu, M., Buck, C. E., Guilderson, T. P., Heaton, T. J., Palmer, J., Reimer, P., Reimer, R. W., Turney, C., & Zimmerman, S. H. H. (2013). SHCAL13 Southern Hemisphere Calibration, 0-50,000 years cal BP. *Radiocarbon, 55*, 1–15.
Iriarte, J. (2003). *Mid-Holocene emergent complexity and landscape transformation: The social construction of early formative communities in Uruguay, La Plata basin*. PhD. Thesis, University of Kentucky.
Iriondo, M. (1999). Climatic changes in the South American plains: Records of a continent-scale oscillation. *Quaternary International, 57*, 83–86.
López Mazz, J. M. (2013). Early human occupation of Uruguay: Radiocarbon database and archaeological implications. *Quaternary International, 301*, 94–103.
López Mazz, J. M., Gascue, A., Villamarzo, V., De León, N., Sotelo, M., & Alzugaray, S. (2009a). La organización de la tecnología lítica de los Pobladores Tempranos del Este de Uruguay. *Intersecciones Antropología, 10*, 63–73.
López Mazz, J. M., Moreno, F., Villarmarzo, E., & Gascue, A. (2009b). Apuntes para una Arqueología costera and del Cabo Polonio. In J. López Mazz & A. Gascue (Eds.), *Arqueología Prehistórica Uruguay en el siglo XXI* (pp. 39–65). Montevideo: Biblioteca Nacional.
Loponte, D., Carbonera, M., & Silvestre, R. (2015). Fishtail projectile points from South America: The Brazilian record. *Archaeological Discovery, 3*, 85–103.
Martínez, S., Veroslavsky, G., & Cabrera, F. (2015). Calizas del Queguay: un enfoque hacia la arqueología. *Revista de Antropología del Museo de Entre Ríos, 1*(2), 1–10.
MEC (Ministerio Educación y Cultura). (1989). *Misión de Rescate Arqueológico de Salto Grande, Tomo 3*. Montevideo: Ministerio Educación y Cultura.
Meneghin, U. (2015). *Secuencia cronoestratigráfica de Urupez II. Nuevas dataciones radiométricas. Orígenes 13*. Montevideo: Fundación Arqueología Uruguay.
Miller, E. T. (1987). Pesquisas arqueológicas Paleoindígenas no Brasil occidental. *Estudios Atacameños, 8*, 37–61.
Miotti, L. L. (2006). La Fachada Atlántica como puerta de ingreso alternativa de la colonización de América del Sur durante la Transición Pleistoceno/Holoceno. In J. C. Jiménez (Ed.), *II Simposio Internacional El Hombre Temprano en América* (pp. 156–188). Mexico City: INAH.
Miotti, L. L., & Terranova, E. (2015). A hill of points in terra incognita from Patagonia: Notes and reflections for discussing the way and tempo of initial peopling. *PaleoAmerica, 1*(2), 181–196.
Nami, H. (2007). Research in the middle Río Negro basin (Uruguay) and the Paleoindian occupation of the Southern Cone. *Current Anthropology, 48*(1), 164–171.
Nami, H. (2009). Crystal quartz and Fishtail projectile points: Considerations on raw-material selection by Paleo South Americans. *Current Research in the Pleistocene, 26*, 9–10.
Nelson, M. C. (1991). The study of technological organization. In M. B. Schiffer (Ed.), *Archaeological method and theory* (Vol. 3, pp. 57–100). Tucson: University of Arizona Press.
Politis, G. (1991). Fishtail projectile points in the Southern Cone of South America: An overview. In R. Bonnichsen & K. Turnmire (Eds.), *Clovis: Origins and adaptations* (pp. 287–301). Corvalis: Oregon State University, Center for the Study of the First Americans.
Politis, G., & Steele, J. (2014). Cronología radiocarbónica. In G. Politis, M. A. Gutiérrez, & C. Scabuzzo (Eds.), *Estado Actual de las investigaciones en el sitio Arroyo Seco 2* (pp. 57–66). Buenos Aires: Universidad Centro de la Provincia INCUAPA.
Prieto, A. (2000). Vegetational history of the Late Glacial-Holocene transition in the grassland of eastern Argentina. *Palaeogeograph, Palaeoclimatology, Palaeocology, 157*, 167–188.
Rabassa, J., & Clapperton, C. M. (1990). Quaternary glaciations of the southern Andes. *Quaternary Science Reviews, 9*, 153–174.
Rabassa, J., Coronato, A., Bujalesky, G., Salemme, M., Roig, C., Meglioli, A., Heusser, C., Gordillo, S., Roig, F., Borromei, A., & Quattrocchio, M. (2000). Quaternary of Tierra del Fuego, Southernmost South America: An updated review. *Quaternary International, 98-71*, 217–214.

Rabassa, J., Coronato, A., & Salemme, M. (2005). Chronology of the Late Cenozoic Patagonian glaciations and their correlation with biostratigraphic units of the Pampean region (Argentina). *Journal of South American Earth Sciences, 20*, 81–103.
Sellet, F. (2004). Beyond the point: Projectile manufacture and behavioral inference. *Journal of Archaeological Science, 31*, 1553–1566.
Sellet, F. (2013). Anticipated mobility and its archaeological signature: A case study of Folsom retooling strategies. *Journal of Anthropological Archaeology, 32*, 383–396.
Shott, M. J. (1986). Technological organization and settlement mobility: An ethnographic examination. *Journal of Anthropological Research, 41*, 15–51.
Shott, M. J., & Seeman, M. (2015). Curation and recycling: Estimating Paleoindian endscraper curation rates at Nobles Pond, Ohio, USA. *Quaternary International, 361*, 316–331.
Stanford, D. (1999). Paleoindian archaeology and Late Pleistocene environments in the Plains and Southwest United States. In R. Bonnichsen & K. L. Turnmire (Eds.), *Ice age peoples of North America* (pp. 281–339). Corvallis: Oregon State University, Center for the Study of First Americans.
Suárez, R. (2003). Paleoindian components of northern Uruguay: New data for early human occupations of the Late Pleistocene and Early Holocene. In L. Miotti, M. Salemme, & N. Flegenheimer (Eds.), *Where the South winds blow: Ancient evidences from Paleo South Americans* (pp. 29–36). College Station: Texas A&M University Press, Center for the Study of the First Americans.
Suárez, R. (2010). *Arqueología Prehistórica En la Localidad Arroyo Catalán Chico. Investigaciones Pasadas, Replanteo y Avances Recientes*. Montevideo: Universidad de la República, Depto. Publicaciones CSIC.
Suárez, R. (2011a). *Arqueología durante la Transición Pleistoceno Holoceno: Componentes Paleoindios, Organización de la Tecnología y Movilidad de los Primeros Americanos en Uruguay, Archaeopress British archaeological reports 2220*. Oxford: Archaeopress Available from Hadrian Books.
Suárez, R. (2011b). Movilidad, acceso y uso de ágata traslúcida por los cazadores-recolectores tempranos durante la transición Pleistoceno Holoceno en el Norte de Uruguay (ca. 11,000-8500 AP). *Latin American Antiquity, 22*(3), 359–383.
Suárez, R. (2014). Pre-Fishtail settlement in the Southern Cone ca. 15,000-13,100 yr cal. BP: Synthesis, evaluation, and discussion of the evidence. In D. Stanford & A. Stanger (Eds.), *Pre Clovis in Americas: International science conference proceedings* (pp. 153–191). Washington D. C: Smithsonian Institution.
Suárez, R. (2015a). The Paleoamerican occupation of the plains of Uruguay: Technology, adaptations, and mobility. *PaleoAmerica, 1*(1), 88–104.
Suárez, R. (2015b). Tecnología Lítica y Conjunto de Artefactos Utilizados Durante el Poblamiento Temprano de Uruguay. *Chungará, 47*, 67–84.
Taddei, A. (1969). Un yacimiento de cazadores superiores del Medio Río Negro, Uruguay. *Pesquisas Antropológicas, 20*, 57–78.
Taddei, A. (1987). Algunos aspectos de la arqueología prehistórica en el Uruguay. *Estudios Atacameños, 8*, 62–93.
Waters, M. R., Amorosi, T., & Stafford, T. (2015). Redating Fell's Cave, Chile and the chronological placement of the Fishtail projectile point. *American Antiquity, 80*, 376–386.

Chapter 14
Changes in the Technological Organization and Human Use of Space in Southern Patagonia (Argentina) During the Late Holocene

Nora Viviana Franco, Luis Alberto Borrero, George A. Brook, and María Virginia Mancini

Introduction

Hunter-gatherers use technology to solve problems posed by the natural and social environment (e.g., Nelson 1991; Bousman 1993). The purpose of this paper is to document the available information on technological and behavioral changes in the southern part of the Upper Santa Cruz River Basin during the Late Holocene and to examine their possible relationships with environmental changes.

To simplify our task, we have defined five periods that seem to differ in both climate and the characteristics of human occupation. For clarity, we will discuss each period in turn from the oldest to the most recent. Periods I and II were characterized by relatively high precipitation in the Andes, but they differ in that there is evidence of human activity at the Rio Bote 1 (RB1) archaeological site on the steppe during Period I but no evidence whatever that this site was used during Period II. Periods III and V constitute what appear to be the driest intervals in the Andes during the last 6000 years (ca. 300 and 200 mm/year precipitation, respectively) and have only limited evidence of human activity. Period IV has evidence that humans used both the Andes and steppe at this time and differs from the other four periods in that the climate fluctuated dramatically between wetter and drier conditions much more than at other times (from ca. 300–400 mm/year over periods of ca. 200–300 years).

N.V. Franco (✉) • L.A. Borrero
CONICET-University of Buenos Aires, Buenos Aires, Argentina
e-mail: nvfranco2008@gmail.com; laborrero2003@yahoo.com

G.A. Brook
Department of Geography, University of Georgia, Athens, GA, USA
e-mail: gabrook@uga.edu

M.V. Mancini
University of Mar del Plata, Mar del Plata, Argentina
e-mail: mvmancin@gmail.com

E. Robinson, F. Sellet (eds.), *Lithic Technological Organization and Paleoenvironmental Change*, Studies in Human Ecology and Adaptation 9,
DOI 10.1007/978-3-319-64407-3_14

Our discussion will emphasize possible relationships between human activities and climate in these different periods, and we will compare the periods with each other.

Background and Methodology

The findings presented here are the outcome of research projects that emphasized the importance of knowledge about the natural and social environment as a way of understanding human behavior in the past (e.g., Borrero and Carballo Marina 1998; Franco 2008). In order to understand paleoenvironmental variations in the past, sediment studies were undertaken at archaeological sites and in ancient fluvial and aeolian dune deposits. Pollen analysis of archaeological sites, peat bogs, and lagoons provided information on past vegetation, while charcoal analysis indicated variations in fire activity. Vegetation changes and climatic interpretations of pollen data were compared with recent pollen-vegetation relationships; these showed that shrubs (Asteraceae subf. Asteroideae) and dwarf shrubs (*Ephedra*, *Nassauvia*) indicate drier conditions, while higher levels of grasses and herbs indicate wetter conditions (Mancini et al. 2013).

Although humans may have started many fires in the Late Holocene, drier-than-average climate conditions may still have been needed for high fire activity (Whitlock et al. 2007). In the case of lakeshore dunes, we presume conditions at the time of dune formation were dry and windy and the level of Lake Argentino was lower so that exposed nearshore sediments could be reworked by waves and wind to form beach ridges and coastal dunes. Periods of soil development within the deposits most likely record higher precipitation and higher lake levels.

Radiocarbon ages in ^{14}C years BP (^{14}C yr BP) were calibrated using CALIB 7.1 (Stuiver and Reimer 1993) and the Southern Hemisphere (SHcal13) atmospheric calibration curve of Hogg et al. (2013). Calibrated ages at the 2σ level are given in calendar years BP (cal BP). OSL ages were initially estimated in years before the measurement year. Here they are expressed in years BP so that they can be compared with calibrated radiocarbon ages.

To understand human strategies of raw material utilization, it was necessary to document the regional availability of lithic resources (sensu Ericson 1984), including their location and characteristics. This is complex in the Upper Santa Cruz River Basin, where secondary raw material sources (sensu Luedtke 1979) are abundant. Our regional sampling of potential lithic sources coupled with geochemical analysis revealed a good correlation between dacite geochemistry and location and has also allowed us to recognize general source areas for different varieties of chalcedony and opal (Franco and Aragón 2004). Design variables (Nelson 1991) and ethnoarchaeological information were used to generate expectations for the lithic record of the exploration of new spaces compared with "effective occupation" of a space (sensu Borrero 1994–95). While the prioritization of versatility, the use of raw materials of inferior quality rather than the ones locally available, and the presence of complete artifacts can be expected during the exploration of new environments, during effective occupation, the tool fragmentation index should increase, and better-quality raw materials should be used because of a better knowledge of the area (Franco 2004a, 2012).

In our environmental research, for the Andes, we have borrowed heavily from the precipitation model of Tonello et al. (2009) based on pollen data from Cerro Frias (CF) south of Lake Argentino and the works of Moreno et al. (2009), Aniya (2013), and Strelin et al. (2014), who have documented periods of glacier advance in the Southern Patagonia Ice Field (SPIF). For the steppe east of the Andes, we have utilized data in Ohlendorf et al. (2014) on Laguna Cháltel. We have assumed following Garreaud et al. (2009) that strong southern westerly winds (SWW) bring more humid conditions to the Andes Mountains and drier conditions to extra-Andean sites. This means that strong winds bring humid conditions to localities immediately east of the Andes, such as Cerro Frías (CF), but drier conditions further east on the steppe at sites like Rio Bote 1 (RB1) and Laguna Cháltel. Some sites, such as Chorrillo Malo 2 (ChM2), are in an intermediate location between the mountains and the steppe so that sometimes precipitation trends match those in the mountains but at other times they match conditions in the steppe. In contrast to strong winds, weaker SWW lessen the west-to-east precipitation gradient allowing humid air masses to reach eastward into the steppe next to the Andes (Moy et al. 2009). An intensification of the SWW and a steepening in the west-to-east precipitation gradient can account for differences in vegetation patterns between the lower eastern slopes of the Andes and sites only several kilometers to the east (Garreaud et al. 2009).

The Area

The study area is very close to the Andean range where the Southern Patagonia Ice Field (SPIF), more than 60 km wide, extends to the Pacific Ocean and limits population movements to the west. The Pacific coast can be reached through the Baguales Range, with altitudes up to ca. 2000 masl (meters above sea level). The Upper Santa Cruz Basin includes lower-elevation plateaus to the east at ca. 200 masl and high plateaus and mountains to the west and south, with altitudes between 500 and 3000 masl (Figs. 14.1 and 14.2). The climate is continental and is strongly influenced by the Pacific anticyclone. It is cold and wet in the west, with a transition to an arid climate in the steppe eastward. Precipitation over this region is mostly produced by disturbances embedded in the westerly flow and is strongly modified by the austral Andes. Uplift on the windward side leads to hyper-humid conditions along the Pacific coast and the western slope of the Andes; in contrast, downslope subsidence dries the eastern plains leading to arid, highly evaporative conditions (Garreaud et al. 2013).

The first evidence of humans in the area dates to ca. 9700 ^{14}C yr BP (ca. 10,800–11,200 cal BP) (Franco and Borrero 2003), while the most recent evidence of hunter-gatherers dates to ca. 350 ^{14}C yr BP (ca. 300–450 cal BP). During this long period, there were changes in precipitation and temperature, including periods of extreme cold (e.g., Mercer and Ager 1983; Pendall et al. 2001; Mancini 2002, 2009; Glasser et al. 2004; Markgraf and Huber 2010). Evidence of human occupation is sparse until ca. 4300 ^{14}C yr BP (ca. 4650–4870 cal BP) and variable afterward.

Fig. 14.1 The Santa Cruz River Basin showing the locations of Lake Argentino, Cerro Frías, and Laguna Cháltel

Fig. 14.2 Map showing the locations of sites mentioned in the text. AB, Alero del Bosque; ChM, Chorrillo Malo 2; CVk1, Cerro Verlika 1; Al, Alice 1 and 2; LN, Laguna Nimez; Huy, Huyliche 1; PB, Punta Bonita 2; CL, Campo del Lago; RA, Rincón Amigo; ChF, Charles Fuhr 2; RB1, Río Bote 1 and 2

Results

Period I: 5300–3700 cal BP

Tonello et al. (2009) have shown that the period ca. 5500–3000 cal BP was the longest, more or less continuous period of increased moisture in the Andes during the Holocene. It was accompanied by lower South Atlantic SSTs (sea surface temperatures), more persistent Antarctic sea ice and multiple periods of glacier advance in the SPIF (Hodell et al. 2001; Moreno et al. 2009; Aniya 2013; Strelin et al. 2014) (Fig. 14.3e, h, j). The most abundant evidence of human occupation at ChM2 (Mehl

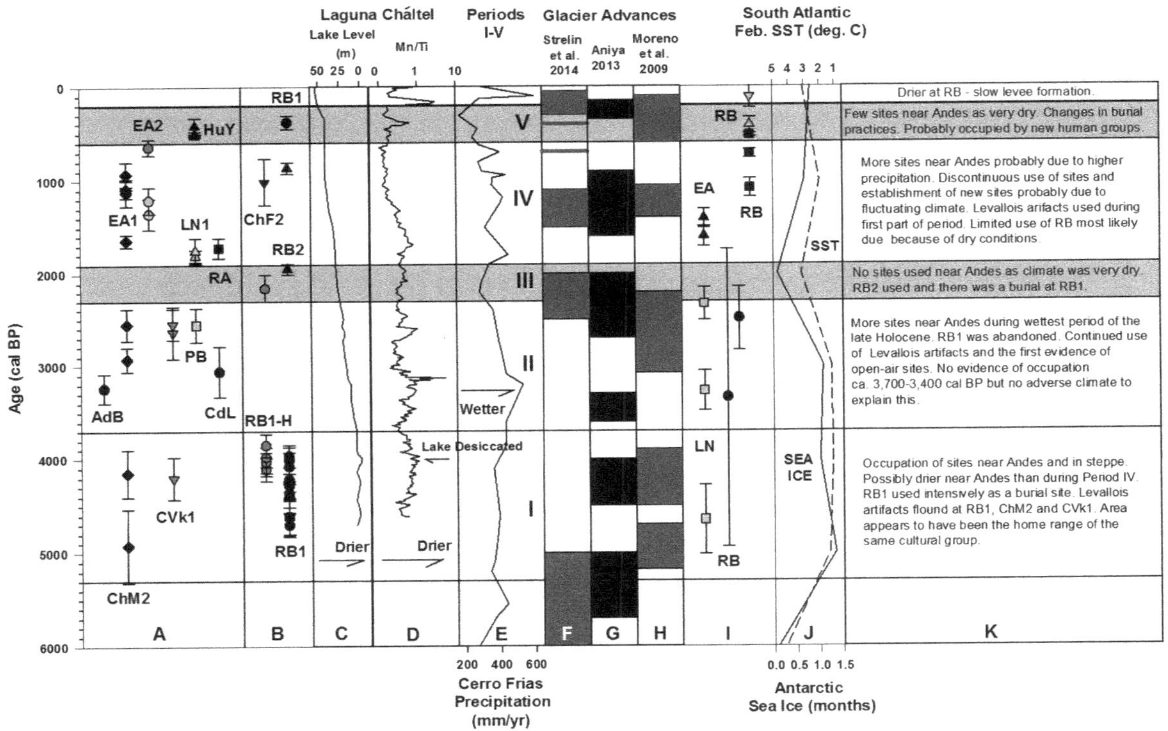

Fig. 14.3 Human occupation of the Upper Santa Cruz River Basin during the Late Holocene and the possible effects of changing environmental conditions. *A* Dates from sites in the Lake Argentino area; *B* dates from the Bote River valley and Charles Fuhr; *C* and *D* lake level and Mn/Ti ratio changes at Laguna Cháltel, Santa Cruz; *E* precipitation model based on pollen data from Cerro Frías (Tonello et al. 2009) and Periods I–V; *F*, *G*, *H* glacier advances in the Lake Argentino area (Strelin et al. 2014) and the SPIF (Moreno et al. 2009; Aniya 2013); *I* OSL and radiocarbon ages for dune and river deposits (*EA* Estancia Alice, *LN* Laguna Nimez, *RB* Rio Bote—*black circles* show OSL ages of RB fluvial deposits indicating dry conditions, *gray squares* and *black triangles* show OSL ages of lakeshore dunes, *black squares* show ages of organic material in buried channels, and gray upright triangle shows the age of buried organics in a cutoff meander and gray inverted triangle the age of the modern levee at 70 cm depth); *J* Antarctic sea ice extent and SST from South Atlantic core TTN057-13-PC4 based on diatom assemblages (after Hodell et al. 2001); *K* major aspects of human occupation and cultural characteristics during Periods I–V. Periods III and V are shaded *gray*, as these were the driest in the Andes in the last 6000 years. Archaeological sites in *A* and *B* are shown from west (*left*) to east (*right*) and abbreviations are given in the text

and Franco 2009) has ages between ca. 5300 and 2300 cal BP (Table 14.1). Pollen indicates dense forest and low fire frequency at CF (Sottile et al. 2012) and grass-shrub steppe at ChM2 (Fig. 14.4) and at Cerro Verlika1 (CVk1) (Mancini 2002). A feldspar OSL age of 4650 ± 370 years BP (UGAOSL Ar10-4) for basal sediments in a coastal dune at Laguna Nimez (LN), Lake Argentino, records low lake levels in the interval between separate periods of glacier advance in Southern Patagonia at ca. 5180–4700 and 4500–3900 cal BP (Moreno et al. 2009; Aniya 2013). Conditions in

Table 14.1 Radiocarbon ages for archaeological sites in the Upper Santa Cruz River Basin. Burial ages are in italics

Archaeological site	Radiocarbon age (^{14}C yr BP)	2σ calibrated age (cal BP)	Laboratory ID or reference
	Period I		
Chorrillo Malo 2	4380 ± 140	4529–5314	LP-2453
Río Bote 1	4200 ± 25	4577–4828	UGAMS-9107
Río Bote 1	4130 ± 25	4444–4810	UGAMS-9104
Río Bote 1	4120 ± 30	4438–4809	UGAMS-9105
Río Bote 1	4100 ± 30	4424–4799	UGAMS-9109
Río Bote 1	4030 ± 25	4319–4566	UGAMS-11064
Río Bote 1	3990 ± 65	4155–4573	UGAMS-9108
Río Bote 1	3980 ± 35	4246–4516	UGAMS-9106
Río Bote 1	3980 ± 25	4250–4514	UGAMS-14195
Río Bote 1	3900 + 25	4157–4410	UGAMS-7534
Río Bote 1	3860 + 25	4091–4400	UGAMS-21776
Cerro Verlika 1	3860 ± 80	3976–4428	Franco et al. (1999)
Río Bote 1	3850 ± 20	4089–4345	UGAMS-14194
Río Bote 1, Individual A	*3800 ± 25*	*3989–4235*	*UGAMS-5916*
Chorrillo Malo 2	3790 ± 80	3895–4405	Franco and Borrero (2003)
Río Bote 1	3768 ± 39	3928–4231	AA-83485
Río Bote 1, Individual G	*3750 ± 25*	*3931–4150*	*UGAMS-5917*
Río Bote 1, Individual J	*3741 ± 54*	*3876–4180*	*Franco* (2008)
Río Bote 1	3727 ± 47	3872–4154	AA-83486
Río Bote 1, Individual E	*3690 ± 25*	*3875–4084*	*UGAMS-7533*
Río Bote 1	3684 ± 39	3847–4086	AA-83487
Río Bote 1, Individual C	*3620 ± 25*	*3729–3977*	*UGAMS-7535*
	Period II		
Alero del Bosque	3110 ± 50	3080–3392	Franco et al. (1999)
Campo del Lago 2	2940 ± 90	2794–3327	Carballo Marina et al. (1999)
Chorrillo Malo 2	2860 ± 35	2798–3058	Franco et al. (2007)
Cerro Verlika 1	2640 ± 110	2356–2923	Franco et al. (1999)
Punta Bonita 2	2540 ± 70	2376–2742	Carballo Marina et al. (1999)
Cerro Verlika 1	2520 ± 20	2381–2721	UGAMS-15764
Chorrillo Malo 2	2525 ± 35	2379–2726	Franco et al. (2007)

(continued)

Table 14.1 (continued)

Archaeological site	Radiocarbon age (^{14}C yr BP)	2σ calibrated age (cal BP)	Laboratory ID or reference
	Period III		
Río Bote 1, Individual B	*2174 ± 43*	*2007–2305*	*AA-83484*
Río Bote 2	2030 ± 25	1891–2003	UGAMS-9110
	Period IV		
Laguna Nimez	1910 ± 20	1738-1873	UGAMS-15767
Laguna Nimez	1877 ± 47	1614–1888	Franco (2008)
Rincón Amigo	1840 ± 40	1610–1827	Beta-138991
Chorrillo Malo 2	1775 ± 30	1571–1713	Ua-32921
Cerro Verlika 1	1685 ± 70	1378–1705	Franco (2002)
Alice 1	1480 ± 70	1186–1520	Borrero et al. (1998–1999)
Alice 1	1370 ± 70	1072–1349	Borrero et al. (1998–1999)
Chorrillo Malo 2	1250 ± 60	979–1269	LP-2463
Chorrillo Malo 2	1240 ± 25	1000–1182	Mehl and Franco (2009)
Charles Fuhr 2	1120 ± 110	768–1264	Carballo Marina et al. (1999)
Chorrillo Malo 2	1070 ± 60	797–1058	LP-2465
Río Bote 2	1010 ± 25	801–927	UGAMS-9114
	Period V		
Alice 2	740 ± 60	557-725	Franco et al. (2004)
Huyliche 1, Individual 3	*520 ± 20*	*500–535*	*UGAMS-14192*
Huyliche 1, Individual 5	*480 ± 20*	*475–522*	*UGAMS-14193*
Huyliche 1, Individual 1	*430 ± 25*	*330–505*	*UGAMS-7536*
Río Bote 1	350 ± 20	309–451	UGAMS-7539

Fig. 14.4 The Chorrillo Malo 2 archaeological site

the Andes were relatively wet throughout Period I but may have been a little drier ca. 4500 cal BP, as suggested by lower Mn/Ti values at Laguna Cháltel, indicating slightly wetter conditions in the steppe and therefore somewhat drier conditions in the Andes (Fig. 14.3d). Pollen records from extra-Andean Patagonia show vegetation changes related to greater environmental heterogeneity, with the increase in shrubs and low grass values indicating lower moisture availability. Reduced moisture levels are also apparent in sediments at Laguna Cháltel in Santa Cruz where carbonate deposits dating to ca. 4040 cal BP record desiccation of the lake basin (Ohlendorf et al. 2014). An ash deposit in sediments at the RB1 archaeological site (Fig. 14.5), from an eruption of the Aguilera volcano in Chile (Stern, 2008, "personal communication"), bracketed by ages of 4030 ± 25 and 3860 ± 25 ^{14}C yr BP (4319–4566 to 4091–4400 cal BP), may have affected humans in the area.

There is sparse evidence of human occupation in the first several hundred years of Period I despite higher precipitation in the Andes. However, after ca 4300 ^{14}C yr BP (ca. 4650–4870 cal BP), evidence of occupation in rock-shelters Río Bote 1 (RB1) in the steppe, and Chorrillo Malo 2 (CHM2) in the Andean foothills ecotone, is more abundant (Table 14.1). At higher elevation, a grass-shrub steppe developed

Fig. 14.5 The Río Bote 1 sediment sequence when the site was first discovered. The white volcanic ash deposit in the upper part of the sequence is cut by the cultural pits, which contained most of the human burials. The *black line* indicates the edge of the deepest excavated pit

at Cerro Verlika 1 (CVk1), located in the Baguales Range, at ca. 1100 masl, between ca. 4500 and 3600 ^{14}C yr BP (4900–3700 cal BP). Toward 3600 ^{14}C yr BP (ca. 3850 cal BP), the vegetation changed to a shrub steppe indicating less moisture. In fact, the first evidence of human presence in the highlands (ca. 3680 ^{14}C yr BP or ca. 3900–4400 cal BP) dates to this relatively dry interval (Franco 2002). This may be related to the fact that highlands and lowlands provide different seasonal opportunities for hunter-gatherers, as guanacos (*Lama guanicoe*), the main prey, move into the highlands during summer, where there is better grazing available.

Although time resolution is low, lowland sites begin to have artifacts made of raw materials such as opals and sedimentary varieties of chalcedony, probably coming from the Baguales Range (Franco 2004b; Franco et al. 2007), as well as burials. At RB1, the burials were multiple; eight individuals—including adults and subadults—were buried at different times between ca. 3800 and 3620 ^{14}C yr BP (ca 4000–3700 cal BP), in some cases, with burial goods (Franco et al. 2011). According to Moraga et al. (2009), two of the skeletons at RB1 belong to haplogroup B. The ages of the burials are close to the age of the Aguilera eruption but also close to the time when Laguna Cháltel in the steppe was desiccated (Fig. 14.3d). It is possible that ash from the eruption only made things worse for hunter-gatherers in the steppe, who were coping already with drought conditions. The regular utilization of the same burial place through time would indicate that its location and significance were passed from one generation to another. Similarities between these burials and others of about the same age near the Magellan Strait suggest regular social contact between peoples of these areas (Franco et al. 2011).

Levallois-flaked artifacts (Boëda 1993), including cores, knives, and sidescrapers, are found for the first time at ChM2 and RB1. Artifacts made using this flaking method were found at RB1 in deposits of the same age as the multiple burials at the site. The Levallois flaking method was identified in cores, knives, and sidescrapers in locally available dacite and basalt. Levallois-flaked artifacts were also identified in the forest area, at Alero del Bosque, in deposits dated a little older than 3300 cal BP. Other flaking methods were used at the same time. Partial data from Chorrillo Malo 2 indicate that, in the case of cores and knives, only 25% were flaked using this method, both on green dacite and basalt (Figs. 14.6 and 14.7). The same flaking method was also recognized in a knife made of gray dacite recovered from Alero del Bosque, which was probably introduced to the site already manufactured. No cores produced by this method were recovered from this site.

Levallois-flaked artifacts were probably also used at Cerro Verlika 1 (CVk1) between ca. 3800 and 2500 ^{14}C yr BP (4000–2700 cal BP). The analysis of lithic artifact frequencies, artifact classes, and characteristics—size and discard tool angle—suggests direct acquisition of raw materials available in highland areas and transport into the lowlands (Franco 2004a). Use of raw materials, most likely from the Baguales Range, suggests that the area was part of the home range of a single population (Franco 2004b). The Levallois flaking method was also used at Cerro Castillo, south of the Baguales Range at about the same time, although other flaking methods were also used (Langlais and Morello 2009).

Fig. 14.6 Knife made from a Levallois flake

Fig. 14.7 Levallois core

Period II: 3700–2300 cal BP

To date we have found no evidence of human presence near Lake Argentino or in the Bote River valley during the first ca. 300 years of Period II, and there is no indication of a climatic extreme that might explain this (Fig. 14.3). There was low fire activity at CF (Sottile et al. 2012), and the highest *Nothofagus* pollen and precipitation values of the entire Holocene at ca. 3500 cal BP (Tonello et al. 2009) are coincident with a period of glacier advance ca. 3600–3300 cal BP (Aniya 2013). The abandonment of RB1 during Period II coincides with a significant, prolonged increase in modeled precipitation at CF (360–530 mm/year) (Fig. 14.3e). Based on Laguna Cháltel Mn/Ti ratios, this increase in precipitation in the Andes was matched by extremely dry conditions in the steppe. However, we don't know the reasons for the abandonment of RB1, as this could also be related to cultural practices due to the death of relatives, as was observed among groups living in Patagonia at the time of European contact. After peak precipitation in the Andean region early in Period II (Tonello et al. 2009), palynological data show that conditions became progressively drier (CF, ChM2, and CVk1).

Feldspar OSL ages of 3270 ± 210 years BP (UGAOSL Ar10–1) and 2330 ± 180 years BP (UGAOSL Ar10-2) for dune sediments at Laguna Nimez (LN) are evidence of strong winds and low lake conditions near the end of Period II, possibly when Mn/Ti values in Laguna Cháltel sediments were low (indicating dry conditions at CF) at ca. 3400 and 3000 cal BP and ca. 2600–2400 cal BP (Fig. 14.5d). Both dune ages fall between periods of glacier advance in the SPIF at ca. 3600–3300 and 3100–2000 cal BP and, considering the age uncertainties, could date within dry intervals in the CF precipitation record (Fig. 14.3e–i). The dune OSL ages also correlate with evidence of increased fire frequency, and with peaks in charcoal centered at ca. 3100 and 2500 cal BP (Sottile et al. 2012). Quartz OSL ages of 3340 ± 1600 years BP (UGAOSL Ar11-2) and 2490 ± 350 years BP (UGAOSL Ar11-3) for a major fluvial sediment deposit upstream of RB1 are evidence of drier conditions in the steppe. The very large age uncertainties make correlations with other data sets difficult, but it is possible that they correspond with dry periods at Laguna Cháltel at 3500 and 3200 cal BP and ca. 2600 cal BP indicated by higher Mn/Ti values (Fig. 14.3d, i). At these times the volume of loose sediment in the Bote River valley was far in excess of what could be transported by the stream, and so much of it was stored in the valley floor.

After ca. 3500 cal BP, there was an increase in the number of sites utilized on the eastern slopes of the Andes and the first evidence of utilization of open-air sites (Campo del Lago 2) (Franco et al. 1999). Some sites continued to be used (ChM2, CVk1) but as mentioned earlier RB1 was abandoned. Levallois-flaked tools continue to be present, and in the case of Chorrillo Malo 2, they account for 30% of the knives recovered from these archaeological deposits. There is also a continuity in the utilization of siliceous raw materials for end scraper manufacture.

Increasingly dry conditions in the Andes toward the end of Period II may have led human populations to explore or move into other areas or to strengthen ties with nearby areas. Levallois-flaked artifacts dating to the end of this period, found on the Atlantic coast at Puesto Viejo, 300 km away, may be evidence of this process.

Period III: 2300–1900 cal BP

We have not recovered evidence of human occupation in the Lake Argentino area during what appears to have been a very prominent 400-year dry period in the Andes. Aniya (2013) and Strelin et al. (2014) suggest glacier advances during the first 300 years of this period (Fig. 14.3f, g), but there is no evidence of wetter conditions or human settlement in western areas at this time. In contrast, Moreno et al. (2009) suggest that the period of glacier advances late in Period II ended around 2200 cal BP. We think this likely also, because southern Atlantic Ocean SSTs were high and Antarctic sea ice significantly reduced during Period III, and these conditions are normally associated with less precipitation in the Andes because of more southerly and weaker SWW (Fig. 14.3j; Hodell et al. 2001). In addition, pollen evidence for the highlands (CVk1) indicates drier conditions with increased fire frequency at CF after ca. 2000 cal BP and shrub steppe indicating drier conditions at ChM2. A new site in the steppe, Rio Bote 2 (RB2), located close to RB1, was used occasionally around 1891–2003 cal BP, and there was a single burial, of Individual B, at RB1 at 2007–2305 cal BP (Franco et al. 2010a), possibly related to increased precipitation in the steppe as indicated by low Mn/Ti values in sediments at Laguna Cháltel (Fig. 14.3d; Ohlendorf et al. 2014). Compared with the multiple burials of Period I, this isolated burial of an individual also belonging to haplogroup B (Moraga et al. 2009) implies a change in the way humans were burying their dead. This change could be related to a change in burial practices, and/or it could signal an end to the conditions that caused the deaths of several individuals over a short time interval during Period I. The use of RB1 again as a burial place, despite no evidence of use during the 1400 years of Period II, suggests cultural continuity because clearly humans knew the location of the site.

Period IV: 1900–600 cal BP

The precipitation model for CF and the Laguna Cháltel Mn/Ti moisture record are inversely correlated during Period IV. The CF record shows high precipitation during the first ca. 300 years, while at Laguna Cháltel higher Mn/Ti ratios suggest slightly drier conditions. Precipitation at CF then declined for the next ca. 300 years with the opposite occurring at Laguna Cháltel. After ca. 1300 cal BP, there is no clear relationship between the CF and Laguna Cháltel records. The evidence from CF shows a fluctuating climate with a precipitation peak centered at ca. 1200 cal BP, while the Laguna Cháltel record shows continued high lake levels (Fig. 14.5c–e). Given evidence of a major period of glacier advance from ca. 1600–900 cal BP, we feel that the CF record is, in this case, more reliable than the Laguna Cháltel data, suggesting that there was probably less moisture on the steppe ca. 1300–1200 cal BP. After the peak in precipitation at CF, the climate becomes progressively drier toward the end of the period, while the level of Laguna Cháltel continues to rise.

Some rock-shelters in both the highlands and lowlands continue to be used during Period IV, but there is less evidence of occupation, and the use of open-air sites seems to be discontinuous. Although the number of ages is still modest, the distribution of ages for sites near Lake Argentino suggests two main periods of occupation at ca. 1900–1600 and 1500–900 cal BP during marked wet intervals at CF, the latter associated with glacier advances in the mountains. Levallois-flaked artifacts dating to the first of these wet periods were found at LN1, a site located on a lakeshore dune of Lake Argentino dating to ca. 1888–1614 cal BP. They were found with hearths and guanaco bones with cut marks but were not present in younger deposits at other sites (Fig. 14.5; Table 14.1).

At Estancia Alice (EA) on the south shore of Lake Argentino, feldspar OSL ages of 1410 ± 100 years BP (UGAOSL Ar10-5) and 1600 ± 110 years BP (UGAOSL Ar10-3) for dune sand over lake sediments date to a period of low precipitation at CF centered on 1500 cal BP. This period of dune development corresponds with a peak in fire frequency at CF centered on 1400 cal BP (Sottile et al. 2012). At Estancia Alice 1 (EA1) in the same area, guanaco bones with cut marks and lithic artifacts dating to 1520–1186 and 1349–1072 cal BP were recovered from a mollisol in aeolian deposits, indicating less windy and relatively moist conditions that limited sand movement (Favier Dubois 2003). The formation of the mollisol corresponds with a period of higher precipitation and minimum fire occurrence at CF, which lasted from ca. 1400–1000 cal BP, and with a period of glacier advances lasting from ca. 1600–900 cal BP. Importantly, this is also a period with evidence of human occupation dating from ca. 1520–1072 cal BP (Fig.14.5; Table 14.1).

At EA, dry, windy conditions followed the wetter conditions that produced the mollisol, burying it with aeolian sand. In fact, a low level of Lake Argentino is recorded by dead, submerged trees in Catalan Bay. The outer layers of one tree date the lake transgression that killed the trees to 899–724 cal BP (AD 1051–1226). As some dead trees were more than a century old when they died, the low lake stand that allowed their growth must have begun before ca. 900 cal BP (Stine 1994). Reconstructed precipitation for CF documents a short but marked dry period, about 200 years long, centered at about 900 cal BP. This was followed by a sharp increase in precipitation around 750 cal BP at the approximate time of the lake transgression (Stine 1994; Tonello et al. 2009). These observations suggest that in the Andes, the Medieval Climate Anomaly (MCA), dating approximately to 1000–700 cal BP, was relatively dry at CF until after about 800 cal BP but became wetter than today at ca. 775–650 cal BP (Tonello et al. 2009; Fig. 14.5e). The pollen record for ChM2 also indicates a dry period between 1200 and 800 cal BP. Archaeological data show that humans occupied EA2 from ca. 725–557 cal BP during this late MCA wet period (Table 14.1). This site is unique in the area as mostly guanaco skulls and mandibles with cut marks were recovered from it, suggesting that its utilization was probably not related to subsistence activities.

Just east of Lake Argentino, pollen from Charles Fuhr 2 (ChF2) records a change to grass steppe and increased moisture (Mancini 2002; Fig. 14.2), which agrees with evidence from Laguna Cháltel of steadily increasing moisture during Period IV. In the Bote River valley, stream discharge varied over time, presumably reflecting the variability in precipitation at CF and therefore in the steppe. From 1182–501 cal BP (AD 768–1449), sediment deposition on the floodplain near RB1 was periodically interrupted by significant floods that cut channels across the floodplain that were later buried by sand. Organic matter from the buried channels defines two periods of flooding, the first prior to 1182–984 cal BP (AD 768–966) and the second before 767–680 cal BP (AD 1183–1270). This pattern of sediment accumulation and erosion suggests higher levels of moisture than during Period II, but it also indicates a fluctuating climate. To date, we have found little evidence that the Bote River valley was occupied, even though there is evidence of increasing wetness in the steppe during Period IV (Fig. 14.5c, d). However, there is some evidence for use of the RB2 rock-shelter at 927–801 cal BP, at the time of a major dry interval at CF ca. 900 cal BP and wetter conditions at Laguna Cháltel.

The frequent and significant climate fluctuations that characterized Period IV may be the reason why humans occupied so many different sites in the area. These changes may also have enabled them to continue the exploration of nearby areas or even move into them when the climate was favorable. Levallois-flaked tools were used in the area only during the first part of this period, but they were identified in younger deposits outside of the area, at the mouth of the Santa Cruz River, at Cañadón de los Mejillones (ca. 1220 ^{14}C yr BP, AD 790–970; cf. Franco et al. 2010b). During this time period, there is also some evidence of interaction with peoples in other areas, such as the utilization of green obsidian from Otway Sound, recovered at ChF2 (Franco 2004a).

Evidence of increased flooding of the Bote River area before ca. 542–328 cal BP (AD 1408–1622) suggests higher and/or more intense rainfall or snowmelt in the steppe region at a time of significantly reduced precipitation at CF (Tonello et al. 2009).

Period V: 600–200 cal BP

This period corresponds to the LIA in Patagonia, and overall it was significantly drier than today in the Lake Argentino area and moister in the steppe. Increased moisture in the steppe is indicated by pollen spectra with higher levels of grass pollen for sites like ChF2 (Mancini 2002) and by deep water and low Mn/Ti ratios at Laguna Cháltel (Ohlendorf et al. 2014). Slightly wetter conditions in the Andes and drier conditions in the steppe at ca. 450 cal BP correlate with evidence of a minor glacier advance at this time (Fig. 14.5e, f). An age of 31–257 cal BP at 70 cm depth in the modern levee of the Bote River records modest floods with vertical accretion over the last ca. 200 years when precipitation at CF increased suggesting drier conditions at RB1, as confirmed by higher Mn/Ti values at Laguna Cháltel (Fig. 14.7d, e, i). This pattern may be related to evidence presented by Aniya (2013) and Strelin et al. (2014) of glacier advances in Southern Patagonia beginning ca. 400/300 cal BP, near the very end of Period V, as this would have increased precipitation near Lake Argentino but reduced it in the steppe to the east. Moreno et al. (2009) suggest glacier advances throughout the LIA, but the later ages of Aniya (2013) and Strelin et al. (2014) are more in agreement with the lake and pollen data.

There is much less evidence of human occupation in western spaces during the LIA. Not only are there fewer sites but the sites that are known have yielded fewer artifacts. The youngest deposits at ChM2 and CVk1, which have very few remains and have not been dated, may belong to this interval. Data from Huyliche 1, a multiple burial site, suggest a change in burial inhumation practices from use of rockshelters to open-air sites, where human remains were covered with rocks, locally called "chenques" (cf. Franco et al. 2010a). Tabular deformation practices were recorded, and initial data suggest morphological differences in the individuals compared to earlier periods (García Guráieb 2010, "personal communication"). Genetic studies are under way. Lithic artifacts include gray-green banded obsidian, which according to geological and archaeological evidence available today comes from the Baguales Range (Stern and Franco 2000). The length of one of the bone tools recovered suggests it was made from the bone of a marine animal (Fernández, P. 2013, "personal communication"), which would imply contacts with easterly or, more likely, southerly spaces.

Raw materials used for artifacts come mostly from the immediate or, in some cases, local area, which, following Meltzer's (1989) review of ethnographical information, implies distances of less than 40 km. The lack of raw materials from more distant sources could reflect the small numbers of artifacts recovered. Most raw materials are of only good or average quality. There are few tools, most of them unbroken; end scrapers were only recovered from ChM2. This body of information along with the existence of few sites, one a burial, suggests that hunter-gatherer groups were not staying for long periods in the area. Changes in technology can be related to a change in the way an area is used or to its utilization by new human groups. The observed changes in burial practices, as well as the introduction of new deformation practices, support the latter hypothesis, but additional information will be needed to test this.

Discussion and Conclusions

A number of studies have suggested that environmental variability influenced hunter-gatherer behavior, especially in desert environments (e.g., Smith et al. 2005; Veth 2005). Annual precipitation in the Patagonian steppe is quite low today, usually below 200 mm/year, and there were very few sources of year-round freshwater during the Holocene (Mayr et al. 2007). For these reasons it has been suggested that in Patagonia, water was a crucial resource that influenced human mobility and settlement significantly (e.g., Borrero and Franco 2000; Goñi et al. 2000–2002).

Although the forces driving changes in climate during the Late Holocene were certainly complex, variations in Antarctic sea ice and SSTs in the South Atlantic that mirror the CF precipitation record were clearly influential as these increased/decreased temperatures and latitudinal temperature gradients in southern South America and thus influenced the position and intensity of the SWW that are so influential in bringing precipitation to southwest Patagonia (Fig. 14.5j; Hodell et al. 2001).

In this analysis we have shown a relationship between human use of space in the Upper Santa Cruz River Basin and long-term trends in precipitation. This relationship is emphasized by the inverse correlation between precipitation near the Andes and that in the steppe further east.

In the Upper Santa Cruz River Basin, the onset of the neoglacial ca. 5000 cal BP saw a significant increase in precipitation in the Andes. The precipitation model for CF suggests that Period I was relatively wet near the Andes with drier periods centered at ca. 5200 and 4100 cal BP. Evidence of human occupation is sparse until ca. 4800 cal BP, when artifact technology and raw materials show that different spaces near the Andes (ChM2) and in the steppe (RB1) were used by the same cultural group, who also buried their dead (multiple individuals) repeatedly at the same site. The initial occupation of the forest probably dates to the end of this period. A similar flaking method was used at least in the lowlands in these different spaces. Highland sites were also used during Period I suggesting that climatic conditions did not limit human use of either the mountains or the steppe for long intervals, perhaps because conditions were never too dry throughout the region or because humans chose to bury their dead in places that were not the best for human occupation (see Guichón et al.2001).

The abandonment of RB1 at the end of Period I coincides with a significant, prolonged increase in modeled precipitation at CF near the Andes starting ca. 4100 cal BP in Period I and continuing until ca. 3200 cal BP in Period II (Fig. 14.5e). This increase in precipitation in the Andes is matched by a marked reduction in precipitation at Laguna Cháltel and at RB1 in the steppe, but the reasons for the abandonment of RB1 as a burial place are not clear and could be linked to cultural practices.

Period II (3700–2300 cal BP) was the wettest near the Andes during the Late Holocene. Some sites in western spaces, like ChM2 and CVk1, continued to be occupied, while an increasing number of new sites—including open-air ones—were used over time. There is a technological continuity with Period I, suggesting that at least western spaces were used by the same cultural group.

During Period III, which was much drier at CF, there is no evidence of human occupation of westerly spaces, but RB2 and RB1 in the steppe were used. The reuse of RB1 as a burial place would imply that human populations knew its function. However, in this case, we are dealing with an isolated human burial, suggesting a change in burial techniques and/or the end of the conditions that led to multiple deaths during a brief interval of Period I.

Climate during Period IV was extremely variable, fluctuating between drier and wetter, with evidence of human presence near the Andes less common during the dry intervals. Early in Period IV, there is evidence of Levallois-flaked artifacts, probably implying that the same population was still using the area. There is no evidence of such artifacts after ca. 1900 cal BP. The lack of sites during drier periods may be related to a more ephemeral occupation of sites, which probably left a very small archaeological signal. The persistence of climatic variability over a prolonged period may also have changed how the Lake Argentino area was used or may even have led to the abandonment of the area.

During Period V, the LIA, it is possible that people moved back into previously occupied spaces and sites or that new hunter-gatherer groups moved into the Lake Argentino area. Changes in technology and burial practices seem to suggest the second option, but further research is needed to confirm this.

Acknowledgments Funding was provided by UBACyT 20020130100664BA and W1/0404, National Geographic Society 8397-08, and Heinz Latin-American (2005) grants. The Universities of Georgia and Arizona assisted with dating. We wish to thank the reviewers, whose comments helped in improving the manuscript. We are grateful to the owners and workers of the Chorrillo Malo and Río Bote ranches, as well as to local inhabitants, particularly Gerardo Povazsan, for their help and support, and to all those who helped us in the field. Assistance was also provided by the Asociación de Guías de Calafate and Mil Outdoor. Austral Construcciones helped with the construction of the scaffold necessary to excavate Rio Bote 1. We also wish to thank Arq. Silvia Mirelman, Dirección de Patrimonio de la provincia de Santa Cruz, and Dirección de Cultura de Calafate.

References

Aniya, M. (2013). Holocene glaciations of Hielo Patagónico (Patagonia Icefield), South America: A brief review. *Geochemical Journal, 47*, 97–105.

Boëda, E. (1993). Le débitage discoid et le débitage Levallois recurrent centripete. *Bulletin de la Societé Préhistorique Francaise, 90*(5), 392–404.

Borrero, L. A. (1994–95). Arqueología de la Patagonia. *Palimpsesto. Revista de Arqueología, 4*, 9–69.

Borrero, L. A., & Carballo Marina, F. (1998). Proyecto Magallania: La cuenca superior del río Santa Cruz. In L. Borrero (Ed.), *Arqueología de la Patagonia Meridional (Proyecto "Magallania")* (pp. 11–27). Concepción del Uruguay: Ed. Búsqueda de Ayllu.

Borrero, L. A., & Franco, N. V. (2000). Cuenca superior del río Santa Cruz: Perspectivas temporales. In J. B. Belardi, F. Carballo Marina & S. Espinosa (Eds.), *Desde el país de los gigantes. Perspectivas arqueológicas en Patagonia* (Vol. II, pp. 345–356). Río Gallegos: Unidad Académica Río Gallegos de la Universidad Nacional de la Patagonia Austral.

Borrero, L. A., Franco, N. V., Carballo Marina, F., & Martin, F. M. (1998–1999). Arqueología de Estancia Alice. *Lake Argentino. Cuadernos del Instituto de Antropología y Pensamiento Latinoamericano, 18*, 31–48.

Bousman, C. B. (1993). Hunter-gatherer adaptations, economic risk and tool design. *Lithic Technology, 18*(1/2), 59–86.

Carballo Marina, F. M., Borrero, L. A., Franco, N. V., Belardi, J. B., Horwitz, V. D., Muñoz, A. S., Campan, P., Martin, F. M., Borella, F., García, M. F., Muñoz, F., Savanti, F., & Lanata, J. L. (1999). Arqueología de la costa de Lake Argentino, río La Leona y pampas altas intermedias. *Praehistoria, 3*, 13–33.

Ericson, J. E. (1984). Toward the analysis of lithic reduction systems. In E. Ericson & B. Purdy (Eds.), *Prehistoric quarries and lithic production* (pp. 11–22). Cambridge: Cambridge University Press.

Favier Dubois, C. M. (2003). Late Holocene climatic fluctuations and soil genesis in southern Patagonia: Effects on the archaeological record. *Journal of Archaeological Science, 30*, 1657–1664.

Franco, N. V. (2002). *Estrategias de utilización de recursos líticos en la cuenca superior del río Santa Cruz*. Buenos Aires: Unpublished PhD dissertation. University of Buenos Aires.

Franco, N. V. (2004a). La organización tecnológica y el uso de escalas espaciales amplias. El caso del sur y oeste de Lake Argentino. In A. Acosta, D. Loponte, & M. Ramos (Eds.), *Temas de Arqueología, Análisis Lítico* (pp. 101–144). Luján: Universidad Nacional de Luján.

Franco, N. V. (2004b). Rangos de acción, materias primas y núcleos preparados al sur de Lake Argentino. In M. T. Civalero, P. Fernández, & A. G. Guráieb (Eds.), *Contra Viento y Marea. Arqueología de la Patagonia* (pp. 105–116). Buenos Aires: Instituto Nacional de Antropología y Pensamiento Latinoamericano and Sociedad Argentina de Antropología.

Franco, N. V. (2008). La estructura tecnológica regional y la comprensión de la movilidad humana: Tendencias para la cuenca del río Santa Cruz. In L. A. Borrero & N. V. Franco (Eds.), *Arqueología del extremo sur del continente americano. Resultados de nuevos proyectos* (pp. 119–154). Buenos Aires: Instituto Multidisciplinario de Historia y Ciencias Humanas (CONICET).

Franco, N. V. (2012). Binford and ethnoarchaeology, a view from the South. Reflections on his contributions to hunter-gatherer archaeology and lithic analysis. *Ethnoarchaeology Journal of Archaeological, Ethnographic, and Experimental Studies, 4*(1), 79–99.

Franco, N. V., & Aragón, E. (2004). Variabilidad en fuentes secundarias de aprovisionamiento lítico: El caso del sur de Lake Argentino (Santa Cruz, Argentina). *Estudios Atacameños, 28*, 71–85.

Franco, N. V., & Borrero, L. A. (2003). Chorrillo Malo 2: Initial peopling of the upper Santa Cruz Basin. In R. Bonnichsen, L. Miotti, M. Salemme, & N. Flegenheimer (Eds.), *Where the south winds blow. Ancient evidence of Paleo South Americans* (pp. 149–152). College Station: Texas A&M University Press, Center for Studies of the First Americans.

Franco, N. V., Borrero, L. A., Belardi, J. B., Carballo Marina, F., Martin, F. M., Campan, P., Favier Dubois, C., Stadler, N., Hernández, M. I., Cepeda, H., Muñoz, A. S., Borella, F., Muñoz, F., & Cruz, I. (1999). Arqueología del Cordón Baguales y sistema lacustre al sur de Lago Argentino. *Praehistoria, 3*, 65–86.

Franco, N. V., Borrero, L. A., & Mancini, M. V. (2004). Environmental changes and hunter-gatherers in southern Patagonia: Lago Argentino and Cabo Vírgenes (Argentina). *Before Farming: the Archaeology and Anthropology of Hunter-gatherers, 2004*(3), 1–17.

Franco, N. V., Mehl, A., & Otaola, C. (2007). Chorrillo Malo 2 (Upper Santa Cruz Basin, Patagonia, Argentina): New data on its stratigraphic sequence. In M. Kornfeld, S. Vasil'ev & L. Miotti (Eds.), *Proceedings of the XV World Congress (Lisboa, 4–9 September 2006)*, Vol. 14, Session C54 (pp. 141–145). Oxford: Archaeopress British Archaeological Reports S1655.

Franco, N. V., Guarido, A. L., García Guráieb, S., Martucci, M., & Ocampo, M. (2010a). Variabilidad en entierros humanos en la cuenca superior y media del río Santa Cruz (Patagonia, Argentina). In J. Bárcena & H. Chiavazza (Eds.), *Arqueología Argentina en el Bicentenario de la Revolución de Mayo. XVII Congreso Nacional de Arqueología Argentina.*, tomo V, capitulo 35 (pp. 1901–1906). Mendoza: Facultad de Filosofía y Letras (UNCuyo) e Instituto de Ciencias Humanas, Sociales y Ambientales (CONICET).

Franco, N. V., Zubimendi, M. A., Cardillo, M., & Guarido, A. L. (2010b). Relevamiento arqueológico en Cañadón de los Mejillones. *Magallania, 38*(1), 269–280.

Franco, N. V., Cirigliano, N., & Ambrústolo, P. (2011). Semejanzas en tecnologías, diseños y prácticas funerarias al sur de la cuenca superior del río Santa Cruz: algunos ejemplos correspondientes al Holoceno tardío. In L. A. Borrero & K. Borrazzo (Eds.), *Bosques, montañas y cazadores: investigaciones arqueológicas en Patagonia Meridional* (pp. 155–178). Buenos Aires: Instituto Multidisciplinario de Historia y Ciencias Humanas (CONICET).

Garreaud, R. D., Vuille, M., Compagnucci, R., & Marengo, J. (2009). Present-day South American climate. *Palaeogeography, Palaeoclimatology, Palaeoecology Special Issue (LOTRED South America), 281*(3–4), 180–195.

Garreaud, R., Lopez, P., Minvielle, M., & Rojas, M. (2013). Large scale control on the Patagonia climate. *Journal of Climate, 26*, 215–230.

Glasser, N. F., Harrison, S., Winchester, V., & Aniya, M. (2004). Late Pleistocene and Holocene palaeoclimate and glacier fluctuations in Patagonia. *Global and Planetary Change, 43*, 79–101.

Goñi, R., Barrientos, G., & Cassiodoro, G. (2000-2002). Condiciones previas a la extinción de las poblaciones humanas del sur de Patagonia: una discusión a partir del análisis del registro arqueológico de la cuenca del Lago Salitroso. *Cuadernos del Instituto Nacional de Antropología y Pensamiento Latinoamericano, 19*, 249–266.

Guichón, R., Barberena, G., & Borrero, L. A. (2001). ¿Dónde y cómo aparecen los restos óseos humanos en Patagonia austral? *Anales del Instituto de la Patagonia, 29*, 103–117.

Hodell, D. A., Kanfoush, S. L., Shemesh, A., Crosta, X., & Guilderson, T. P. (2001). Abrupt cooling of Antarctic surface waters and sea ice expansion in the South Atlantic sector of the Southern Ocean at 5000 cal yr B.P. *Quaternary Research, 56*, 191–198.

Hogg, A. G., Hua, Q., Blackwell, P. J., Buck, C. E., Guilderson, T. P., Heaton, T. J., Niu, M., Palmer, J. G., Reimer, P. J., Reimer, R. W., Turney, C. S. M., & Zimmerman, S. R. H. (2013). SHCal13: Southern hemisphere calibration, 0-50,000 years cal BP. *Radiocarbon, 55*(4). doi:10.2458/azu_js_rc.55.16783.

Langlais, M., & Morello, F. (2009). Estudio tecno-económico de la industria lítica de Cerro Castillo (Provincia de Última Esperanza, Chile). *Magallania, 37*(1), 61–83.

Luedtke, B. E. (1979). The identification of sources of chert artifacts. *American Antiquity, 44*, 744–756.

Mancini, M. V. (2002). Vegetation and climate during the Holocene in Southwest Patagonia, Argentina. *Review of Paleobotany and Palynology, 122*, 101–115.

Mancini, M. V. (2009). Holocene vegetation and climate changes from a peat record of the forest-steppe ecotone, Southwest of Patagonia (Argentina). *Quaternary Science Reviews, 28*, 1490–1497.

Mancini, M. V., Franco, N. V., & Brook, G. A. (2013). Palaeoenvironment and early human occupation of southernmost South America (South Patagonia, Argentina). *Quaternary International, 299*, 13–22.

Markgraf, V., & Huber, U. M. (2010). Late and postglacial vegetation and fire history in Southern Patagonia and Tierra del Fuego. *Palaeogeography, Palaeoclimatology, Palaeoecology, 97*, 351–366.

Mayr, C., Lücke, A., Stichler, W., Trimborn, P., Ercolano, B., Oliva, G., Ohlendorf, C., Soto, J., Fey, M., Haberzettl, T., Janssen, S., Schabitz, F., Schleser, G. H., Wille, M., & Zolitschka, B. (2007). Precipitation origin and evaporation of lakes in semi-arid Patagonia (Argentina) inferred from stable isotopes (d18O, d2H). *Journal of Hydrology, 334*, 53–63.

Mehl, A., & Franco, N. V. (2009). Cambios en la morfología de los reparos rocosos: el caso de los sitios arqueológicos Chorrillo Malo 2 y Río Bote 1 (pcia. de Santa Cruz, Argentina). In M. Salemme, F. Santiago, M. Álvarez, E. Piana, M. Vazquez & M. Mansur (Eds.), *Arqueología de Patagonia: Una Mirada Desde el Ultimo Confín* (Vol. 2, pp. 893–900). Ushuaia: Editorial Utopias.

Meltzer, D. J. (1989). Was stone exchanged among Eastern North American Paleoindians? In C. J. Ellis & J. Lothrop (Eds.), *Eastern Paleoindian lithic resource use* (pp. 11–39). Boulder: Westview Press.

Mercer, J., & Ager, T. (1983). Glacial and floral changes in South Argentina since 14,000 years ago. *National Geographic Society Research Reports, 15*, 457–477.

Moraga, M., Mena, F., Reyes, O., Barrientos, G., Goñi, R., Franco, N., & Borrero, L. (2009). Linajes mitocondriales fundadores en restos humanos prehistóricos de Patagonia y Tierra del Fuego. In *Actas de las Novenas Jornadas Nacionales de Antropología Biológica* (p. 42). Puerto Madryn: Asociación de Antropología Biológica Argentina.

Moreno, P. I., François, J. P., Villa-Martínez, R. P., & Moy, C. M. (2009). Millennial-scale variability in southern hemisphere westerly wind activity over the last 5000 years in SW Patagonia. *Quaternary Science Reviews, 28*, 25–38.

Moy, C. M., Moreno, P. I., Dunbar, R. B., Kaplan, M. R., Francois, J.-P., Villalba, R., & Haberzettl, T. (2009). Climate change in Southern South America during the last two millennia. In F. Vimeux, F. Sylvestre, & M. Khodri (Eds.), *Past climate variability in South America and surrounding regions, from the last glacial maximum to the Holocene* (pp. 353–394). New York: Springer.

Nelson, M. C. (1991). The study of technological organization. *Archaeological Method and Theory, 3*, 57–100. Tucson: University of Arizona Press.

Ohlendorf, C., Fey, M., Massaferro, J., Haberzettl, T., Laprida, C., Lücke, A., Maidana, N., Mayr, C., Oehlerich, M., Ramón Mercau, J., Wille, M., Corbella, H., St-Onge, G., Schäbitz, F., & Zolitschka, B. (2014). Late Holocene hydrology inferred from lacustrine sediments of Laguna Cháltel (southeastern Argentina). *Palaeogeography, Palaeoclimatology, Palaeoecology, 411*, 229–248.

Pendall, E., Markgraf, V., White, J., & Drier, M. (2001). Multiproxy record of late Pleistocene-Holocene climate and vegetation changes from a peat bog in Patagonia. *Quaternary Research, 55*, 168–178.

Smith, M., Veth, P., Hiscock, P., & Wallis, L. A. (2005). Global deserts in perspective. In P. Veth, M. Smith, & P. Hiscock (Eds.), *Desert peoples, archaeological perspectives* (pp. 1–13). Oxford: Blackwell Publishing Ltd..

Sottile, G. D., Bamonte, F. P., Mancini, M. V., & Bianchi, M. M. (2012). Insights into Holocene vegetation and climate changes at the southeastern side of the Andes: *Nothofagus* Forest and Patagonian Steppe fire records. *The Holocene, 22*(11), 1201–1214.

Stern, C. R., & Franco, N. V. (2000). Obsidiana gris verdosa veteada de la cuenca superior del río Santa Cruz, extremo sur de Patagonia. *Anales del Instituto de la Patagonia, 28*, 265–273.

Stine, S. (1994). Extreme and persistent drought in California and Patagonia during medieval time. *Nature, 369*, 546–549.

Strelin, J. A., Kaplan, M. R., Vandergoes, M. J., Denton, G. H., & Schaefer, J. M. (2014). Holocene glacier history of the Lago Argentino basin, Southern Patagonian Icefield. *Quaternary Science Reviews, 101*, 124–145.

Stuiver, M., & Reimer, P. J. (1993). Extended 14C database and revised CALIB radiocarbon calibration program. *Radiocarbon, 35*, 215–230.

Tonello, M. S., Mancini, M. V., & Seppä, H. (2009). Quantitative reconstruction of Holocene precipitation changes in Southern Patagonia. *Quaternary Research, 72*(3), 410–420.

Veth, P. (2005). Cycles of aridity and human mobility: Risk minimization among late Pleistocene foragers of the Western Desert, Australia. In P. Veth, M. Smith, & P. Hiscock (Eds.), *Desert peoples: Archaeological perspectives* (pp. 100–115). Oxford: Blackwell Publishing Ltd..

Whitlock, C., Moreno, P. I., & Bartlein, P. (2007). Climate controls of Holocene fire patterns in southern South America. *Quaternary Research, 68*, 28–36.

Chapter 15
The Costs and Benefits of Technological Organization: Hunter-Gatherer Lithic Industries and Beyond

Michael J. Shott

Introduction

Technological organization is a set of ideas that archaeologists use to interpret the material record. It is about how the design and character of ancient material cultures, mostly tools and industrial debris, were determined not just narrowly by function or tradition but broadly by the cultural and adaptive context in which that use or tradition was embedded. Contextual factors that technological organization commonly invokes include distribution, abundance and quality of toolstone, settlement mobility broadly defined, and risk reduction. Essentially, technological organization is about having what one needs and no more, having it when and where needed, and minimizing the costs of having while maximizing the benefits of using.

As a research program, that is, technological organization starts from the assumption that ancient people optimized their technologies. Costs include toolstone acquisition and processing, the transport of partly finished and finished tools, and tool maintenance. Costs also include those associated with the use of the resulting tools, including search and pursuit time and other acquisition costs. Risks of toolstone acquisition and processing include the prospect of exhausting tool supplies far from sources of the raw materials needed to replenish them and the probability of failure in production. Risks of tool use include the prospect of tool failure or depletion or ineffectiveness in use and the failure in hunting and other tasks that follow. Benefits sought include high success and acquisition rates, efficiency in use, effort minimization, and perhaps prestige.

Traditionally, technological organization has been considered uniquely archaeological theory. Viewed in a broader perspective, however, it is a special domain of behavioral ecology theory that pertains to the human use of stone and other raw

M.J. Shott (✉)
Department of Anthropology and Classical Studies, University of Akron, Akron, OH, USA
e-mail: shott@uakron.edu

E. Robinson, F. Sellet (eds.), *Lithic Technological Organization and Paleoenvironmental Change*, Studies in Human Ecology and Adaptation 9,
DOI 10.1007/978-3-319-64407-3_15

materials to make and use the things that achieve human goals (Shott 1986; Nelson 1991; Kuhn and Miller 2015). Like behavioral ecology, then, technological organization is a set of ideas about how people solve problems, informed by cost-benefit analysis. Sellet's use of the marginal value theorem, related to the field processing model (Beck et al. 2002; Kuhn and Miller 2015; Shott 2015), illustrates this quality of technological organization and helps locate it within the larger body of behavioral ecology theory.

The value of the technological-organization concept can be gauged in part by its scope and the quality of its applications. As this volume shows, technological organization is a productive interpretive approach for the study of lithic assemblages in a wide range of Paleolithic, Paleoindian, and other cultural contexts. Chapters illustrate its application to a wide range of lithic-analytic domains from toolstone acquisition and processing to strategies for tool production, use, and maintenance. They form a set of case studies of impressive quality and considerable rigor. As a collection, the volume exemplifies the current state of thought and practice in technological organization.

Yet because technological organization is a variant of behavioral ecology, its scope of application in archaeology mirrors behavioral ecology in general. Technological-organizational explanations are rare in the record of later prehistoric complex polities and to materials other than stone. In practice but not in principle, therefore, technological organization is largely confined to the analysis and interpretation of hunter-gatherer lithic assemblages and industries. Illustrating these emphases, chapters cover an impressive range of global Paleolithic contexts and industries accessible to Western archaeologists. Among them, the only conspicuous omission is East African Lower Paleolithic industries. Otherwise, everything from Old-World Middle Paleolithic to New-World Paleoindian is represented, along with some Holocene industries.

Archaeology's reluctance to apply technological organization to other culture types and materials may owe in part to a misapprehension that it somehow concerns stone tools only. There is no reason that technological organization should not pertain just as well to, say, ceramic technology, although examples of its use there are rare. But more than a problem of material, technological organization suffers from the larger misapprehension that such behavioral theory somehow applies only to hunter-gatherer material adaptation as it registers in lithic industries. Yet technological organization can model social constraints in lithic technology (Wiessner 1983; Nelson 1991: 88). Ancient people confronted problems in all cultural contexts, some of which could be constructively reduced to the costs and benefits of various solutions. Yet even passing familiarity with the archaeological literature on complex societies shows that traditions of explanation there differ dramatically. Apparently, archaeologists believe that hunter-gatherers can be understood as are animals except that they made and used sophisticated stone tools and that otherwise only in more complex societies did ancient people become uniquely human.

It goes without saying that hunter-gatherers were and are human (but perhaps bears emphasizing in this context!), no less nor more than anyone else. Everyone alive today can trace ancestry to a hunter-gatherer past that is not distant by either evolutionary or archaeological standards. In their individual perceptions, motives, and actions, of course hunter-gatherers were no less complex than the most manipulative chief or the most aggrandizing emperor. Technological organization and the larger body of behavioral ecology theory from which it derives do not reduce or simplify those human qualities, but are useful explanations when cultural context is taken into account.

Technological Organization's Origin and Development

The technological-organization concept traces back to Binford (Spry and Stern 2016). In two influential papers, Binford (1977, 1979) characteristically did not formally define the concept, instead introducing it by application to the particulars of his cases. In this way, he provided a sense more than a detailed, formal exposition of what technological organization was. Binford used a set of concepts to explain how hunter-gatherers obtained, processed, and used materials and, finally, why tool proportions in discard do not necessarily match proportions in use. Throughout, he located these behaviors in the broader context of land-use scale and mobility and the risks, both opportunity and failure, that confronted tool makers and users. Later Nelson defined technological organization as "the selection and integration of strategies for making, using, transporting, and discarding tools and the materials needed for their manufacture and maintenance" (1991:57).

Binford's development of the technological-organization concept owed, like so much of his thought, to his experiences in interpreting the cause and structure of Mousterian assemblage variation in the Dordogne (Binford and Binford 1966). In contrast to Bordes's view that Mousterian variation was the product of interdigitating technotypological social groups like Quina Mousterian (Clarke 1973:10) Binford, following Thomson (1939), favored a functional view that accommodated much wider and more structured variation in a culture's archaeological assemblages than did Bordes's culture-historical or identity view. That is, to Bordes and other archaeologists steeped in the traditions of mid-twentieth century typological systematics, all significant assemblage variation was normative. Binford and Thomson instead viewed cultures in more varied and systematic, if narrowly adaptive or material, terms. One culture could produce a range of tool and assemblage types depending upon conditions to which it adapted. To this extent, technological organization comprises one facet of processual archaeology's reaction to the manifest shortcomings of culture history.

Technological organization thus arose during a time of ferment and dissatisfaction with traditional approaches to the study of lithic assemblages and forager cultures. Among other things, its consideration of toolstone selection and the full range of reduction processes and products reflected broader trends in archaeology at the time. Whereas previously lithic analysts focused almost exclusively upon finished tools and interpreted their assemblages in culture-historical or crude functional terms, by the 1970s their treatment encompassed entire lithic industries and their by-products as well. Although cores may have cultural or chronological meaning depending upon the context, few types of debitage (if debitage types truly exist) possessed those qualities. Yet the size of and technological character of debitage assemblages could reveal production processes and strategies. Although tools, by their size and form, reflected age and coarse function, their design, functional range and specificity, and patterns and degrees of use also were informative and therefore deserved study. In sum, to view lithic assemblages and industries as integral parts of their encompassing cultures required interpretive frameworks beyond culture history.

Following Binford's lead, the technological-organization concept soon was applied to a range of interpretive problems, most or all involving archaeological study of the adaptations of hunter-gatherers that also concerned Binford and that, as above, has become typical of the approach. Torrence (1983) interpreted the design of high-latitude toolkits as a response to time stress, the availability of critical resources in very short intervals that pay a premium to reliability. Bleed (1986) distinguished reliable from maintainable weapons, although the qualities are not mutually exclusive. Shott (1986) related the number of tool types and their functional range to mobility, particularly mobility magnitude. These contributions concerned the nature and design of only finished tools; Nelson's (1991) early synthesis of technological organization encompassed cores and debitage as well.

The Curation Concept

In initial formulation of a theory of technological organization, among Binford's most important ideas was the concept of curation. Elaborated in greatest detail in his celebrated interpretation of 47 Nunamiut hunting trips, curation may have been described ambiguously, but its importance was clear. Curation played a central role in determining the size, composition, and structure of tool assemblages. With curation, tool "replacement rates are directly proportional to the life span or utility of the tool under maintenance care, and may bear no direct relationship to the frequency of activity performance involving tool use" (1977:34). Essentially, the curation concept is assimilated to assemblage formation via various, often complex, permutations of tool: task mapping relations and tool discard rates (Ammerman and Feldman 1974; Shott 1986:18–19).

Because Binford described the curation concept variously and ambiguously, it remained somewhat protean in meaning. Later curation was redefined as a continuous, quantitative relationship between tools' realized and maximum utility (Shott 1996), a view assimilated in some research traditions today (e.g., Andrefsky 2009:70–71). In this view, curation is a variable like length that pertains to individual tools, not a qualitative state to which assemblages or industries conform. It stands in opposition to ill-defined notions like "expediency" no more than length has an opposite unless we arbitrarily bifurcate it, reducing its inherent continuity to ordinal "shortness" and "longness." Conceiving curation as a variable gives access to methods of tool analysis that can measure continuous degree of variation in the quantity within and between assemblages and identify the failure processes that form those assemblages (e.g., Morales 2016; Shott and Seeman 2017). For its correlation with use life, curation also is crucial to identifying, measuring, and controlling for the dependence between assemblage size and typological richness that must precede behavioral interpretation from assemblage composition (Shott 2003).

But the ambiguity that inhered in Binford's formulation, noted earlier (Shott 1996; Attenbrow 2004:229), persists somewhat in this volume. Here, curation is a qualitative condition opposed to expediency (Riel-Salvatore and Negrino), a state conferred on tools by virtue of being transported (Morisaki et al., Sellet, Wragg Sykes) and/or retouched (Morisaki et al., Riel-Salvatore and Negrino, Wragg Sykes), or a property of assemblages rather than tools (Riel-Salvatore and Negrino). It is a condition to identify, not a variable to measure. Yet Morisaki et al. and Suarez also perceive curation as a continuous property of tools, suggesting some convergence between the research traditions represented here and the conception of curation as continuous variable.

However valuable as an analytical concept, curation also can be viewed as a variable in a technological-organizational perspective. For instance, it is a way to manage toolstone supply relative to technological requirements. Because curation rate also varies with tool-task mapping relations, it is subject to manipulation in response to organizational constraints like mobility magnitude and the transport costs it imposes (Shott 1986). Because curation and behavioral ecology models share the concept of utility, stone tools themselves can be treated as, for instance, resource patches whose kind and particularly degree of use can be modeled as optimization problems (e.g., Kuhn 1994; Surovell 2009: 142–170; Clarkson et al. 2015:124–128; Kuhn and Miller 2015: 178–184). Curation is at once a concept that links archaeological formation theory to general theory and a variable at the disposal of tool users as they manage the design of their reduction processes, their tools relative to the range of tasks they must perform subject to sometimes rigorous performance criteria, and the discard rates of those tools.

Curation also bears upon the measurement of occupational intensity, essentially by artifact density, in several chapters. Directly, artifact density measures discard rate, not length of occupation or size of the occupying group. Discard rate itself depends upon curation rate and also, in archaeological deposits, sedimentation rates. All else equal, density and apparent occupational intensity can rise merely with a corresponding decline in curation (or vice versa). Therefore, a quantity used in modeling technological organization—occupational intensity—itself is a variable that depends upon organizational parameters like curation rate. It is as much a consequence of technological organization as it is a constraint upon it.

Problems

Technological organization arose indirectly in response to dissatisfaction with the culture-historical paradigm's emphasis upon the construction of static, normative taxa and its sterile cataloguing of their changing fortunes over time. Technological-organizational studies here and generally in the past 30 years demonstrate its ability to reveal the complex, often subtle, ways that ancient people integrated stone tools (and, as above, other materials) into their cultures, documenting the great variety and effectiveness of problem-solving of which our ancestors were capable. Its accomplishments are undeniable, but in important respects technological organization has not fully exploited its analytical potential. Some areas of improvement in the application of technological organization are readily apparent in this volume, less as criticism of contributions than as rueful acknowledgment of the collective shortcomings of a still-imperfect perspective to which many of us have contributed and for which all of us are responsible. In ascending order of generality and importance, they include the following.

Scope of Application to Lithic Assemblages

Despite the scope of lithic analysis expanding from finished tools to entire industries and reduction sequences that technological organization exemplified, even now detailed analyses of debitage assemblages are not often attempted in organizational studies (in different ways, Riel-Salvatore and Negrino and Goebel et al. here study them to some degree). They should enable, for instance, inference regarding the kinds and numbers of cobbles reduced or tools made or allocate them to one or more segments of complex reduction sequences, all relevant to technological organization and behavioral theory. Instead, organizational analyses continue largely to disregard the most abundant portion of lithic assemblages. For instance, the staging of tool production that Sellet integrates into his comparison of Clovis and Folsom technologies can be documented and analyzed in debitage as in tools themselves (e.g., Beck et al. 2002; Shott 2015; Shott and Habtzghi 2016).

Tools are distinguished from flakes in part by type, degree, and pattern of retouch. Flakes that lack retouch often, probably usually, were not used. But some flakes were used without benefit of retouch as integral components of organized lithic technologies. As elusive as such practices may seem, not to mention the challenge that measurement of their prevalence and pattern poses, assemblage analysis grounded in technological perspective can identify them and account for their role in organized industries (e.g., Holdaway et al. 2015). Organizational studies must take more seriously the potential use of unretouched flakes. Conceding the frequent use of unretouched flakes, it does not follow that core reduction to produce tools was uncommon or that reduction for tool production is overemphasized. It does, however, acknowledge that complex technological practices could involve apparently simple tools.

Debitage assemblages are generated more in tool production than in use. Tool function involves which parts of tools were used on what materials and the kinetics of their use, areas that are the domain of use-wear studies. Because it is time- and labor-intensive and because it reveals only the last uses to which tools were put, use-wear analysis has obvious but limited value in archaeology. One value is to document the functional range or versatility of various tool types, which is implicated in theory (Ammerman and Feldman 1974; Shott 1986) and in practice here and elsewhere in organizational studies. Yet use-wear analysis remains uncommon in this research tradition.

Assemblage Analysis and Size-Composition Dependence

Lithic assemblages are characterized by their size—number of specimens of relevance to analysis, whether finished tools or not—and composition, the distribution and proportion of specimens across types however defined. Analysis of assemblage composition is common in chapters here. Binford's (1977) foundational paper on Nunamiut technology was among the first to demonstrate organizational effects upon assemblage size and composition, not least by explaining how curation mediated the equation between rate of use and rate of discard. Owing to complex relationships between the functional range of tool types, rates of use, and rates of curation (specified in Ammerman and Feldman's (1974) classic model, itself anticipated by Clarke's (1972: 26–28) "interdigitation" argument to interpret Perigordian Middle Paleolithic industries and elaborated in Schiffer's (1975) theoretical treatment of the complex relationships between occupation span, use lives, and assemblage structure), assemblage composition often varies systematically as size increases. This size-composition dependence that formation theory explains is demonstrated in Paleolithic and Paleoindian data that are among the most popular subjects of analysis in this volume and elsewhere (e.g., Shott 2003). Variation in assemblage size is interpreted in several chapters here (e.g., Wragg Sykes notes that larger Late Middle Paleolithic assemblages are more technologically diverse than smaller ones, which may be less a behavioral difference than a formation effect), but differences in composition that may be a function of assemblage size rather than organizational structure or behavioral adaptation are not always considered. Probably size-composition effects would not alter their conclusions, but these studies would only be strengthened by controlling for such effects before attributing to adaptation assemblage differences arising from formation processes.

Scales of Observation and Strategies of Inference

Like its culture-historical predecessor, technological organization often involves comparison between strata or time units and industries or assemblages that emphasizes qualitative difference as much as continuous variation within and between them.

As Kuhn and Miller (2015: 173; see also Kuhn 1994: 428) put it, technological-organizational studies tend to be "narrative" rather than predictive. In this volume and in earlier studies including my own (e.g., Shott 1986), analysis often starts by charting environmental differences between study units. Differences in technological organization between units then are accommodated to environmental conditions, often in qualitative terms that involve differences in toolstone, core type and reduction practices, or tool blanks and types.

However, useful that an organizational approach is in explaining technological adaptations and their archaeological expressions, therefore, these chapters demonstrate that the approach remains mostly comparative and, to a considerable extent, qualitative. Sellet, for instance, uses the marginal value theorem and distinguishes Clovis and Folsom technological organization partly in continuous terms, yet in analysis qualitatively contrasts Folsom staged biface-reduction sequences to, presumably, Clovis quarry production. Earliest tests of biface staging in terms of that model were, perhaps necessarily, qualitative, but more recent research shows how the model accommodates complex, continuous metric and assemblage variation (e.g., Kuhn and Miller 2015; Shott 2015) that might resolve Clovis and Folsom biface-staging differences of degree, not merely kind.

Organization may concern technological properties like toolstone selection, reduction practices and technologies, and the kind, number, design, functional versatility (Shott 1986:19, exemplified here in Hiscock's "diverse craft activities"), and curation rate of finished tools, on the one hand, and constraints like toolstone distribution and abundance, mobility magnitude and frequency (sensu Kelly 1983, Shott 1986), and landscape structure including patch size and productivity. In practice, however, any of the technological properties might be linked to any of the constraints, and the degree and direction of their covariation are themselves free to vary. This can create interpretive ambiguity, as here, for instance, when Mackay and Hallinan associate microblades with low residential mobility and Morisaki et al. with high mobility. In their own contexts, probably both interpretations are valid. But the underdeveloped state of technological organization fosters ambiguity and unresolved equifinality, where problems and inferred solutions are freely matched to conform to patterns in archaeological evidence.

Ambiguity owing to technological organization's common practice of narrative accommodation to evidence arises in other contexts as well. As noted previously (Shott 2013:152–155), archaeologists have interpreted nearly the same evidence to reach opposing conclusions about the historical relationship between Beringian Nenana and Denali industries or traditions, on the one hand, and Clovis on the other. Whatever these differences imply about the limitations of culture-historical units is less important in this context than a third possibility, which Cole and Graf acknowledge but reject: that they comprise distinct industrial and presumably behavioral modes of a single broader adaptive pattern (e.g., Bever 2006; Odess and Rasic 2007). In this way, the Beringian debate reproduces the Bordes-Binford functional argument. That this argument remains unresolved a half century after its emergence suggests a failure of archaeological method and theory, notably the theory of assemblage formation whose foundation Ammerman and Feldman laid and that links technological organization to its archaeological correlates, as well as the tenacity of culture-historical habits of thought.

Whatever the merits of competing arguments, an irony of the debate in Beringian context is that the toolkit thesis is grounded at least as strongly in technological-organization theory as is its opposite. Thus, Bever (2006:607) invoked technological organization and the impetus to increasing technological diversity and complexity with latitude that he felt it documented, not historical traditions like Denali, to explain the occurrence of microblades. Similarly, the curation concept and the interpretation of tool caches counseled Odess and Rasic against the definition of static or "archetypal" toolkits (2007:711). Instead, they perceived complex assemblage variation that culture-historical taxa cannot accommodate and that requires consideration of formation theory (e.g., to calibrate archaeological frequencies of types to varying use rates and use lives and to control for size-composition interactions) and then behavioral technological-organization theory to explain the occurrence and distribution of technological solutions to adaptive problems. Glossing the assemblage-formation theory that articulates closely with technological organization, studies here that invoke culture-historical taxa risk the equivalent of, in Clarke's memorable phrasing, interpreting "the French Mousterian sequence, of more than 30,000 years duration, in terms of the acrobatic manoeuverings of...typological tribes" (1973:10).

Commensurateness of Time Scales

As a version of behavioral ecology theory, technological organization explains adaptive behavior in particular contexts. Understandably, most chapters here carefully describe the environmental contexts in which technological organization guides solutions to problems. In the process, they link particular solutions to particular conditions. Yet technological organization as a theory of behavior is at once timeless and focused on short-term behavior. Unavoidably, the time resolution of paleoenvironmental inference tends to be coarse. In contrast, adaptive practices modeled by technological organization can vary subtly and over short spans in response to environmental variation that is not captured by the customary scale of paleoenvironmental inference.

As a result, time periods that vary greatly in scale often are treated as homogeneous units. Contributors here do not argue that study assemblages preserve a complete record of adaptation during a period of, say, 5000 years to which they are assigned. Nevertheless, periods that range from <500 to >25,000 years tend to be characterized in essentialist terms corresponding to the character of their technological organizations. In this way, short-term behavioral inference guided by technological organization is projected to variable spans, some quite long. Put differently, the time scale of projection may not be commensurate with the scale of behavior inferred. Either remarkable behavioral and organizational stasis prevailed in these cases, or study assemblages as time-averaged accumulations elide complex combinations of behavioral responses that are collapsed into single period-long adaptations, or there was only partial sampling of relevant archaeological evidence.

Theoretical Specification

Technological organization is less a body of mature theory than a set of interpretive norms. Lacking is a more developed body of theory that specifies in both qualitative and quantitative detail the strength and direction of relationships between technological properties and organizational constraints. Among the many questions begged, a few stand out. What qualities of environments, toolstone, land-use practices, and task applications are the efficient causes to which technological properties are adapted? Should we expect incremental or abrupt, dramatic changes in the use of local versus exotic toolstone or in curation rates as either environments or land-use scale change? That is, what is the shape and scale of the response function linking the variables? Should tool production shift qualitatively from, for instance, blade to biface production or by proportion between the technologies? When and why should toolstone distribution and abundance impose greater organizational constraints than mobility parameters? How does type and degree of risk interact with toolstone supply and mobility constraints? Knowing that any environmental parameter declined by x over time, how should organized technologies respond by what kind, degree, and direction of change y? Today questions the nature of these are idle because the complete theory that would establish their relevance and ground their answers remains undeveloped. Current practice's comparison between cases and conditions and accommodation of organizational expectations to myriad patterns of empirical variation is useful to a degree, but technological-organization theory remains badly underdeveloped. Some relevant research already exists (e.g., Clarkson et al. 2015; Kuhn 1994; Kuhn and Miller 2015; Shott 1986; Surovell 2009). Only when fully specified in these and other ways can technological organization achieve the predictive value that scientific theories possess.

Until that level of theoretical specification is reached, at least comparative organizational studies should be conducted. As examples, should two different cases facing similar environmental conditions and toolstone supply show similar technological responses? If not, why not? Should two cases facing different environmental and toolstone conditions show different responses? If so, can the degree or kind of difference be correlated with degree or kind of difference in conditions or supply?

History and Adaptation

However much technological organization urgently requires better theoretical specification, and it also must account for the role of history. This will not involve the normative culture-historical taxa long since found wanting, but instead recent research demonstrates the remarkably long persistence, even to some extent in defiance of environmental trends, of sets of transmitted cultural practices or traits (e.g., Mathew and Perreault 2015). Acknowledgment of history's role does not mean that past cultures faced no adaptive constraints. Instead, we must study how adaptation and historical transmission interact, identify the conditions under which either is prevalent, and integrate both into more detailed and better contextualized accounts of the cultural past from the archaeological record.

Like all adaptive or behavioral theory, technological organization must contribute to this integration. Merely as one suggestion, technological organization can encompass preexisting cultural practices and traits into its definition of the environment of adaptation. In the process, it can invoke canalization as a factor that limits the effective range of organization responses to changing conditions. At the broadest scale, for instance, canalization may help explain why chapters here interpret very different technological emphases (e.g., microliths [Clarkson et al., Hiscock, Jochim, Mackay and Hallinan, Marwick, Morisaki et al.] versus bifaces [Franco et al., Morisaki et al., Sellet]) as solutions to similar problems (e.g., risk reduction, mobility constraints, and transport costs). Microlith or blade industries are common in Europe and Asia; in the admittedly shorter North American record, such technological products and tools are rare. Instead, North America is biface central, a continent in which biface technologies were adopted early on and thereafter comprised a major component of almost all lithic industries. The many forms of bifaces both beautiful and wonderful that followed colonization may represent, in a broad sense, one set or sequence of solutions to problems that in many Old-World contexts that followed different canalization trajectories were met using other technologies.

But it is essential to the healthy development of technological organization as a body of theory that canalization not merely become another narrative (Kuhn and Miller 2015) trope invoked as post hoc explanation for patterns found in archaeological evidence. Instead, just as technological organization must become generally more systematic, fully specified, and predictive in application in order to transcend its current phase of inductive accommodation to evidence and emerge as a genuine body of scientific theory, this imperative must encompass both adaptation and historically transmitted tradition. It will achieve this integration by the use, in part, of the canalization concept.

Conclusion

As chapters demonstrate, technological organization is a productive approach to the study of hunter-gatherer lithic technologies. The range—in time, space, adaptation, and cultural context—of applications presented here attests to the concept's breadth and analytical vigor. Yet like an archaeology that remains mired in a prescientific state, technological organization is burdened with its own self-imposed limitations. It must be expanded in scope and applications to become truly general. More important, it must build upon its relationship to behavioral ecology to become a fully specified body of theory. This volume illustrates at once the potential and inherent limitations of a technological organization focused upon hunter-gatherer adaptions and narrative interpretations. It is an exceptional assay of the current state of an interpretive art that, if developed along lines that include those suggested here, may yet emerge as a deductive science.

References

Ammerman, A. J., & Feldman, M. W. (1974). On the 'making' of an assemblage of stone tools. *American Antiquity, 39*, 610–616.

Andrefsky, W. (2009). The analysis of stone tool procurement, production, and maintenance. *Journal of Archaeological Research, 17*, 65–103.

Attenbrow, V. (2004). *What's changing? Population size or land–use patterns? The archaeology of Upper Mangrove Creek, Sydney Basin.* Terra Australis No. 21. Canberra: Pandanus Press, Australian National University.

Beck, C., Taylor, A., Jones, G. T., Fadem, C., Cook, C., & Millward, S. (2002). Rocks are heavy: Transport costs and Paleoarchaic quarry behavior in the Great Basin. *Journal of Anthropological Archaeology, 21*, 481–507.

Bever, M. (2006). Too little, too late? The radiocarbon chronology of Alaska and the peopling of the New World. *American Antiquity, 71*, 595–620.

Binford, L. R. (1977). Forty-seven trips: A case study in the character of archaeological formation processes. In R. V. Wright (Ed.), *Stone tools as cultural markers: Change, evolution and complexity* (pp. 24–36). Canberra: Australian Institute of Aboriginal Studies.

Binford, L. R. (1979). Organization and formation processes: Looking at curated technologies. *Journal of Anthropological Research, 35*, 255–273.

Binford, L. R., & Binford, S. R. (1966). A preliminary analysis of functional variability in the Mousterian of Levallois facies. *American Anthropologist, 68*, 238–295.

Bleed, P. (1986). The optimal design of hunting weapons: Maintainability or reliability. *American Antiquity, 51*, 737–747.

Clarke, D. L. (1972). Models and paradigms in contemporary archaeology. In D. L. Clarke (Ed.), *Models in archaeology* (pp. 1–60). London: Methuen.

Clarke, D. L. (1973). Archaeology: The loss of innocence. *Antiquity, 47*, 6–18.

Clarkson, C., Haslam, M., & Harris, C. (2015). When to retouch, haft, or discard? Modeling optimal use/maintenance schedules in lithic tool use. In N. Goodale & W. Andrefsky (Eds.), *Lithic technological systems and evolutionary theory* (pp. 117–138). Cambridge: Cambridge University Press.

Holdaway, S., Dougless, M., & Phillipps, R. (2015). Flake selection, assemblage variability, and technological organization. In M. Shott (Ed.), *Works in stone: Contemporary perspectives on lithic analysis* (pp. 46–62). Salt Lake City: University of Utah Press.

Kelly, R. L. (1983). Hunter-gatherer mobility strategies. *Journal of Anthropological Research, 39*, 277–306.

Kuhn, S. L. (1994). A formal approach to the design and assembly of mobile toolkits. *American Antiquity, 59*, 426–442.

Kuhn, S. L., & Miller, D. S. (2015). Artifacts as patches: The marginal value theorem and stone tool life histories. In N. Goodale & W. Andrefsky (Eds.), *Lithic technological systems and evolutionary theory* (pp. 172–197). Cambridge: Cambridge University Press.

Mathew, S., & Perreault, C. (2015). Behavioural variation in 172 small-scale societies indicates that social learning is the main mode of human adaptation. *Proceedings of the Royal Society Series B, 282*, 20150061.

Morales, J. I. (2016). Distribution patterns of stone-tool reduction: Establishing frames of reference to approximate occupational features and formation processes in Paleolithic societies. *Journal of Anthropological Archaeology, 41*, 231–245.

Nelson, M. C. (1991). The study of technological organization. *Archaeological Method and Theory, 3*, 57–100.

Odess, D., & Rasic, J. T. (2007). Toolkit composition and assemblage variability: The implications of Nogahabara I, northern Alaska. *American Antiquity, 72*, 691–718.

Schiffer, M. B. (1975). The effects of occupation span on site content. In M. Schiffer & J. House (Eds.), *The Cache River Archeological project: An experiment in contract Archeology* (pp. 265–269). Fayetteville, AR: Arkansas Archeological Survey Research Series No. 8.

Shott, M. J. (1986). Technological organization and settlement mobility: An ethnographic examination. *Journal of Anthropological Research, 42*, 15–51.

Shott, M. J. (1996). An exegesis of the curation concept. *Journal of Anthropological Research, 52*, 259–280.

Shott, M. J. (2003). Size as a factor in assemblage variation: The European Middle Palaeolithic viewed from a North American perspective. In N. Moloney & M. Shott (Eds.), *Lithic analysis at the millennium* (pp. 137–149). London: Archtype.

Shott, M. J. (2013). Human colonization and late Pleistocene lithic industries of the Americas. *Quaternary International, 285*, 150–160.

Shott, M. J. (2015). Glass is heavy too: Testing the field-processing model at the Modena obsidian quarry, Lincoln County, southeastern Nevada. *American Antiquity, 80*, 548–570.

Shott, M. J., & Habtzghi, D. (2016). Toward disentangling stages in mixed assemblages of flake debris from biface reduction: An experimental approach. *Journal of Archaeological Science, 70*, 172–180.

Shott, M. J., & Seeman, M. F. (2017). Use and multifactorial reconciliation of uniface reduction measures: A pilot study at the Nobles Pond Paleoindian site. *American Antiquity, 81*(4.) October.

Spry, C., & Stern, N. (2016). Technological organization. In J. L. Jackson (Ed.), *Oxford bibliographies in anthropology*. New York: Oxford University Press.

Surovell, T. A. (2009). *Toward a behavioral ecology of lithic technology: Cases from Paleoindian archaeology*. Tucson, AZ: University of Arizona Press.

Thomson, D. (1939). The seasonal factor in human culture illustrated from the life of a contemporary nomadic group. *Proceedings of the Prehistoric Society, 5*, 209–221.

Torrence, R. (1983). Time budgeting and hunter-gatherer technology. In G. Bailey (Ed.), *Hunters-gatherer economy in prehistory: A European perspective* (pp. 11–22). Cambridge: Cambridge University Press.

Wiessner, P. (1983). Style and social information in Kalahari San projectile points. *American Antiquity, 48*, 253–276.

Index

A
Alaska. *See* Owl Ridge
Archaeological method and theory, 328, 330, 331
Archaeological site formation processes, 2
Archaic period, 236, 237
Assemblage-formation theory, 329
Australian lithic technology. *See also* Backed artifacts. *See also* Environmental change
  NCT, 79
  paleoenvironmental change, 79
  terminal Pleistocene and Holocene, 79

B
Backed artifacts
  archaeological examinations, 85
  craft goods, 83
  defined, 80
  degree of craft specialization, 82
  economic and social contexts, 87
  extractive tools, 87
  game hunting and individual pursuit strategies, 83
  geographical variation, 80, 81
  hafting positions, 82
  low chronological resolution, 85
  Mangrove Creek catchment, 83
  master-apprentice relationships, 84
  process of cultural changes, 88
  production, 81
  production behaviors, 84
  public signaling, 88
  resin residues and stains, 82
  scrutinizing truthfulness of signal, 84
  size and shape, 81
  specimens identification, 80
  territory and group composition, 87
  utilitarian kinds, 83
Behavioral ecology theory, 321, 323
BER. *See* Bonneville estates rockshelter (BER)
Bifacial technology, 281
Blade/microblade systems, 7
Bonneville estates rockshelter (BER)
  early Holocene sediments, 246
  lithic assemblages, 245
  map of, 243
  paleoindian occupation, 244
  radiocarbon dates, 245
  raw material procurement, 246–250
  raw material selection, 252
  stratigraphic profile, 244
  technological activities, 250, 251
  tools production, 251, 252
Bordes-Binford functional argument, 328
British LMP. *See* Middle palaeolithic in Britain

C
Cerro Verlika 1 (CV), 309
Chorrillo Malo 2, 307
Chronological resolution, archeological records, 4, 5
CLIMAP project, 2
Clovis-Folsom transition, 274
Clovis site, 266
Colonization, 124, 139, 141–145, 281, 282, 294, 297

E. Robinson, F. Sellet (eds.), *Lithic Technological Organization and Paleoenvironmental Change*, Studies in Human Ecology and Adaptation 9,
DOI 10.1007/978-3-319-64407-3

Colonization, South Asia, 39, 40
Curated technologies, 1, 7
Curation, 1, 3, 7, 8, 325, 329

D
Debitage, 1, 8
Dispersal model, 40

E
Eastern Beringia, 211–213
  archaeological context, 204–206
  archaeological investigations, 203
  colonization, 203
  environmental change (*see* Environmental change)
  Owl Ridge, 208, 210
  paleoenvironmental, 206–208
  Pleistocene-Holocene transition, 228
  technological strategies, 203
Eastern Great Basin. *See* Environmental change
El Niño-Southern Oscillation (ENSO) system, 85
Environmental change, 189–193, 237–242
  Australian lithic technology, 85–87
  eastern Beringia, 211–213
  eastern Great Basin, 237
    alluvial histories, 240
    cave records, 241
    Gunnison shoreline, 239
    intermountain region, 237
    Lake Bonneville, geographic extent, 238
    paleobotanical studies, 240
    Pleistocene large-mammal extinctions, 241
    pollen record, 240
    small-mammal assemblages, 241
    transgressive/regressive cycle, 239
    in western Bonneville basin, 241, 242
  southern Germany
    animal species, 190
    archaeological entity, 192
    boreal forest patterns, hunters, 192
    carnivore species, 192
    fish productivity and diversity, 193
    foraging technology, 190
    habitat structure, 190
    late and postglacial periods, 190
    and lithic technology, 189
    lowlands along migratory routes, 191
    Magdalenian and late Mesolithic, 189
    oldest Dryas period, 192
    patchiness, 190
    rapid warming, 191
    temperate deciduous forests, 193
    vegetational communities, 191
Estancia Alice (EA), 313
Estancia Alice 1 (EA1), 313
Expediency, 1, 7

F
Feldspar OSL ages, 311
Field processing model, 322
Final paleolithic
  archaeological entity, 192
  faunal assemblages, 194
  high and low spatial predictability, 196
  late, 191
Fine-grained igneous (FGI) toolstones, 219
Fire activity, 302, 311
Fishtail points
  morphological variations, 287
  PA, 288, 289
Folsom
  and Clovis stone tool systems, 263
  hunters, 267
  tool kits, 272
Formal *vs.* informal technologies, 219, 221
Fynbos taxa, 14

G
Good/bad environment, 7
Great Plains, 268, 273

H
Hoabinhian, southeast Asia
  artifact makers, 73
  assemblages, at Tham Lod and Jerimalai, 71
  bamboo hypothesis, 72, 73
  east Asia, 67, 68
  hominin taxonomic affiliations, 72
  Levallois technology, in Tham Lod, 72
  lithic elements, 65
  microlithic technology, 71
  southeast Asia, 66, 68–70
  stone artifacts, 63
  sumatraliths, 72
Hokkaido region, 111–112
Holocene
  lower-ranked foods, 85
  probabilities of radiometric dates, 87
  technological change, Australia, 79

Homo sapiens, 164, 182
Howiesons Poort
artifacts and discard rates, 21
classic/later, 16
earlier Howiesons Poort, 15, 29
fairly intensive occupation, 29
post-Howiesons Poort, 17, 30
and Still Bay, 23
Human adaptive responses
archaeological data, 92
archaeological features, 106, 108, 110
archaeological sites, 105–197
climatic amelioration, 114
climatic fluctuations, 93
data, 96
environmental change, 91
fauna, 96
flora, 95–96
lithic technology, 113, 115
microblade industries, 111
radiocarbon dates, 106
southeastern Hokkaido, 111
stone weaponry systems, 106
technology, 92
Unoki-minami site, 111
Human behavioral ecology, 268
Human mobility patterns, 268
Human occupation, 305, 308, 315, 316

I
Interpretive ambiguity, 4, 7

J
Jake Bluff Clovis site, 266, 268
Jomon lithic technology, 114
Jomon pottery types, 108

K
Kanto/Chubu region, 109–110
Kyushu region, 108

L
Last glacial maximum (LGM), 94, 283
Late glacial (LG), 92
Late Holocene, 305, 316
Levallois-flaked artifacts, 309, 312, 313, 317
Lithic analysis, 7, 8, 41, 263, 294
application scope, 326, 327
artifacts, 315
assemblages, 8
food and people, 265
Hoabhinhian, 65
hominin taxonomic affiliations, 72
and paleoenvironments, in east and southeast Asia, 74
in peninsula Thailand, 68
observation and strategies, 327, 328
recording strategies, 7
resources, 302
size-composition dependence, 327
south Asia (*see* South Asia, India)
types, 274
in Vietnam, 69

M
Magdalenian
assemblages, 193
food resources, 194
high and low spatial predictability, 196
hunters, 191
lakeshore sites, 194
steppe-tundra landscape, 189
Marginal value theorem, 269
Marine isotope stage (MIS) 4, 14
Mesolithic
archaeological entity, 192
early and late, 190, 191
with faunal assemblages, 194
geometric microlithics, 194
locations, 193
low–moderate predictability, 196
tool manufacture, 197
Microlithic, in south Asia
adaptive model, 41, 54
blade size and elongation, 53
dispersal model, 40, 53
Jurreru River valley, 43, 44, 46
from middle paleolithic, 41, 42
Middle palaeolithic in Britain
climatic and environmental contexts, 123, 137–139
colonization, exploration and risk, 141–145
continental records, 123
cultural adaptation, 124
cultural isolation and technological homogeneity, 140, 141
cyclical landscape exploitation, 145–147
dated assemblages, 134
distinctiveness, 139, 140
geographical situation, 123
geographic location, 123, 132
raw materials, 131

Middle palaeolithic in Britain (*cont.*)
- sites and chronology, 124–134
- technological characterization, 134–137
- UF radiocarbon, 133

Middle stone age (MSA), 13
- SA's, winter rainfall zone (*see* Southern Africa, WRZ)

Milankovitch hypothesis, 2

N

Neanderthal re-colonization of Britain (Wragg-Sykes), 6
Neanderthals
- dated assemblages, 134
- geographic location and species, 123
- geological resources across entire landscapes, 135
- mobility, techno-economic system, 145
- subarctic/artic contexts, 142

Niche construction Balzi Rossi site complex, 163
Niche construction technological patterns, 163, 164, 183
Niche construction theory (NCT), 79
Nunamiut, 1, 2

O

Owl Ridge
- artifact types, 213, 214
- cultural components, 209
- debitage, 213
- lithic assemblages, 210
- Nenana and Teklanika River valleys, 208
- Pleistocene-Holocene boundary, 210
- sample of tools, 213, 215
- stratigraphic profile, 209

P

Paleoclimatic data, 264
Paleoenvironment, 91–96, 302, . *See also* Australian lithic technology
- BER, 236
- eastern Beringia, 206–208

Paleoenvironmental change, 2–4, 6, 8, 268
Paleoenvironmental conditions
- Brazil, 283
- landscape features, 283
- LGM, 283
- Uruguay, 284

Paleo-Honshu island, 96
Paleoindian
- archeology, 274
- hunters, 267
- period, 236, 244, 256, 263–265, 272, 274
- points, 289, 293, 294, 297

Patagonia
- LIA, 314
- pollen records, 308
- SWW, 316

Pay Paso 1, 281, 282, 284
Pay Paso point, 292, 293
Penetration angle (PA), 288, 295
Period I
- 5300–3700 cal BP
  - CVk1, 309
  - Holocene, 304
  - human occupation, 305
  - radiocarbon ages, 306, 307
  - RB1, 309
  - reduced moisture levels, 308

Period II:3700–2300 cal BP, 311, 312
Period III:2300–1900 cal BP, 312
Period IV:1900–600 cal BP, 312–314
Period V:600–200 cal BP, 314, 315
Phlegrean fields, 182
Pleistocene, 267
- glaciations, 283
- Holocene transition, 91, 111, 240, 241, 281–283, 285, 295, 296

Pollen analysis, 302
Primary reduction activities, 218
Projectile point designs, 271, 294
Proto-Aurignacian behavioral package, 169
Proto-Aurignacian lithic assemblages
- adaptations, 181
- allochtonous material, 177
- archaeological signature, 164, 170
- behavioral strategies, 170
- behavioral variability and environmental variability, 181, 182
- bladelets, 177, 178, 181
- Bombrini, 168, 176, 183
- *chaînes opératoires*, 172
- climatic factors, 183
- coastal plains, 183
- discarding pieces, 178
- equids and rhinoceros, 168
- excavation, 165
- hunter-gatherer adaptations, 163
- landscape, 183
- levels, 175
- makers, 177
- occupants, 166
- paleoenvironmental datasets, 164
- proto-Aurignacian levels, 167

radiocarbon dates, 167
raw material, 170
resources, 183
retooling, 178
retouched and unretouched subsets, 178
Riparo Bombrini, 164, 169–171
technology and typology, 171–173
typological inventories, 172
typological inventory, 173
unretouched bladelets, 178
variability, 181–184
volumetric densities and frequencies, 174
Provisioning
artifacts, in Howiesons Poort assemblages, 18
definition, 17
humid conditions, 18, 19
of individuals, 17
in late Pleistocene of southern Africa, 18
mobility, 18
of places, 17, 18
systems, 17

Q
Quartz OSL ages, 311

R
Radiocarbon ages, 306, 307
Raw material procurement, 175
lithic landscape, 214–216
variability, 216–218
Raw materials, Southern Africa's WRZ, 22, 23
Región Arqueológica Catalanes Nacientes Arapey (RACNA), 285
Rio Bote 1 (RB1), 301, 308
Riparo Bombrini, 164, 181
location, 165
profile, 166
Riparo Mochi, 181

S
Seasonal constraints, 5
Secondary reduction activities, 219
South America, 287, 289
Fishtail point
morphological variations, 287
PA, 287, 289
Pampa and Patagonia, 287
paleoindian points, 289
Pay Paso point, 292, 293
Tigre points, 290, 291
South Asia, India
chopper, pebble and small tool, 38
chronology, 38
colonization, 39
demographic expansions, 38
environmental settings, 42
Jurreru River Valley, 45, 46
archaeological excavations, 43
in the blades, 45
lithic sequence, 44
raw materials, 45
middle paleolithic assemblages, 43
microlithic adaptation model, 38
microlithic origins (*see* Microlithic, in South Asia)
middle paleolithic, to microlithic transition, 41, 42
Middle Son Valley
blade characteristics, 51
changes, in core scar, 50
combined technological sequence, 50
deposits, 49
Dhaba locality, 49
location and border, 47
stone artifacts, 49
sub-continents, 38
timing, *Homo sapiens*' arrival, in India, 39, 40
Southeast Asia, 63
archaeology, 63
Hoabinhian, 65 (*see also* Hoabinhian, Southeast Asia)
locations, 63, 64
Southern Africa, WRZ, 19, 21, 24–28
analytic methods, for shelter sites, 20
cortex data, 23
Diepkloof, 15
on factors controlling precipitation, 14
fine-grained rocks, 15
fynbos, 14
Howiesons Poort, 21 (*see also* Howiesons Poort)
implements, 21
mobility, 18
MSA archaeological sequence, 15
Olifants River valley, 19
open site data
Clanwilliam Dam surveys, 24, 25
Doring River surveys, 26–28
Knersvlakte surveys, 28
population, 18

Southern Africa, WRZ (*cont.*)
  provisioning changes, 17, 18, (*see also* Provisioning)
  quartzite and quartz, 19
  raw material proportions, 22, 23
  retouched flakes and adjusted residuals, 22
  rock- shelters, 20
  silcrete cores, 23, 24
  technological systems design theory, 15
Southern Germany, 190–193
  environmental change (*see* Environmental change)
  faunal assemblages, 194
  food resources, 194
  geometric microlithics, 194
  heat-tempered artifacts, 195, 196
  host of factors, 189
  lakes channeled animal movements, 194
  lithic technology, 189
  Magdalenian assemblages, 189, 193
  microlithic barbs, 195
  raw materials, 195 (*see also* Technological organization (TO))
  vegetational changes, 194
Southern Patagonia Ice Field (SPIF), 303
Southern westerly winds (SWW), 303
Stemmed triangular with wings (STW), 290, 291
Still Bay
  bifacial points, 20
  Clanwilliam Dam Surveys, 25
  at Diepkloof, 22
  discard rates, 21, 22, 29
  of fine-grained rocks, 15
  and Howiesons Poort, 20
  Knersvlakte surveys, 28
  termination, at rock- shelter sites, 15
  WRZ site, Diepkloof, 15
Stone tools, 295
Sumatraliths, 65, 69, 72
Summer rainfall zone (SRZ), southern Africa, 14, 15

T

Technological organization (TO), 92, 196–199, 326–328
  behavioral ecology, 322
  Binford's development, 323
  bison oriented systems, 270
  by-products, tool production, 1, 2
  chronological resolution, 4–6
  CLIMAP project, 2
  Clovis and Folsom groups, 270, 273, 328
  cost, 321
  debitage, 8
  defined, 321
  environmental conditions, 4
  flute, 274, 295
  history and adaptation, 330, 331
  human mobility patterns, 268
  hunter-gatherer material adaptation, 322
  lithic assemblages
    application scope, 326, 327
    observation and strategies, 327, 328
    size-composition dependence, 327
  marginal value theorem, 269
  Nunamiut, 2
  and organizational constraints, 6, 7
  origin, 324
  paleoclimate forcing mechanisms, 4
  and paleoenvironmental change, 2, 3
  Paleoindian hunter, 270
  problem-solving, 326
  projectile point design, 271
  reporting strategies, 7
  southern Germany
    blades in assemblages by period, 197
    cores and blades, local raw materials, 199
    debitage, 198, 199
    faceted core platforms, 197, 198
    high and low predictability, 196
    stone tool typology, 196
    tools, anticipatory production, 197
  theoretical specification, 330
  time scale, 329
  time-series data, 8
  tool-caching behavior, 271
  tool forms, 272
  toolstone acquisition and processing, 321
Techno-typological uniformity, 173
Thailand, 65, 68
Tigre points, 290, 291
Time scales, 329–330
Time-series arrays, 8
Tohoku region, 110–111
Tool-Catching behavior, 271–272
Toolstone, 330
Translucent agate, 296

U

Ultrafiltrated (UF) radiocarbon, 133
Unifacial and bifacial technologies, 221–223
Uruguay
  bifacial artifacts, 286

Paleoenvironmental conditions, 284
and Paraná Rivers, 281
RACNA, 285
Tigre site, 285
Urupez 2 and K87, 285

V
Vietnam, 65, 69

W
Winter rainfall zone (WRZ), 13
southern Africa, 14, (*see also* Southern Africa, WRZ)

Y
Year-round rainfall zone (YRZ), southern Africa, 14, 15
Younger Dryas (YD) dynamics
bison population, 266
climatic event, 265
Clovis sites, 266
Folsom hunters, 267
paleoindian tradition, 265

GPSR Compliance
The European Union's (EU) General Product Safety Regulation (GPSR) is a set of rules that requires consumer products to be safe and our obligations to ensure this.

If you have any concerns about our products, you can contact us on

ProductSafety@springernature.com

In case Publisher is established outside the EU, the EU authorized representative is:

Springer Nature Customer Service Center GmbH
Europaplatz 3
69115 Heidelberg, Germany

www.ingramcontent.com/pod-product-compliance
Lightning Source LLC
Chambersburg PA
CBHW060743150926
51501CB00003B/79
9783319644059